Sustainable Materials for Engineering Applications

Sustainable Materials for Engineering Applications

Guest Editors

Abdul Aabid
Muneer Baig

MDPI

Basel • Beijing • Wuhan • Barcelona • Belgrade • Novi Sad • Cluj • Manchester

Guest Editors

Abdul Aabid
Engineering Management
Prince Sultan University
Riyadh
Saudi Arabia

Muneer Baig
Engineering Management
Prince Sultan University
Riyadh
Saudi Arabia

Editorial Office
MDPI AG
Grosspeteranlage 5
4052 Basel, Switzerland

This is a reprint of the Special Issue, published open access by the journal *Materials* (ISSN 1996-1944), freely accessible at: www.mdpi.com/journal/materials/special_issues/54D7438D47.

For citation purposes, cite each article independently as indicated on the article page online and as indicated below:

Lastname, A.A.; Lastname, B.B. Article Title. *Journal Name* **Year**, *Volume Number*, Page Range.

ISBN 978-3-7258-4098-4 (Hbk)
ISBN 978-3-7258-4097-7 (PDF)
https://doi.org/10.3390/books978-3-7258-4097-7

Contents

About the Editors

Abdul Aabid

Dr. Abdul Aabid is currently working as a Researcher at the College of Engineering, Prince Sultan University, in the Kingdom of Saudi Arabia. He earned his PhD in Mechanical and Aerospace Engineering from the International Islamic University Malaysia in 2020. He completed a master's degree in Aerospace Engineering at Jawaharlal Nehru Technological University, Hyderabad, in 2014 and his undergraduate degree in Aeronautical Engineering from Visvesvaraya Technological University in 2010. Dr. Aabid has gained valuable teaching experience, serving as an Assistant Professor at the Mangalore Institute of Technology and Engineering from 2014 to 2016 and as a Lecturer at KBN College of Engineering from 2010 to 2012. Dr. Aabid has been recognized among the world's top 2% of scientists by the Stanford and Elsevier list for his research contributions. His research interests span a broad spectrum, including solid mechanics and structures, fluid dynamics and control, as well as material science. His achievements have been recognized with various awards throughout his career, including outstanding student, outstanding researcher, and best paper presentation awards. Beyond his academic and research commitments, Dr. Aabid is also actively involved in professional societies around the world. He is a full member of the Saudi Arabia Society for Composite Materials and the International Association of Engineers (IAENG) in Hong Kong (since 2024), and he previously served as a member of the RAeS in the UK, Sigma Xi, and ASME in the USA. Moreover, he contributes to the academic community as an editorial board member for the Journal of *Fracture and Structural Integrity* at the Italian Group of Fracture in Italy and *Fluid Dynamics & Materials Processing at Tech Science Press* in the USA and as the Chief Guest Editor for the two Special Issues of the *Materials* Journal at MDPI in Switzerland.

Muneer Baig

Dr. Muneer Baig holds a Ph.D. in Mechanical Engineering from the University of Maryland, Baltimore, USA. He earned his M.S. in Mechanical Engineering from the University of New Orleans, USA, and his B.E. in Production Engineering specializing in Mechanical Engineering from Osmania University, Hyderabad, India. Currently, Dr. Baig is an Associate Professor in the Engineering Management Department at Prince Sultan University, Saudi Arabia, where he joined in 2017 as an Assistant Professor and was promoted to Associate Professor in 2019. Previously, he served as an Assistant Professor at the Center of Excellence for Research in Engineering Materials at King Saud University from 2010 to 2017, leading significant projects funded by the university and KACST. His research areas include nanocrystalline materials, severe plastic deformation, polymer composites, material modelling, and dynamic characterization. Dr. Baig is actively involved in professional societies globally, serving as a full member of the Saudi Arabia Society for Composite Materials and the International Association of Engineers in Hong Kong. He has served as an editor for two Special Issues of the *Materials* MDPI Journal and two Special Issues of *Fluid Dynamics & Materials Processing Tech Science Press*, USA. Dr. Baig received the Research Excellence Award from PSU (2022–2024), holds editorial roles in two journals, reviews for numerous journals, and has patented innovations in mechanical tooling mechanisms, demonstrating his significant academic and industrial contributions.

Preface

This reprint is a culmination of carefully curated research contributions previously published in the *Materials* journal under the Special Issue titled "Sustainable Materials for Engineering Applications". The collection presented here highlights cutting-edge advancements and interdisciplinary approaches toward sustainable material development, characterization, and applications in diverse engineering domains.

The growing urgency to address environmental degradation, resource scarcity, and climate change has led to a paradigm shift in engineering practices—one that prioritizes sustainability and circular economy principles. With this backdrop, this reprint integrates a wide range of studies—from innovative uses of industrial by-products and recycled materials to novel biopolymers and energy-efficient manufacturing techniques. These chapters explore both theoretical frameworks and experimental validations, reflecting the multidisciplinary nature of modern materials science.

Our objective in compiling these works is to provide researchers, engineers, scholars, and policy-makers with a comprehensive resource that not only demonstrates successful sustainable material applications but also inspires further innovation. The research contributions included span sectors such as civil infrastructure, energy storage, water treatment, and green construction, demonstrating real-world relevance and measurable impact.

This compilation is also a testament to the collaborative effort between academia and industry in addressing the global sustainability challenge. We believe the insights presented in this volume will serve as both a reference and a catalyst for future research and practical implementations in sustainable engineering.

We extend our deepest appreciation to the contributing authors, reviewers, and the editorial team at *Materials* for their valuable inputs and support in making this initiative a success.

Abdul Aabid and Muneer Baig
Guest Editors

materials MDPI

Editorial

Sustainable Materials for Engineering Applications

Abdul Aabid * and Muneer Baig

Department of Engineering Management, College of Engineering, Prince Sultan University, P.O. Box 66833, Riyadh 11586, Saudi Arabia; mbaig@psu.edu.sa
* Correspondence: aaabid@psu.edu.sa

This newly added Special Issue (SI) of the *Materials* journal, titled "Sustainable Materials for Engineering Applications", focuses on the foundations, characterizations, and applications of several sustainable materials. In recent years, there has been an increasing global focus on sustainability and the urgent need to transition towards environmental practices in various industries. Engineering, as a key sector driving innovation and development, plays a crucial role in shaping a sustainable future. This Special Issue aims to explore the latest advancements and challenges in the field of sustainable materials for engineering applications. Sustainable materials encompass a wide range of materials and technologies that minimize environmental impact while maintaining or even improving performance. This Special Issue welcomes original research, review articles, case studies, and perspectives that shed light on sustainable materials' applications across different engineering disciplines. The objective is to provide a comprehensive platform for researchers, scientists, engineers, and practitioners to share their insights, exchange knowledge, and present cutting-edge solutions to address sustainability challenges.

The exploration and development of environmentally friendly materials involves a wide array of strategies, ranging from bio-based and recycled materials to those designed for improved end-of-life management. Extensive studies delve into the life cycles of these sustainable materials, yielding methodologies, tools, and practical case studies that rigorously assess their environmental impact and resource consumption. Innovative techniques are pivotal in sustainable material processing, with a focus on energy-efficient manufacturing, waste reduction, and pollution prevention, not only in improving the environmental aspects of material production but also by enhancing sustainability across industries. These sustainable material advancements extend to structural applications, exemplified by eco-friendly high-performance concrete, sustainable metals, composites, and polymers, striking a balance between impressive strength-to-weight ratios and environmental conscientiousness.

Renewable energy technologies greatly benefit from these materials, powering components in solar panels, wind turbines, energy storage, and fuel cells, and ultimately boosting efficiency and promoting cleaner energy sources. Sustainable materials are also transforming transportation through lightweight materials, electric components, and alternative fuels, significantly reducing the sector's environmental impact. Additionally, in water treatment, tailored sustainable materials address water scarcity and pollution challenges, significantly enhancing the environmental performance of vital water management systems. The successful integration of these sustainable materials into engineering projects serves as a showcase of their potential, providing valuable insights into their benefits, challenges, and real-world applications. As industries increasingly prioritize sustainability, the continuous exploration and integration of such materials are vital for shaping an eco-conscious future.

At present, there is an increased interest in replacing common engineering materials with sustainable materials due to the environmental challenges that our planet is exposed to. These challenges, encountered due to the use of common engineering materials, include (but are not limited to) changes in the climatic conditions, a decrease in available resources, and an increase in pollution. There has been a significant increase in the use of sustainable materials in the construction industry, as it poses a significant impact on the environment [1,2]. Studies have been performed to develop sustainable concrete materials to improve thermal insulation properties [3–6], and these materials are considered to be environmentally friendly and sustainable. Additionally, biopolymers are being considered as a replacement for polymers. These polymers are biodegradable and are considered to be sustainable materials in many engineering applications. Moreover, biopolymers obtained naturally have been gaining importance recently as they exist in abundance and are available from natural sources, which minimizes the impact on the environment and lowers costs [7,8].

It is well known that materials play a significantly critical role throughout their life span, in any product. However, their production poses several challenges. The traditional manufacturing processes are responsible for increasing carbon emissions, leading to a higher global warming concern. The presence of several studies in the literature related to the selection of materials highlights the importance of challenges involved with the proper selection of materials [9–11]. Currently, one of the most formidable challenges confronting humanity is the issue of global energy sustainability for the future. Concerns surrounding our environment, the depletion of finite natural resources, challenges in energy storage, environmental hazards, and the specter of natural disasters have collectively intensified our focus on the remarkable potential of sustainable energy materials. This pertains to their development, production, distribution, and packaging. As the demand for energy continues to surge, there is a pressing need to minimize adverse environmental effects while optimizing the utilization of thermal, electrical, optical, and chemical energy inherent in these sustainable energy materials [12]. Extensive research is being carried out in Lithium (Li) to develop Li-ion batteries as an energy storage device. In addition, in recent studies, Lithium-ion batteries have proven to be effective energy storage solutions for portable electronic devices and as energy sources in electric vehicles [13–15].

By bringing together diverse perspectives, this Special Issue seeks to foster interdisciplinary collaborations and promote the adoption of sustainable materials in engineering applications. It will serve as a valuable resource for researchers, scientists, professors, engineers, representatives, and participants committed to driving sustainable development and creating a more environmentally conscious future. Papers from the whole sustainable materials and manufacturing process, with high performance and energy efficiency, and referencing the experimental work, numerical simulation, mathematical expression, and optimizations techniques, with a focus on the anticipated and achieved results, are greatly welcomed.

Author Contributions: Conceptualization, A.A. and methodology, A.A.; formal analysis, M.B.; investigation, A.A.; resources, A.A.; data curation, A.A.; writing—original draft preparation, A.A.; writing—review and editing, M.B.; supervision, M.B.; project administration, A.A. and M.B.; funding acquisition, M.B. All authors have read and agreed to the published version of the manuscript.

Acknowledgments: The authors acknowledge the support of the Structures and Materials (S&M) Research Lab of Prince Sultan University.

Conflicts of Interest: The authors declare no conflict of interest.

References

1. Sharma, S.; Sharma, N.K. Advanced materials contribution towards sustainable development and its construction for green buildings. *Mater. Today Proc.* **2022**, *68*, 968–973. [CrossRef]
2. Sangmesh, B.; Patil, N.; Jaiswal, K.K.; Gowrishankar, T.P.; Selvakumar, K.K.; Jyothi, M.S.; Jyothilakshmi, R.; Kumar, S. Development of sustainable alternative materials for the construction of green buildings using agricultural residues: A review. *Constr. Build. Mater.* **2023**, *368*, 130457. [CrossRef]
3. Wei, S.; Yiqiang, C.; Yunsheng, Z.; Jones, M. Characterization and simulation of microstructure and thermal properties of foamed concrete. *Constr. Build. Mater.* **2013**, *47*, 1278–1291. [CrossRef]

4. Gencel, O.; Bayraktar, O.Y.; Kaplan, G.; Arslan, O.; Nodehi, M.; Benli, A.; Gholampour, A.; Ozbakkaloglu, T. Lightweight foam concrete containing expanded perlite and glass sand: Physico-mechanical, durability, and insulation properties. *Constr. Build. Mater.* **2022**, *320*, 126187. [CrossRef]
5. Yu, H.; Zahidi, I.; Liang, D. Sustainable porous-insulation concrete (SPIC) material: Recycling aggregates from mine solid waste, white waste and construction waste. *J. Mater. Res. Technol.* **2023**, *23*, 5733–5745. [CrossRef]
6. Al Shouny, A.; Issa, U.H.; Miky, Y.; Sharaky, I.A. Evaluating and selecting the best sustainable concrete mixes based on recycled waste materials. *Case Stud. Constr. Mater.* **2023**, *19*, e02382. [CrossRef]
7. Biswas, M.C.; Jony, B.; Nandy, P.K.; Chowdhury, R.A.; Halder, S.; Kumar, D.; Ramakrishna, S.; Hassan, M.; Ahsan, A.; Hoque, E.; et al. Recent Advancement of Biopolymers and Their Potential Biomedical Applications. *J. Polym. Environ.* **2022**, *30*, 51–74. [CrossRef]
8. Román-Hidalgo, C.; Barreiros, L.; Villar-Navarro, M.; López-Pérez, G.; Martín-Valero, M.J.; Segundo, M.A. Electromembrane extraction based on biodegradable materials: Biopolymers as sustainable alternatives to plastics. *TrAC Trends Anal. Chem.* **2023**, *162*, 117048. [CrossRef]
9. Mousavi-Nasab, S.H.; Sotoudeh-Anvari, A. A comprehensive MCDM-based approach using TOPSIS, COPRAS and DEA as an auxiliary tool for material selection problems. *Mater. Des.* **2017**, *121*, 237–253. [CrossRef]
10. Atici, K.B.; Podinovski, V.V. Using data envelopment analysis for the assessment of technical efficiency of units with different specialisations: An application to agriculture. *Omega* **2015**, *54*, 72–83. [CrossRef]
11. Cicek, K.; Celik, M.; Topcu, Y.I. An integrated decision aid extension to material selection problem. *Mater. Des.* **2010**, *31*, 4398–4402. [CrossRef]
12. Robinson, J.; Kumari, N.; Srivastava, V.K.; Taskaeva, N.; Mohan, C. Sustainable and environmental friendly energy materials. *Mater. Today Proc.* **2022**, *69*, 494–498. [CrossRef]
13. Huang, B.; Li, X.; Wang, Z.; Guo, H.; Shen, L.; Wang, J. A comprehensive study on electrochemical performance of Mn-surface-modified LiNi0.8Co0.15Al0.05O2 synthesized by an in situ oxidizing-coating method. *J. Power Sources* **2014**, *252*, 200–207. [CrossRef]
14. Benveniste, G.; Sánchez, A.; Rallo, H.; Corchero, C.; Amante, B. Comparative life cycle assessment of Li-Sulphur and Li-ion batteries for electric vehicles. *Resour. Conserv. Recycl. Adv.* **2022**, *15*, 200086. [CrossRef]
15. Xu, P.; Tan, D.H.; Gao, H.; Rose, S.; Chen, Z. Recycling of Li-Ion Batteries for Electric Vehicles. *Encycl. Energy Storage* **2021**, *4*, 98–107. [CrossRef]

materials

MDPI

Article

Pyrolyzed Agro-Food By-Products: A Sustainable Alternative to Coal

Lukáš Jeníček *, Jan Malaťák, Jan Velebil and Michal Neškudla

Faculty of Engineering, Czech University of Life Sciences Prague, Kamýcká 129, 165 00 Prague, Czech Republic; malatak@tf.czu.cz (J.M.); velebil@tf.czu.cz (J.V.); neskudla@tf.czu.cz (M.N.)
* Correspondence: jenicekl@tf.czu.cz

Abstract: This study investigates the potential use of biochar derived from residues—such as spruce wood, spent coffee grounds, tea waste, and nutshells—as a sustainable coal substitute—to enhance the decarbonization of European energetic systems and decrease the dependence on fossil fuels. The biomasses were pyrolyzed at 250–550 °C, analyzed for calorific value and composition, and evaluated for energy retention and mass loss. The results show significant energy density improvements, with optimal temperatures varying by material (e.g., spruce wood reached 31.56 MJ·kg^{-1} at 550 °C, retaining 21.84% of its mass; spent coffee grounds peaked at 31.26 MJ·kg^{-1} at 350 °C, retaining 37.53%). Economic analysis confirmed pyrolyzed biomass as a cost-effective alternative to coal, especially considering emission allowance costs. Integrating biomass pyrolysis into regional energy systems supports decarbonization, reduces emissions, and advances us towards a circular, economy,

Keywords: biomass; biochar; calorific value; energy residual; spruce wood; spent coffee ground; tea waste; nut shells

1. Introduction

The dwindling reserves of fossil fuels, coupled with Europe's goal to reduce dependency on them [1], have spurred interest in sustainable alternatives like biomass waste within a circular economy framework. Renewable energy sources such as solar, hydro, wind, and biomass are crucial for sustaining economic growth while protecting the environment [2]. Biomasses from agricultural and forestry by-products such as sawdust, bagasse, coffee husk, and nutshells are already commonly used as solid fuels in developing regions [3–5] and have the potential to replace fossil fuels in developed countries [6,7]. Biomass can also be co-combusted with coal in existing coal power plants in order to achieve reduced greenhouse gas emissions and reduced solid waste from the combustion process [8]. The results show that blends with 70 wt. %, 50 wt. % and 30 wt. % biomass could reduce the non-renewable CO_2 emissions per produced unit of energy by about 75%, 57% and 34%, respectively [9]. Purohit and Chaturvedi [10] estimated that associated carbon dioxide mitigation potential in India would be 205 Mt in 2030–2031 if the entire biomass surplus is to be diverted for power generation. Coal remains the primary global fuel for heat and power production, but its use significantly contributes to greenhouse gas emissions [11]. Major CO_2 emitters include China, the USA, India, and European nations such as Germany and Poland [12]. Finding viable alternatives to fossil fuels remains a pressing challenge [13–15]; however, pyrolysis technologies utilizing renewable biological resources show promise in producing substitutes for coal [16].

Recent advancements in biochar research have further reinforced its potential use as a sustainable energy resource. For example, a study from Prabha [17] highlighted that biochar could significantly contribute to improving soil health and mitigating climate change. This is due to its carbon sequestration capabilities and its ability to act as a clean-burning fuel source. Other studies, such as one by Rajput [18], have emphasized the role of optimized pyrolysis processes in tailoring biochar properties to enhance its calorific value and material applications, ranging from renewable energy to industrial uses. These findings underline the critical importance of refining production methods to maximize both efficiency and versatility, paving the way for broader applications across sectors.

A bibliometric analysis of biochar trends published by Wu [19] demonstrated a growing focus on the economic feasibility of biochar applications in agricultural, environmental, and energy sectors. Such advances underscore the importance of improving biochar's production efficiency and scalability, ensuring that its benefits extend beyond niche uses to become a cornerstone of renewable energy strategies.

In the Czech Republic, energy policy prioritizes the use of alternative biomass resources like damaged timber and logging residues for fuel, aligning with plans to decarbonize the energy sector [20,21]. The region's significant challenges, including extensive forest disturbances [22,23] caused by climate change [24–27] and associated salvage logging [28], provide a substantial supply of by-products suitable for biochar production. These by-products not only offer an environmentally friendly solution for waste management, but also present an economically viable resource for renewable energy [29].

One of the key technologies for enhancing biomass energy properties is low-temperature pyrolysis, often called torrefaction, a process involving the slow heating of organic materials like wood, spent coffee ground [30–32], straw, and nutshells in a low-oxygen environment. This method not only improves the energy properties of biomass, but also makes it a more effective, clean-burning fuel [33–35]. Pyrolysis increases biochar's energy density, making it easier to handle and transport, and reduces overall logistics costs. Furthermore, biochar production supports renewable energy generation and aids in carbon sequestration, which is crucial for mitigating climate change [36,37].

Recent research has demonstrated that pyrolysis can be optimized using machine learning techniques to enhance energy recovery and adapt material properties to specific applications [38,39]. Studies have also explored the co-pyrolysis of multiple biomass types, revealing synergistic effects that enhance yield and quality, a promising area for further exploration [40,41].

Economic evaluations have also underscored the viability of using biochar as a substitute for coal. The economic and environmental benefits of transitioning to biochar-based fuels are supported by their lower emissions [42], reduced waste disposal costs, and ability to create new markets for agricultural by-products [43–46]. Studies, including those examining the use of biochar in decentralized energy systems, emphasize the efficiency of localized production and utilization [47–49], with additional benefits such as soil conditioning for biochar [50]. Decentralized approaches not only reduce the environmental costs associated with transportation, but also increase resilience by enabling energy generation close to biomass sources. These findings further bolster the argument for incorporating pyrolyzed biomass into the Czech Republic's energy mix [51], supporting EU decarbonization goals and promoting circular economy approaches.

In summary, advancements in biochar research and its integration into renewable energy policies suggest that optimizing local waste materials for energy production contributes towards achieving energy security and sustainability in the region. This study aims to evaluate the potential of pyrolyzed biofuels made from residues and waste materials, including spruce wood, spent coffee grounds, tea waste, walnut shells, pistachio shells, and peanut shells, as substitutes for coal. Aside from comparing the proximate and ultimate analysis results with those of other authors, this article also focuses on the economic aspect of biochar usage.

By leveraging recent technological and economic insights, the potential for biochar to transform energy systems and contribute to environmental sustainability is made increasingly apparent.

2. Materials and Methods

To confirm data consistency, results from prior research conducted in 2020–2022 [52–55] are here re-evaluated alongside new analyses performed in 2023. Additional analyses were conducted on the samples to assess the overall energy potential of these sustainable biochar materials as substitutes for fossil resources.

The selected pyrolysis temperatures (250 °C, 300 °C, 350 °C, 450 °C, and 550 °C), along with a non-pyrolyzed sample (designated as 0 °C), were chosen to represent the full range of slow pyrolysis processes, from low to moderately high temperatures. Pyrolysis at 250 °C induced notable changes in material composition for most samples, as illustrated in the Van Krevelen diagram (Section 3.5), with a minimal correlation between temperature and mass loss [56]. In contrast, pyrolysis at 550 °C led to highly active pyrolysis reactions, depending on the nature of the tested materials [57], resulting in the transformation of the material into gaseous and liquid states. These selected temperature points align with those from previous studies, ensuring methodological consistency and enabling reliable comparisons across datasets.

The samples, sourced from renewable energy materials, were selected as by-products from the forestry and food industries. The primary focus was to determine the optimal energy utilization of these samples as substitutes for fossil fuels.

Spruce timber was harvested from the Vysočina region in the Czech Republic in 2023. The samples, sourced from logs cut 12 to 18 months earlier, were processed into wood chips and combined to form a representative composite sample. In 2020, the EU had around 159 million hectares of forests, a 10% increase since 1990. Forest areas grew in most EU countries, with Ireland, Spain, and Malta experiencing notable expansions. In 2019, the EU's wood stocks were 28.4 billion m^3, with Germany, Sweden, and France leading in volume [29]. The Czech Republic's wood volume rose by 15%, representing 2.7% of European wood stocks [58].

Spent coffee grounds (SCG) were obtained from *Coffea arabica*, which constitutes approximately 75% of global coffee production [59,60]. These grounds, collected from a single household (4 kg total), were sun-dried for several days until reaching a stable moisture content of 8.45%. Coffee, as one of the world's most widely consumed beverages, generates substantial amounts of spent coffee grounds, totaling approximately 6.5 megatons annually worldwide [61]. In the Czech Republic, about 3 billion cups of coffee consumed in 2019 produced roughly 24,000 tons of SCG [62]. Rich in cellulose, sugars, minerals, and lipids, coffee beans are valuable for producing biodiesel or fuel pellets, with a calorific value of about 20.9 $MJ \cdot kg^{-1}$ [63–65].

Black tea residues (*Camellia sinensis* L. Kuntze) were gathered from domestic brewing processes over four months and sun-dried to reach a moisture content of 8.31%. Tea is the second most popular non-alcoholic drink globally, with nearly 6 million tons produced annually [66–68]. In Czech Republic, the annual production is approximately 2000 tons of tea waste [62]. Like coffee, tea consumption generates significant waste, with over 90% of the tea material left unused after brewing [69]. Studies have demonstrated tea waste's potential as a fuel pellet resource, showing high calorific values and suitability as a natural additive to other biomass materials [70–74]. However, tea waste-based pellets may contain high ash levels and deposit-forming elements [75,76].

Nutshells, namely, walnut shells harvested from English walnut (*Juglans regia* L.), pistachio shells from *Pistacia vera* L., and peanut shells derived from *Arachis hypogaea* L., were also used. The United States is a leading producer of nuts [77,78], with its industries generating significant by-products. In California, there were 1.5 million tons of almonds, 0.5 million tons of pistachios and 0.4 million tons of walnuts produced [79], while in Czech Republic, 22 tons of nut of waste was generated [62] in 2020. Walnut and almond shells, often discarded, can be converted into biochar, offering enhanced properties and reduced emissions compared to direct combustion. Processing companies report that utilizing nutshells as biochar or biofuel offsets disposal costs and provides an environmentally friendly alternative [80–86].

All samples were ground to a particle size of 1 mm using a Retsch SM 100 mill (Haan, Germany) and then subjected to pyrolysis in a LECO TGA 701 analyzer (St. Joseph, MI, USA) at five distinct temperatures: 250 °C, 300 °C, 350 °C, 450 °C, and 550 °C. Each sample was heated in an inert atmosphere at a rate of 10 °C per minute and held at the target temperature for an additional 30 min. The 30-min residence time was selected based on the available literature, as this duration has been shown to optimize pyrolysis performance. This has been demonstrated in the processing of fruit bio-waste [87], as well as in the pyrolysis of olive pits [88].

2.1. *Proximate and Ultimate Analysis*

Each sample underwent proximate and ultimate analysis to understand its material composition and characteristics. In the proximate analysis, measurements included elemental composition, moisture content, and ash content. The ultimate analysis involved determining the calorific value [89]. All measurements were conducted 5 times, and the average value was used for the following calculations.

A LECO TGA 701 thermogravimetric analyzer was used for measuring moisture and ash contents [90]. To calculate moisture content, 1 g of each sample was dried at 105 °C to a constant weight [91]. For ash content, 1 g of each sample was heated to a constant weight up to 550 °C.

The chemical compositions were analyzed using a LECO CHN628+S elemental analyzer (LECO Corporation, St. Joseph, MI, USA) through the LECO instrumental biomass combustion method. Each 0.1-gram sample was burned in oxygen at 950 °C to measure the amounts of carbon (C), hydrogen (H), and nitrogen (N). The oxygen content was determined by difference to a total of 100% on a dry basis. The LECO elemental analyzer was regularly calibrated using ethylenediaminetetraacetic acid, rice flour, and oatmeal.

To determine the calorific value, 1 g of each sample was placed in a stainless-steel crucible and secured in a cylinder. The pressure increased to 3 MPa at a reference temperature of 28 °C. The sample was then ignited with a cotton thread inside a LECO AC600 isoperbolic calorimeter (LECO Corporation, St. Joseph, MI, USA). Benzoic acid was used to calibrate the instrument. All proximate and ultimate analysis results were reported on a dry basis.

Net calorific value was calculated from the measured gross calorific value using the equation

$$Q_n = Q_v - (24.42 \times ((W) + (8.94 \times H))) \quad (1)$$

where Q_n (kJ·kg^{-1}) is net calorific value, Q_v (kJ·kg^{-1}) is gross calorific value, W (%) is water contain in the sample, H (kg·kg^{-1}) is hydrogen content in the sample, 24.42 is the amount equal to energy needed to evaporate 1% of water at 25 °C and 8.94 is the coefficient for the conversion of hydrogen to water.

2.2. Energy Residual

The energy residual of the original mass was calculated by adjusting the calorific value for the material's enthalpy and accounting for the mass loss.

The enthalpy of the material ($kJ \cdot kg^{-1}$) was determined using the standard for spruce wood. Weight loss was measured with a LECO TGA 701 (LECO Corporation, St. Joseph, MI, USA). During the material's pyrolysis, weight measurements were taken with an accuracy of 10^{-4} g.

2.3. Thermogravimetric Analysis

To express the dynamics of volatile matter loss during heating, LECO TGA 701 was used to perform a TGA analysis. The samples were enclosed in aluminum foil and put in covered crucibles. The temperature program consisted of two phases. In the first one, the samples were dried at 105 °C until constant weight. In the second, the temperature would ramp up at a rate of 5 °C min^{-1} from 105 °C to 580 °C. The results were processed into weight loss curves (TGA) and their derivations (DTG curves). The values were converted to express the loss of combustible matter. The DTG curves were then taken to represent the rate of loss (wt. % min^{-1}) of combustible matter as a function of temperature.

2.4. Economic Assessment of Material Pyrolysis

An economic assessment was performed considering the price of the material and energy residual from the pellet produced.

Pellet production was calculated as a fixed energy cost E_p.

$$Q_P = Q_T - E_p \tag{2}$$

where Q_P ($kJ \cdot kg^{-1}$) is energy residual in the pellet, Q_T ($kJ \cdot kg^{-1}$) is energy residual in the material and E_P = 108 $kJ \cdot kg^{-1}$ is the fixed energy cost of pellet production [92].

The final price of 1 MWh (EUR) for all materials in the form of a pellet was calculated by use of the following equation:

$$P_{MWh} = \frac{P_{kg}}{Q_P * 0.2778} \times 1000 \tag{3}$$

where P_{MWh} (EUR) is a price of 1 MWh, P_{kg} is a price of 1 kg, Q_P ($kJ \cdot kg^{-1}$) is energy residual in the pellet and 0.2778 is a coefficient to display kg in MWh.

2.5. Statistical Analysis

The statistical analysis of the results was conducted using the STATISTICA 14.0.0.15 CZ (StatSoft, Tulsa, OK, USA) software. The data were subjected to a one-way ANOVA.

3. Results and Discussion

3.1. Proximate and Ultimate Analysis

Elemental composition, ash and water content, and calorific value were measured for all samples. Measurements were performed five times for all the samples. The final average results with the given deviation are shown in the Table 1.

The overall results of the proximate and ultimate analysis of the present materials do not fundamentally differ from those reported by other authors, e.g., from the processing of olive pits [88] and other agricultural waste materials [87]. The only notable differences are in moisture and ash content, which are specific to the region where the original samples were collected and have a significant impact on the net calorific value.

Table 1. Proximate and ultimate analysis of all samples.

	Carbon (wt. %)	Oxygen (wt. %)	Hydrogen (wt. %)	Nitrogen (wt. %)	Ash (wt. %)	Water (wt. %)	Gross Calorific Value (MJ·kg^{-1})
Spruce wood	47.01 ± 0.82	38.82	6.41 ± 0.14	0.23 ± 0.01	0.33 ± 0.01	7.19 ± 0.09	20.17 ± 0.20
Spruce wood 250 °C	52.00 ± 0.62	37.79	5.98 ± 0.06	0.28 ± 0.01	0.56 ± 0.01	3.38 ± 0.12	20.41 ± 0.24
Spruce wood 300 °C	57.24 ± 0.92	31.84	5.73 ± 0.07	0.33 ± 0.01	0.71 ± 0.01	4.14 ± 0.15	22.18 ± 0.16
Spruce wood 350 °C	72.88 ± 0.61	16.93	4.59 ± 0.03	0.41 ± 0.01	0.92 ± 0.02	4.26 ± 0.05	28.42 ± 0.18
Spruce wood 450 °C	79.00 ± 0.44	11.14	3.32 ± 0.09	0.46 ± 0.01	1.29 ± 0.07	4.78 ± 0.12	30.32 ± 0.17
Spruce wood 550 °C	84.34 ± 0.74	5.03	2.6 ± 0.14	0.56 ± 0.01	1.63 ± 0.07	5.84 ± 0.05	32.27 ± 0.29
Spent coffee ground	49.38 ± 0.82	31.24	7.18 ± 0.05	2.22 ± 0.03	1.63 ± 0.01	8.35 ± 0.11	21.51 ± 0.22
Spent coffee ground 250 °C	59.27 ± 0.84	28.24	6.90 ± 0.04	2.68 ± 0.02	1.99 ± 0.06	0.92 ± 0.01	25.38 ± 0.23
Spent coffee ground 300 °C	70.71 ± 0.67	14.49	6.88 ± 0.11	3.39 ± 0.02	2.96 ± 0.08	1.56 ± 0.02	30.63 ± 0.24
Spent coffee ground 350 °C	76.35 ± 0.66	7.63	6.20 ± 0.06	3.99 ± 0.01	4.10 ± 0.10	1.73 ± 0.03	32.65 ± 0.28
Spent coffee ground 450 °C	77.52 ± 0.84	5.13	3.83 ± 0.10	4.43 ± 0.04	6.38 ± 0.08	2.71 ± 0.11	30.78 ± 0.23
Spent coffee ground 550 °C	76.32 ± 0.72	3.20	3.21 ± 0.04	4.11 ± 0.04	6.96 ± 0.09	6.20 ± 0.06	28.34 ± 0.15
Tea waste	46.21 ± 0.63	34.83	5.99 ± 0.07	1.82 ± 0.03	4.10 ± 0.05	7.04 ± 0.11	18.5 ± 0.20
Tea waste 250 °C	55.48 ± 0.45	30.64	5.45 ± 0.13	2.37 ± 0.01	5.31 ± 0.10	0.74 ± 0.01	21.37 ± 0.29
Tea waste 300 °C	64.24 ± 0.44	19.31	5.04 ± 0.04	2.59 ± 0.03	7.01 ± 0.07	1.80 ± 0.05	24.59 ± 0.18
Tea waste 350 °C	68.27 ± 0.53	13.38	4.43 ± 0.13	2.56 ± 0.03	8.74 ± 0.09	2.58 ± 0.12	27.22 ± 0.11
Tea waste 450 °C	72.65 ± 0.68	7.63	3.32 ± 0.13	2.49 ± 0.03	11.12 ± 0.16	2.76 ± 0.08	24.80 ± 0.29
Tea waste 550 °C	73.00 ± 0.80	3.93	2.76 ± 0.12	2.23 ± 0.01	11.75 ± 0.13	6.29 ± 0.14	25.22 ± 0.24
Walnut shell	43.53 ± 0.63	39.24	6.23 ± 0.06	0.48 ± 0.01	2.48 ± 0.06	8.05 ± 0.10	18.51 ± 0.16
Walnut shell 250 °C	55.90 ± 0.89	34.54	5.35 ± 0.11	0.52 ± 0.01	3.44 ± 0.05	0.25 ± 0.01	21.21 ± 0.26
Walnut shell 300 °C	66.88 ± 0.87	22.31	5.23 ± 0.03	0.59 ± 0.01	4.54 ± 0.02	0.45 ± 0.01	26.24 ± 0.25
Walnut shell 350 °C	74.60 ± 0.89	12.25	4.36 ± 0.09	0.69 ± 0.01	6.76 ± 0.03	1.35 ± 0.10	26.72 ± 0.19
Walnut shell 450 °C	82.06 ± 0.49	4.14	3.52 ± 0.11	0.74 ± 0.01	8.09 ± 0.05	1.47 ± 0.15	29.13 ± 0.28
Walnut shell 550 °C	83.24 ± 0.74	1.67	2.77 ± 0.11	0.82 ± 0.01	9.84 ± 0.03	1.66 ± 0.05	30.65 ± 0.28
Pistachio shell	44.53 ± 0.73	40.89	6.48 ± 0.11	0.38 ± 0.01	1.00 ± 0.03	6.71 ± 0.13	17.82 ± 0.27
Pistachio shell 250 °C	54.85 ± 0.45	37.10	5.86 ± 0.09	0.44 ± 0.01	1.46 ± 0.04	0.29 ± 0.01	21.39 ± 0.15
Pistachio shell 300 °C	66.01 ± 0.82	25.19	5.30 ± 0.13	0.49 ± 0.01	2.24 ± 0.03	0.77 ± 0.01	25.99 ± 0.18
Pistachio shell 350 °C	75.69 ± 0.93	14.94	4.42 ± 0.06	0.58 ± 0.01	2.74 ± 0.05	1.65 ± 0.08	28.43 ± 0.28
Pistachio shell 450 °C	82.37 ± 0.45	7.53	3.57 ± 0.08	0.62 ± 0.01	4.26 ± 0.09	1.65 ± 0.06	29.99 ± 0.25
Pistachio shell 550 °C	87.34 ± 0.80	2.72	2.83 ± 0.14	0.72 ± 0.01	4.66 ± 0.01	1.72 ± 0.04	31.34 ± 0.29
Peanut shell	48.97 ± 0.92	34.96	6.42 ± 0.14	1.42 ± 0.03	1.75 ± 0.02	6.48 ± 0.06	20.05 ± 0.11
Peanut shell 250 °C	56.43 ± 0.88	32.83	6.06 ± 0.05	1.56 ± 0.03	2.15 ± 0.02	0.97 ± 0.01	21.22 ± 0.24
Peanut shell 300 °C	63.36 ± 0.66	24.77	5.76 ± 0.09	1.73 ± 0.03	2.96 ± 0.09	1.43 ± 0.03	23.09 ± 0.11
Peanut shell 350 °C	72.33 ± 0.73	15.01	4.61 ± 0.15	1.94 ± 0.03	4.24 ± 0.06	1.87 ± 0.04	25.84 ± 0.23
Peanut shell 450 °C	77.12 ± 0.60	9.36	3.45 ± 0.09	1.88 ± 0.02	5.93 ± 0.10	2.26 ± 0.11	27.17 ± 0.24
Peanut shell 550 °C	81.70 ± 0.82	4.87	2.84 ± 0.10	1.96 ± 0.03	6.17 ± 0.01	2.46 ± 0.12	28.79 ± 0.27

The spruce wood samples' results are very close to those of Jenicek et al., Malatak et al. and Arias et al. [53,93,94].

The spent coffee ground proximate analysis results are in line with those of Fermoso and Mašek, Masek et al. and Mayson and Williams [95–97]. The calorific values of SCG are similar to those found by Silva et al. and Colantoni et al. [98,99].

The tea waste results here are similar to those found by Tunklova et al. [54], while the calorific value in the non-torrefied sample here is bit lower than that in Sermyagina et al. [100], whose result was 20.39 MJ·kg^{-1}.

The results for nutshells are very close to those found by Jenicek et al. and Vassilev et al. [55,101] in all aspects of proximate and ultimate analysis.

3.2. Calorific Value and Energy Residual of the Samples

The net calorific value in the dry state (Table 2) was calculated from the calorific value, which was measured using the LECO AC600 isoperbolic calorimeter (Table 1).

Table 2. Net calorific value in the dry state of samples (MJ·kg^{-1}).

	0 °C	250 °C	300 °C	350 °C	450 °C	550 °C
Spruce wood	18.6	19.02	20.83	27.31	29.48	31.56
Spent coffee ground	19.74	23.85	29.09	31.26	29.88	27.49
Tea waste	17.02	20.16	23.45	26.19	24.01	24.47
Walnut shell	16.96	20.03	25.09	25.74	28.33	30.00
Pistachio shell	16.24	20.11	24.81	27.43	29.17	30.68
Peanut shell	18.48	19.87	21.80	24.78	26.36	28.11

The results here are very close to those of Malatak et al. [93], Gendek et al. [102] and Jenicek et al. [42] for spruce wood, with 18.6 MJ·kg^{-1} for dried material and 31.56 MJ·kg^{-1} for the sample pyrolyzed at 550 °C.

Colantoni et al. [99] and Silva et al. [98] reported the calorific values of spent coffee grounds with approximate results of 20.92 MJ·kg^{-1} and 22.36 MJ·kg^{-1}, respectively. Similar results were also achieved by Kovalcik et al., McNutt and He, Nepal et al., Bejenari et al. and Ballesteros et al. [103–107].

Tea waste net calorific values were close to the results derived by Tunklova et al. [54], in the range of 17.04–24.31 MJ·kg^{-1} depending on the level of pyrolysis, but were slightly lower compared to Sermyagina et al. [108], who reported gross calorific values of 20.39 MJ·kg^{-1} in green tea and 20.26 MJ·kg^{-1} in black tea.

Measuring the nutshell's net calorific value, we find results close to those of Jenicek et al. [55], who used the same method, with a walnut net calorific value of 18.66 MJ·kg^{-1} (non-pyrolyzed) to 30.55 MJ·kg^{-1} (550 °C), a pistachio net calorific value of 17.58 MJ·kg^{-1} (non-pyrolyzed) to 31.35 MJ·kg^{-1} (550 °C) and a peanut net calorific value of 19.93 MJ·kg^{-1} (non-pyrolyzed) to 28.78 MJ·kg^{-1} (550 °C).

Material enthalpy for spruce wood was applied for all samples to estimate the energy needed for material pyrolysis. Enthalpy varied from 0.274 MJ·kg^{-1} for the material pyrolyzed at 250 °C to 0.602 MJ·kg^{-1} at 550 °C. The results for net calorific value adjusted by material enthalpy are shown in Table 3.

Table 3. Net calorific values of samples adjusted by material enthalpy (MJ·kg^{-1}).

	0 °C	250 °C	300 °C	350 °C	450 °C	550 °C
Spruce wood	18.600	18.746	20.501	26.927	28.987	30.958
Spent coffee ground	19.740	23.576	28.761	30.877	29.387	26.888
Tea waste	17.020	19.886	23.121	25.807	23.517	23.868
Walnut shell	16.960	19.756	24.761	25.357	27.837	29.398
Pistachio shell	16.240	19.836	24.481	27.047	28.677	30.078
Peanut shell	18.480	19.596	21.471	24.397	25.867	27.508

Another important parameter when assessing the pyrolyzed material is the weight loss under each process temperature. There is an overview in Table 4 of what percentage of original mass remained after the material was pyrolyzed. The highest level of material loss when pyrolyzed at 250 °C was seen in pistachio shells, with only 72.35% of original mass left. The highest loss at 550 °C occurred for spent coffee grounds, which kept only 20.05% of their original mass.

Table 4. Material mass residue (%).

	0 °C	250 °C	300 °C	350 °C	450 °C	550 °C
Spruce wood	100%	83.49%	74.90%	39.56%	27.59%	21.84%
Spent coffee ground	100%	79.31%	52.63%	37.53%	22.87%	20.05%
Tea waste	100%	77.30%	54.35%	41.16%	31.83%	28.82%
Walnut shell	100%	77.49%	52.57%	36.06%	29.19%	25.81%
Pistachio shell	100%	72.35%	45.70%	31.14%	24.53%	21.98%
Peanut shell	100%	83.53%	65.04%	41.19%	31.40%	27.96%

When combining the net calorific value of the pyrolyzed material adjusted by the material enthalpy and weight loss of the pyrolyzed material, we get the energy residual of the original mass weight unit. As we can see in Table 5, for spruce wood, pyrolysis led to a decrease in energy residual from 18.600 MJ·kg^{-1} to 6.762 MJ·kg^{-1}. In other words, the direct combustion of 1 kg of dried spruce wood would lead to the generation of 18.600 MJ·kg^{-1}, or if the same 1 kg of dried spruce wood is first pyrolyzed at 550 °C, the total output would lead to the generation of only 6.762 MJ·kg^{-1}. This energy loss was significantly greater than that measured by Liu et al. [109], who showed bamboo pellet mass losses of 2.12%, 3.31% and 7.45% at 180 °C, 200 °C and 220 °C, respectively. The difference in mass loss is caused by the lower temperature in the experiments of Liu, as well as different materials and different processes of pyrolysis.

Table 5. Energy residual of original mass (MJ·kg^{-1}).

	0 °C	**250 °C**	**300 °C**	**350 °C**	**450 °C**	**550 °C**
Spruce wood	18.600	15.651	15.355	10.653	7.998	6.762
Spent coffee ground	19.740	18.697	15.136	11.587	6.720	5.392
Tea waste	17.020	15.371	12.567	10.623	7.486	6.880
Walnut shell	16.960	15.309	13.018	9.144	8.127	7.587
Pistachio shell	16.240	14.352	11.189	8.422	7.035	6.612
Peanut shell	18.480	16.369	13.965	10.049	8.123	7.691

To summarize the results, Table 2 shows the net calorific value of the sample; the higher pyrolysis of the material leads to higher energy density and therefore higher net calorific value. On contrary, Table 5 shows that the total energy output of the original dried material was lowered by the material enthalpy and the weight loss caused by the pyrolysis. An equilibrium between increasing the material energy density and losing the energy of original mass when modifying the material by pyrolysis needs to be set for each material separately.

3.3. *Calorific Value Density and Weight Loss Comparison*

Spruce wood showed a significant drop in energy residual between pyrolysis temperatures of 300 °C and 350 °C; in the same interval, spruce wood increases its calorific value density from 20.83 MJ·kg^{-1} to 27.31 MJ·kg^{-1} (see Figure 1). The recommended temperature for spruce wood pyrolysis is 300 °C. This is lower in comparison to that found by Jenicek et al. [53], recommending pyrolysis at 350 °C, as the mass loss was not considered in this respective article.

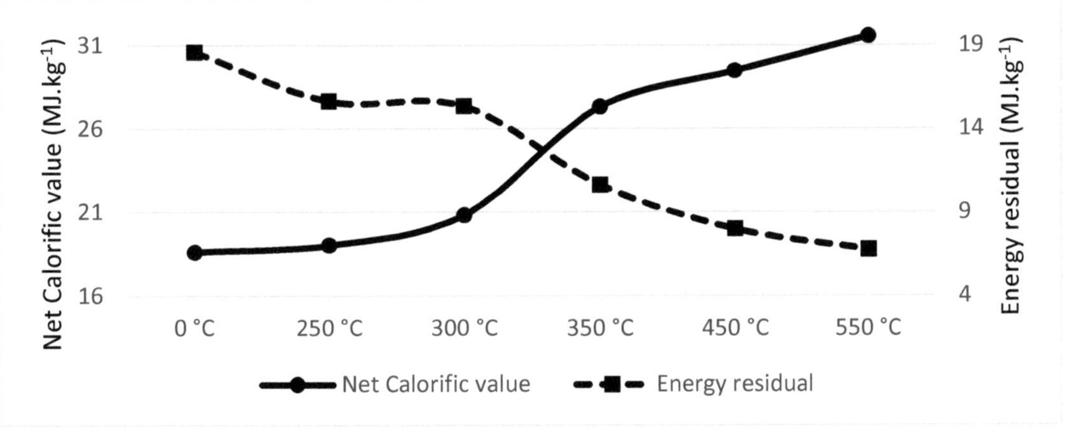

Figure 1. Net calorific value (MJ·kg^{-1}) and energy residual (MJ·kg^{-1}) of Spruce wood.

Spent coffee ground dropped in total energy considerably from the temperature level of 250 °C, as seen in Figure 2. The energy density increased substantially between 250 °C and 300 °C. The recommended pyrolysis temperature was also 300 °C, as it was for spruce wood, which is slightly lower than recommended by Jenicek et al. [52], who recommended 350 °C as an ideal pyrolysis temperature for spent coffee ground, and did not show the mass loss figure estimations in their original article.

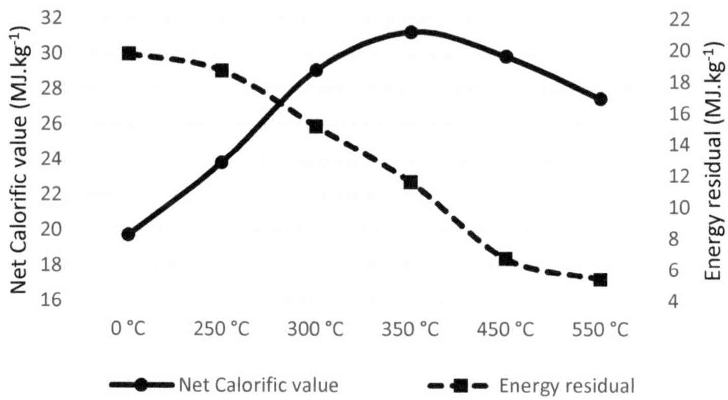

Figure 2. Net calorific value (MJ·kg^{-1}) and energy residual (MJ·kg^{-1}) of spent coffee ground.

Tea waste pyrolysis showed a significant drop in total energy produced above the temperature of 250 °C. From the research of Tunklova et al. [54], it can be confirmed that increasing the share of ash also occurred for tea waste pyrolysis; therefore, its recommended temperature is no more than 250 °C. Detailed curve of net calorific values and energy residuals of the pyrolyzed material are visible on Figure 3.

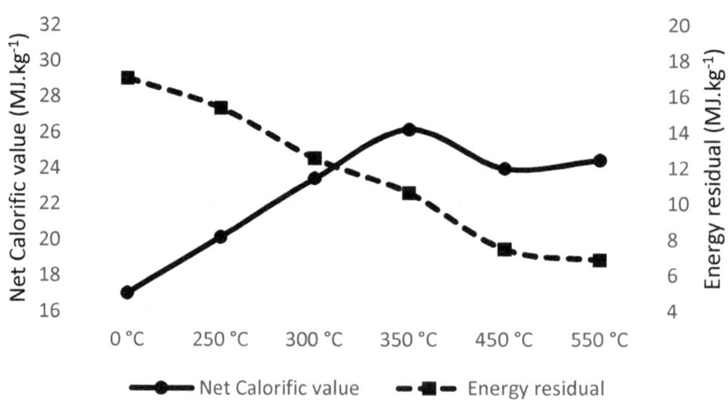

Figure 3. Net calorific value (MJ·kg^{-1}) and energy residual (MJ·kg^{-1}) of Tea waste.

Walnut shell pyrolysis significantly increases the net calorific value at the temperature of 300 °C; for higher temperatures, the net calorific value remains stable or increases indistinctly. On the other hand, the energy residual drops at the temperature of 350 °C, as seen in Figure 4. As such, the recommended temperature for the pyrolysis of walnut shell is 300 °C, the same as measured by Jenicek et al. [55].

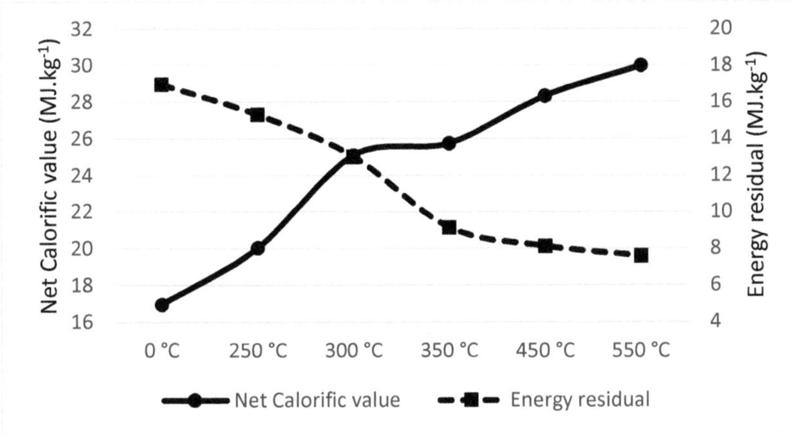

Figure 4. Net calorific value (MJ·kg^{-1}) and energy residual (MJ·kg^{-1}) of Walnut shell.

Pistachio shell shows an almost linear increase in net calorific value and an almost linear decrease in energy residual with pyrolysis temperature increases. To keep at least 14 MJ·kg^{-1} of total residual, the recommended temperature for pyrolysis should not surpass 250 °C. This is slightly lower than Jenicek et al.'s [55] recommendation of 300 °C pyrolysis. Detailed results are shown in Figure 5.

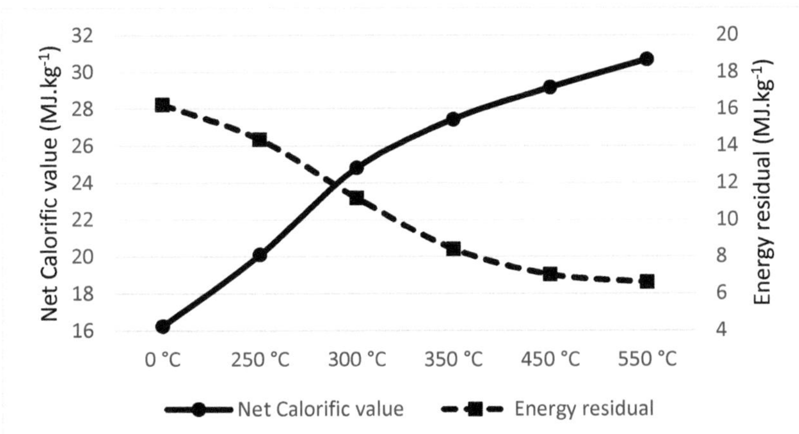

Figure 5. Net calorific value (MJ·kg^{-1}) and energy residual (MJ·kg^{-1}) of Pistachio shell.

Peanut shell shows a similar nearly linear increase in net calorific value and decrease in energy residual with pyrolysis temperature increases. On the contrary, the maximum net calorific value of a peanut shell sample pyrolyzed at 550 °C was 28.11 MJ·kg^{-1} with the energy residual 7.661 MJ·kg^{-1}; meanwhile, the pistachio shell sample showed a higher residual net calorific value of 30.68 MJ·kg^{-1} with a lower energy residual of 6.589 MJ·kg^{-1} in the same laboratory conditions. Detailed results are shown in Figure 6. The recommended temperature for the pyrolysis of peanut shell is 300 °C—far lower than suggested by Jenicek et al., who recommended 550 °C.

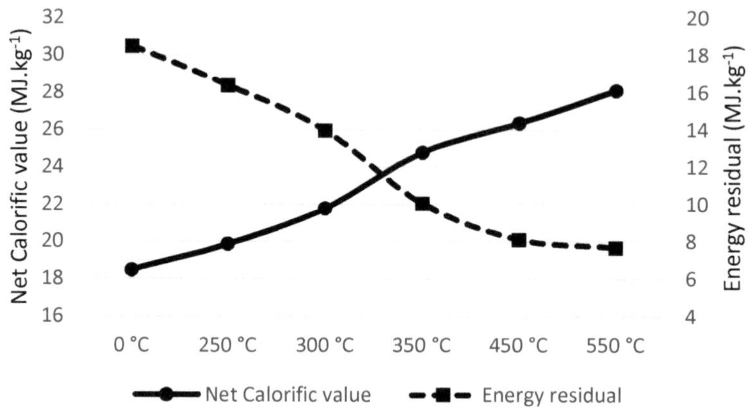

Figure 6. Net calorific value (MJ·kg^{-1}) and energy residual (MJ·kg^{-1}) of Peanut shell.

3.4. Economic Assessment of Material Pyrolysis

In order to assess the economic aspects of pyrolysis, let us first adjust the energy residual of dried material by pellet production losses. Pyrolyzed material is pressed into a form of a pellet of 8 mm in diameter and 15 to 20 mm in length. During the pellet production, approximately 0.108 MJ·kg^{-1} [92] of energy is used. The net calorific value of a pellet shown in original mass weight units is presented in Table 6.

Table 6. Energy residual of original mass—pellet production (MJ·kg^{-1}).

	0 °C	**250 °C**	**300 °C**	**350 °C**	**450 °C**	**550 °C**
Spruce wood	18.492	15.561	15.275	10.610	7.969	6.738
Spent coffee ground	19.632	18.612	15.079	11.547	6.696	5.371
Tea waste	16.912	15.288	12.509	10.578	7.451	6.849
Walnut shell	16.852	15.225	12.961	9.105	8.095	7.559
Pistachio shell	16.132	14.274	11.140	8.388	7.009	6.589
Peanut shell	18.372	16.279	13.895	10.005	8.089	7.661

In real operation, the production of pellets from material pyrolyzed at high temperatures (more than 400 °C) requires additional components to ensure sufficiently high mechanical durability. This fact is omitted for the purpose of the purity of results of this research [110].

The energy residual of original material combined with the price of the original material gives us the result of price per MWh for each analyzed sample. The spruce wood price was set as EUR 50 m^{-3}, which is ca. EUR 106 ton^{-1} of dried spruce wood. The prices of all other materials are very complicated to estimate, as the market is not centralized and not yet well developed; therefore, the prices for spent coffee grounds, tea waste and all nutshells were set as EUR 200 ton^{-1}.

The final material prices are shown in Table 7. The lowest price was derived for naturally dried spruce wood, with EUR 20.71 MWh^{-1}; the highest price is for spent coffee ground pyrolyzed at 550 °C, with EUR 134.06 MWh^{-1}. Recommended pyrolysis conditions are marked in bold letters, with the lowest value of EUR 25.07 MWh^{-1} for spruce wood pyrolyzed at 300 °C, and the highest value of EUR 55.55 MWh^{-1} for walnut shell pyrolyzed at 300 °C.

Table 7. Material price (EUR MWh^{-1}).

	0 °C	250 °C	300 °C	350 °C	450 °C	550 °C
Spruce wood	20.71	24.61	**25.07**	36.10	48.06	56.84
Spent coffee ground	36.67	38.69	**47.75**	62.36	107.53	134.06
Tea waste	42.57	**47.10**	57.56	68.06	96.63	105.13
Walnut shell	42.72	47.29	**55.55**	79.07	88.94	95.25
Pistachio shell	44.63	**50.44**	64.63	85.84	102.73	109.28
Peanut shell	39.19	44.23	**51.82**	71.97	89.01	93.98

For a complete overview, these results need to be compared to those of the standard fossil fuel brown coal, which is today still the most widely used fuel around the world. The average price of brown coal for the last 12 months was EUR 110 ton^{-1}, which is approximately EUR 28.5 MWh^{-1}. Next to the price of coal itself, the price of emission allowances needs to be combined with the 12-month average of EUR 57.58 $t(CO_2)^{-1}$, which is approximately ERU 86.77 MWh^{-1}. In total, this gives over EUR 115 MWh^{-1} in material costs when using brown coal. Due to the high price of emission allowance, any pyrolyzed biomass would be cheaper to use, as it is not burdened by emission allowances. The economic assessment of biomass usage as an alternative to brown coal in municipal power plants was also calculated by Malatak et al. [93], with positive results when considering timber as a more profitable fuel than coal.

Economic evaluations highlight biochar's viability as a coal substitute due to its lower emissions, reduced waste disposal costs, and potential to create new markets for agricultural by-products [54,55,93]. However, the lack of centralized markets for materials like spent coffee grounds, tea waste, and nutshells leads to price variability. Additionally, collection and transportation costs can significantly impact economic feasibility, particularly for dispersed sources.

Decentralized waste collection, co-locating pyrolysis units near waste generation sites, and integrating with existing supply chains can reduce logistical costs. Future analyses should include transportation impact modeling to improve economic assessments. Strategies such as leveraging local supply chains and minimizing transport distances further enhance cost efficiency.

In summary, while biochar's benefits are clear, overcoming logistical and economic barriers is vital for large-scale adoption.

3.5. Van Krevelen Diagram

The van Krevelen diagram (Figures 7 and 8) shows the evolution of the three main components of fuel—carbon, oxygen, and hydrogen. During carbonization, oxygen is released approximately twice as fast as hydrogen until black coal is formed. Further transformation into anthracite decreases the H/C ratio while keeping the oxygen content low.

Figure 7 shows the van Krevelen diagram for biochar produced from spruce wood, spent coffee grounds, and tea waste, where it is evident that increasing process temperatures lead to a reduction in the H/C and O/C ratios. Figure 8 shows similar trends for biochar derived from walnut, pistachio, and peanut shells.

Like fossil fuels, the O/C ratio drops faster than the H/C ratio. Temperatures between 300 and 350 °C cause the greatest decrease in these ratios. For spruce wood, a 100 °C increase lowers the O/C ratio from 0.42 to 0.17, and a further 100 °C increase drops the H/C ratio from 0.17 to 0.11. Similar reductions are seen in other wood biomasses, with the H/C ratio dropping from 1.37 to 0.75 between 250 and 350 °C [111]. Spent coffee grounds and tea waste show ratio changes starting at 250 °C. Walnut, pistachio, and peanut shells display almost identical changes in ratios with rising temperatures.

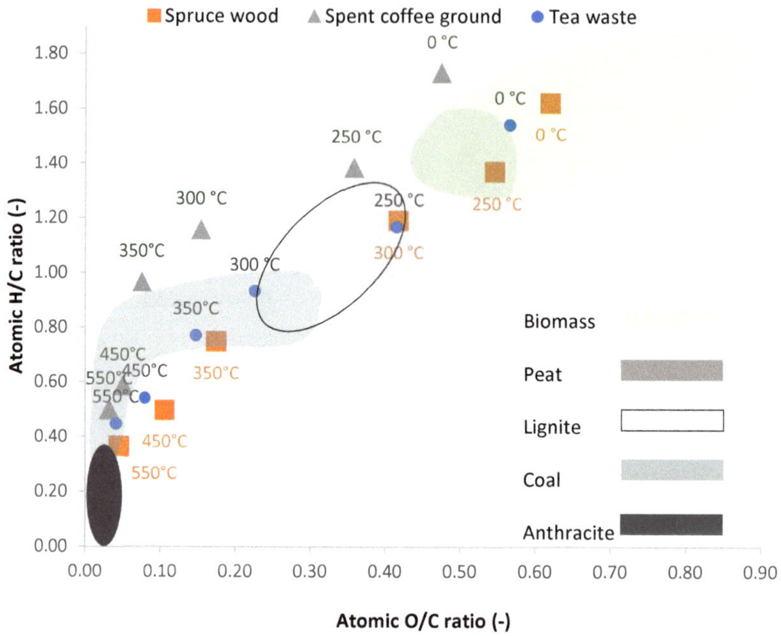

Figure 7. Van Krevelen diagram for spruce wood, spent coffee ground and tea waste ranges for material types according to Van Loo and Koppejan [112].

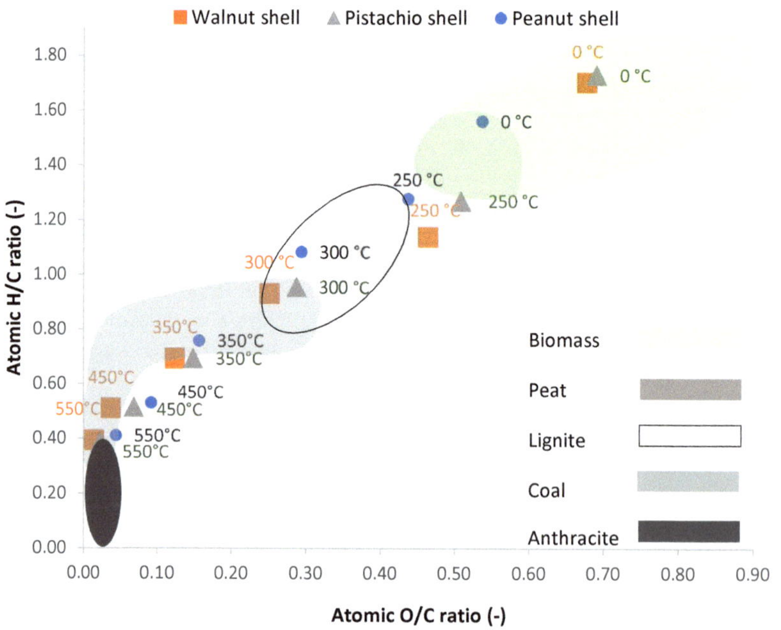

Figure 8. Van Krevelen diagram for walnut shell, pistachio shell and peanut shell ranges for material types according to Van Loo and Koppejan [112].

3.6. Thermogravimetric Curves

To express the differences in the release of volatile matter, e.g., during the first phase of combustion, between the original biomass materials and the biochar, all samples have undergone TGA analysis. The results are expressed as DTG curves of combustible matter in Figures 9–11. In raw biomass materials, the spread of results is relatively broad. The significant release of volatiles starts at 250 °C for all materials; however, the trends do vary, owing to the differences in chemical constitution, between the source biomasses, i.e., different proportions of hemicelluloses, cellulose and lignin, all of which tend to decompose in different temperature ranges [113]. In particular, in all samples apart from SCG, the lignin content is apparent in the second DTG peak around 350 °C.

In biochar produced at 300 °C (Figure 10), significant decomposition starts above 300 °C, with the first DTG peak being almost uniform around 350 °C. Markedly different again is the SCG, with a different combustible release profile, possibly thanks to it already being thermally treated as a part of coffee production. Spruce wood biochar achieved a very high maximum (over 5% min^{-1}) in volatile release. However, in the other biochar, the release was more gradual, with peaks approximately half as significant compared to raw biomass. This would make the combustion process more similar to coal combustion, or better suited for coal-fired boilers [39,45,48].

In the biochar produced at 400 °C, the DTG curves were much lower than in the previous case, with DTG peaks generally below 0.5% min^{-1}. Here, the volatile matter release is not finished by the end of the analysis at 580 °C. In all samples, there is an early peak around 200 °C, which can be inferred to be caused by bound water release. Above this temperature, the profile seems to correlate relatively well with that of, e.g., lignite coal [114,115]. Importantly, the volatile release profiles correlate very well with each other, and therefore are expected to behave similarly in combustion.

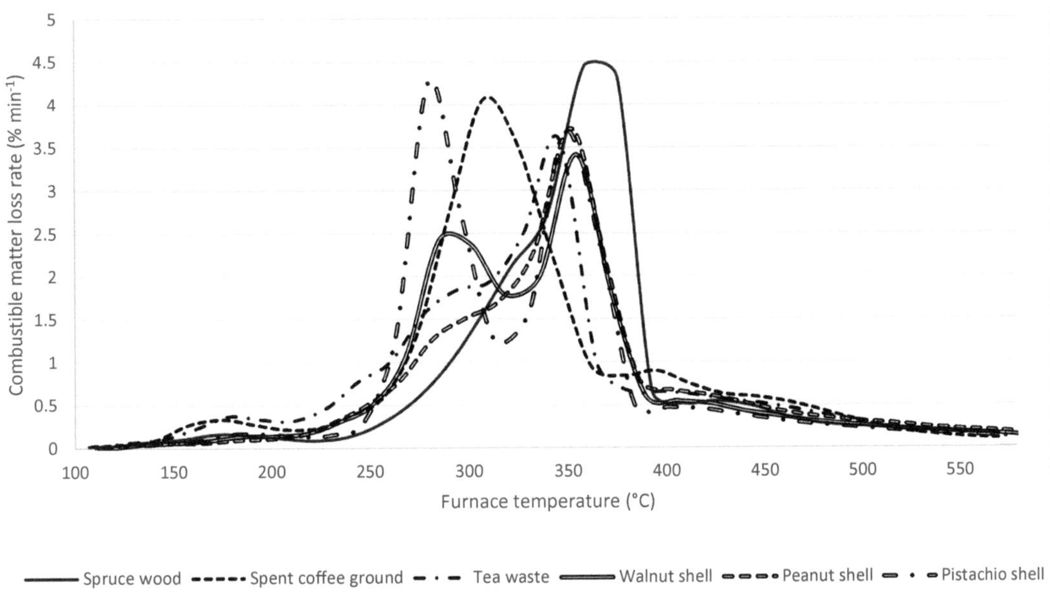

Figure 9. DTG curves of untreated biomass. The curves were converted to express the rate of loss of combustible matter, i.e., dry, ash-free state.

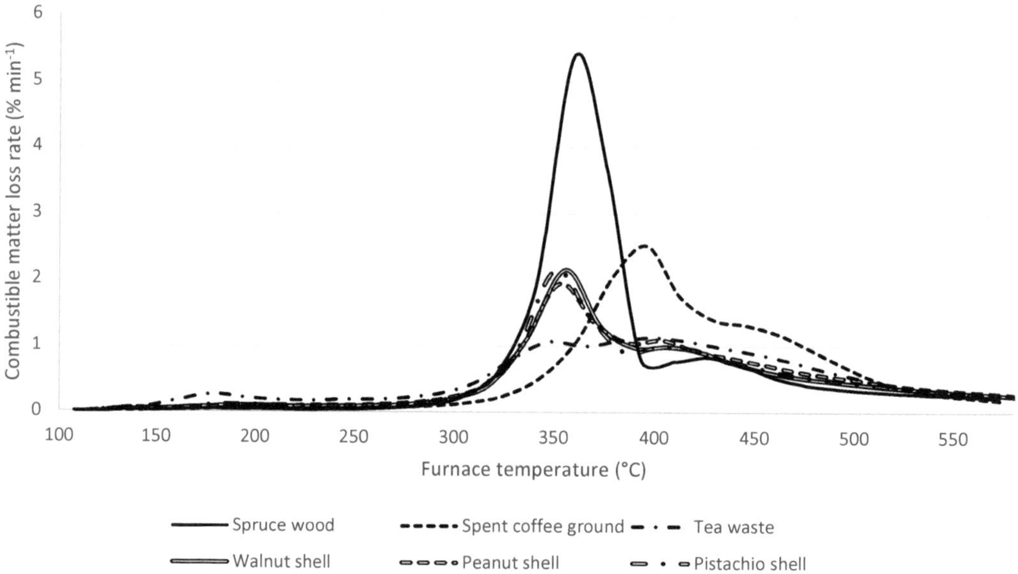

Figure 10. DTG curves of samples treated at 300 °C. The curves were converted to express the rate of loss of combustible matter, i.e., dry, ash-free state.

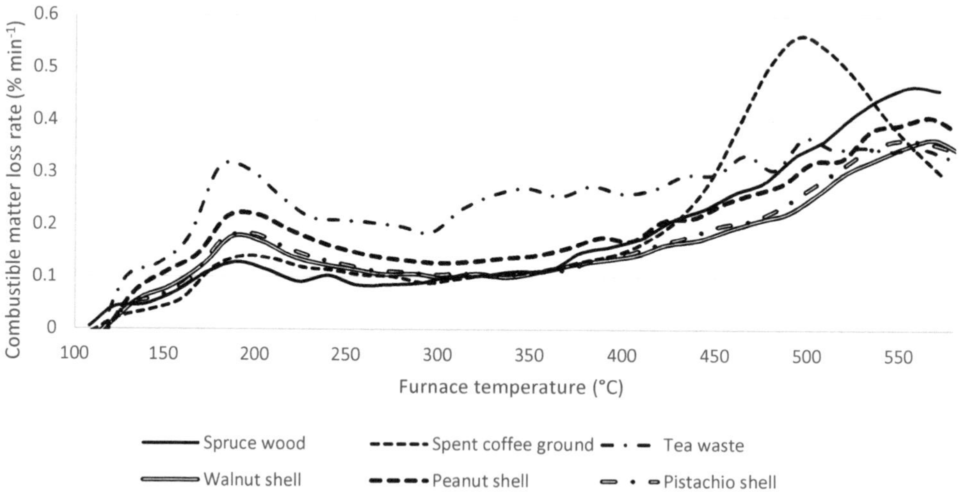

Figure 11. DTG curves of samples treated at 400 °C. The curves were converted to express the rate of loss of combustible matter, i.e., the dry, ash-free state.

4. Conclusions

This study underscores the transformative potential of pyrolyzed biomass derived from daily residues and waste materials, such as spruce wood, spent coffee grounds, tea waste, walnut shells, pistachio shells, and peanut shells, as sustainable and high-calorific-value substitutes for coal. By harnessing these underutilized agro-waste resources, which often end up in mixed municipal waste landfills, this approach not only provides a renewable energy solution, but also promotes circular economy principles. These raw materials demonstrate significant promise for local energy applications, ensuring long-term sustainability while maintaining critical quality metrics, such as calorific value and elemental composition. This work reinforces the viability of integrating pyrolyzed biomass into energy systems to reduce dependence on fossil fuels and meet decarbonization goals effectively.

The results demonstrate that pyrolysis significantly enhances the net calorific value of these materials, with optimal temperatures varying based on the feedstock. For instance, spruce wood achieves a net calorific value of 31.56 $MJ \cdot kg^{-1}$ at 550 °C, while spent coffee grounds reach their peak at 31.26 $MJ \cdot kg^{-1}$ at 350 °C. However, higher temperatures also result in greater mass loss, reducing the overall energy retained per unit of original biomass. For example, spruce wood retains only 21.84% of its original mass at 550 °C, while spent coffee grounds retain 37.53% at 350 °C, highlighting a critical trade-off between maximizing energy density and preserving material mass. These findings emphasize the importance of determining material-specific pyrolysis conditions to optimize both energy yield and resource efficiency.

Economically, pyrolyzed biomass offers a cost-effective alternative to fossil fuels when factoring in emission allowances, which significantly increase the cost of coal. The decentralized production of biochar, particularly near waste generation sites, not only reduces logistical and transportation costs, but also supports local economies and waste management strategies. Furthermore, the use of renewable waste materials aligns with circular economy principles, reducing environmental impacts and diversifying energy sources.

In summary, integrating pyrolyzed biofuels into the Czech Republic's energy framework can play a pivotal role in achieving EU decarbonization targets. By leveraging local biomass resources and prioritizing decentralized energy production, this approach addresses logistical, economic, and environmental challenges, advancing the transition toward a more sustainable and resilient energy system.

Author Contributions: Conceptualization, L.J. and M.N.; methodology, J.M. and J.V.; validation, J.V. and J.M.; formal analysis, L.J.; investigation, L.J. and M.N.; resources, L.J. and M.N.; data curation, L.J.; writing—original draft preparation, L.J.; writing—review and editing, J.M., L.J., J.V. and M.N.; visualization, L.J.; supervision, J.M.; project administration, L.J.; funding acquisition, L.J. All authors have read and agreed to the published version of the manuscript.

Funding: This study was supported by the Internal Grant Agency of the Faculty of Engineering, Czech University of Life Science Prague, grant number: 2022:31170/1312/3104.

Institutional Review Board Statement: Not applicable.

Informed Consent Statement: Not applicable.

Data Availability Statement: The raw data supporting the conclusions of this article will be made available by the authors on request.

Conflicts of Interest: The authors declare no conflict of interest.

References

1. European Environment Agency Closing the Loop—An EU Action Plan for the Circular Economy COM/2015/0614 Final—European Environment Agency. Available online: https://www.eea.europa.eu/policy-documents/com-2015-0614-final (accessed on 28 August 2022).
2. Okedu, K.; Al Senaidi, A.S.; Al Hajri, I.; Al Rashdi, I.; Salmani, W. Al Real Time Dynamic Analysis of Solar PV Integration for Energy Optimization. *Int. J. Smart Grid* **2020**, *4*, 68–79. [CrossRef]
3. Lubwama, M.; Yiga, V.A. Development of Groundnut Shells and Bagasse Briquettes as Sustainable Fuel Sources for Domestic Cooking Applications in Uganda. *Renew Energy* **2017**, *111*, 532–542. [CrossRef]
4. Lubwama, M.; Yiga, V.A. Characteristics of Briquettes Developed from Rice and Coffee Husks for Domestic Cooking Applications in Uganda. *Renew Energy* **2018**, *118*, 43–55. [CrossRef]

5. Ujjinappa, S.; Sreepathi, L.K. Production and Quality Testing of Fuel Briquettes Made from Pongamia and Tamarind Shell. *Sadhana—Acad. Proc. Eng. Sci.* **2018**, *43*, 58. [CrossRef]
6. IEA Technology Roadmap—Delivering Sustainable Bioenergy—Analysis—IEA. Available online: https://www.iea.org/reports/technology-roadmap-delivering-sustainable-bioenergy (accessed on 1 September 2022).
7. Garrido, M.A.; Conesa, J.A.; Garcia, M.D. Characterization and Production of Fuel Briquettes Made from Biomass and Plastic Wastes. *Energies* **2017**, *10*, 850. [CrossRef]
8. Variny, M.; Varga, A.; Rimár, M.; Janošovský, J.; Kizek, J.; Lukáč, L.; Jablonský, G.; Mierka, O. Advances in Biomass Co-Combustion with Fossil Fuels in the European Context: A Review. *Processes* **2021**, *9*, 100. [CrossRef]
9. Vasileiadou, A. Reduction of CO_2 Emissions Through the Co-Combustion of Lignite with Biomass Residues: Renewable and Non-Renewable CO_2 per Produced Megajoule and Fuel Characterization. *Therm. Sci. Eng. Prog.* **2024**, *50*, 102566. [CrossRef]
10. Purohit, P.; Chaturvedi, V. Biomass Pellets for Power Generation in India: A Techno-Economic Evaluation. *Environ. Sci. Pollut. Res.* **2018**, *25*, 29614–29632. [CrossRef]
11. Falup, O.; Mircea, I.; Ivan, R.; Ionel, I. Novel Approach for the Current State of Greenhouse Gases Emissions. *Rom. Case Study. J. Environ. Prot. Ecol.* **2014**, *15*, 807–818.
12. Lee, M.; Zhang, N. Technical Efficiency, Shadow Price of Carbon Dioxide Emissions, and Substitutability for Energy in the Chinese Manufacturing Industries. *Energy Econ* **2012**, *34*, 1492–1497. [CrossRef]
13. Demirbas, A. Potential Applications of Renewable Energy Sources, Biomass Combustion Problems in Boiler Power Systems and Combustion Related Environmental Issues. *Prog Energy Combust. Sci.* **2005**, *31*, 171–192. [CrossRef]
14. Goyal, H.B.; Seal, D.; Saxena, R.C. Bio-Fuels from Thermochemical Conversion of Renewable Resources: A Review. *Renew. Sustain. Energy Rev.* **2008**, *12*, 504–517. [CrossRef]
15. Gürdil, G.A.K.; Selvi, K.C.; Malaták, J.; Pinar, Y. Biomass Utilization for Thermal Energy. *AMA Agric. Mech. Asia Afr. Lat. Am.* **2009**, *20*, 80–85.
16. Mohan, D.; Pittman, C.U.; Steele, P.H. Pyrolysis of Wood/Biomass for Bio-Oil: A Critical Review. *Energy Fuels* **2006**, *20*, 848–889. [CrossRef]
17. Key Advances in Biochar Research from 2024—Biochar Today. Available online: https://biochartoday.com/2024/12/20/key-advances-in-biochar-research-from-2024/?utm_source=chatgpt.com (accessed on 16 January 2025).
18. Rajput, V.; Saini, I.; Parmar, S.; Pundir, V.; Kumar, V.; Kumar, V.; Naik, B.; Rustagi, S. Biochar Production Methods and Their Transformative Potential for Environmental Remediation. *Discov. Appl. Sci.* **2024**, *6*, 408. [CrossRef]
19. Wu, P.; Fu, Y.; Vancov, T.; Wang, H.; Wang, Y.; Chen, W. Analyzing the Trends and Hotspots of Biochar's Applications in Agriculture, Environment, and Energy: A Bibliometrics Study for 2022 and 2023. *Biochar* **2024**, *6*, 78. [CrossRef]
20. Šafařík, D.; Hlaváčková, P.; Michal, J. Potential of Forest Biomass Resources for Renewable Energy Production in the Czech Republic. *Energies* **2022**, *15*, 47. [CrossRef]
21. Rečka, L.; Ščasný, M. Brown Coal and Nuclear Energy Deployment: Effects on Fuel-Mix, Carbon Targets, and External Costs in the Czech Republic up to 2050. *Fuel* **2018**, *216*, 494–502. [CrossRef]
22. Červenka, J.; Bače, R.; Brůna, J.; Wild, J.; Svoboda, M.; Heurich, M. Mapping of Mountain Temperate Forest Recovery After Natural Disturbance: A Large Permanent Plot Established on Czech-German Border. *Silva Gabreta* **2019**, *25*, 492–510.
23. Purwestri, R.C.; Hájek, M.; Hochmalová, M.; Palátová, P.; Huertas-Bernal, D.C.; Garciá-Jácome, S.P.; Jarský, V.; Kašpar, J.; Riedl, M.; Marušák, R. The Role of Bioeconomy in the Czech National Forest Strategy: A Comparison with Sweden. *Int. For. Rev.* **2022**, *23*, 31–41. [CrossRef]
24. Hlásny, T.; Zimová, S.; Merganičová, K.; Štěpánek, P.; Modlinger, R.; Turčáni, M. Devastating Outbreak of Bark Beetles in the Czech Republic: Drivers, Impacts, and Management Implications. *For. Ecol. Manag.* **2021**, *490*, 119075. [CrossRef]
25. Cheng, T.; Veselská, T.; Křížková, B.; Švec, K.; Havlíček, V.; Stadler, M.; Kolařík, M. Insight into the Genomes of Dominant Yeast Symbionts of European Spruce Bark Beetle, *Ips typographus*. *Front. Microbiol.* **2023**, *14*, 1108975. [CrossRef]
26. Blake, M.; Straw, N.; Kendall, T.; Whitham, T.; Manea, I.A.; Inward, D.; Jones, B.; Hazlitt, N.; Ockenden, A.; Deol, A.; et al. Recent Outbreaks of the Spruce Bark Beetle *Ips typographus* in the UK: Discovery, Management, and Implications. *Trees For. People* **2024**, *16*, 100508. [CrossRef]
27. Netherer, S.; Schebeck, M.; Morgante, G.; Rentsch, V.; Kirisits, T. European Spruce Bark Beetle, *Ips typographus* (L.) Males Are Attracted to Bark Cores of Drought-Stressed Norway Spruce Trees with Impaired Defenses in Petri Dish Choice Experiments. *Forests* **2022**, *13*, 537. [CrossRef]
28. Maitah, M.; Toth, D.; Malec, K.; Appiah-Kubi, S.N.K.; Maitah, K.; Paňka, D.; Prus, P.; Janků, J.; Romanowski, R. The Impacts of Calamity Logging on the Sustainable Development of Spruce Fuel Biomass Prices and Spruce Pulp Prices in the Czech Republic. *Forests* **2022**, *13*, 97. [CrossRef]
29. Eurostat Forests, Forestry and Logging—Statistics Explained. Available online: https://ec.europa.eu/eurostat/statistics-explained/index.php?title=Forests,_forestry_and_logging#Forests_in_the_EU (accessed on 30 September 2022).

30. Nosek, R.; Tun, M.M.; Juchelkova, D. Energy Utilization of Spent Coffee Grounds in the Form of Pellets. *Energies* **2020**, *13*, 1235. [CrossRef]
31. Chen, Y.C.; Jhou, S.Y. Integrating Spent Coffee Grounds and Silver Skin as Biofuels Using Torrefaction. *Renew Energy* **2020**, *148*, 275–283. [CrossRef]
32. Cardarelli, A.; Pinzi, S.; Barbanera, M. Effect of Torrefaction Temperature on Spent Coffee Grounds Thermal Behaviour and Kinetics. *Renew Energy* **2022**, *185*, 704–716. [CrossRef]
33. Malaťák, J.; Passian, L. Heat-Emission Analysis of Small Combustion Equipments for Biomass. *Res. Agric. Eng.* **2011**, *57*, 37–50. [CrossRef]
34. Malaták, J.; Velebil, J.; Bradna, J.; Gendek, A.; Tamelová, B. Evaluation of Co and NoxEmissions in Real-Life Operating Conditions of Herbaceous Biomass Briquettes Combustion. *Acta Technol. Agric.* **2020**, *23*, 53–59. [CrossRef]
35. Tamelová, B.; Malaťák, J.; Velebil, J.; Gendek, A.; Aniszewska, M. Impact of Torrefaction on Fuel Properties of Aspiration Cleaning Residues. *Materials* **2022**, *15*, 6949. [CrossRef] [PubMed]
36. Oldfield, T.L.; Sikirica, N.; Mondini, C.; López, G.; Kuikman, P.J.; Holden, N.M. Biochar, Compost and Biochar-Compost Blend as Options to Recover Nutrients and Sequester Carbon. *J. Environ. Manag.* **2018**, *218*, 465–476. [CrossRef] [PubMed]
37. Sang, F.; Yin, Z.; Wang, W.; Almatrafi, E.; Wang, Y.; Zhao, B.; Gong, J.; Zhou, C.; Zhang, C.; Zeng, G.; et al. Degradation of Ciprofloxacin Using Heterogeneous Fenton Catalysts Derived from Natural Pyrite and Rice Straw Biochar. *J. Clean. Prod.* **2022**, *378*, 134459. [CrossRef]
38. Onsree, T.; Tippayawong, N.; Phithakkitnukoon, S.; Lauterbach, J. Interpretable Machine-Learning Model with a Collaborative Game Approach to Predict Yields and Higher Heating Value of Torrefied Biomass. *Energy* **2022**, *249*, 123676. [CrossRef]
39. Ni, Z.; Bi, H.; Jiang, C.; Wang, C.; Tian, J.; Zhou, W.; Sun, H.; Lin, Q. Investigation of the Co-Pyrolysis of Coal Slime and Coffee Industry Residue Based on Machine Learning Methods and TG-FTIR: Synergistic Effect, Kinetics and Thermodynamic. *Fuel* **2021**, *305*, 121527. [CrossRef]
40. Wang, X.; Deng, S.; Tan, H.; Adeosun, A.; Vujanović, M.; Yang, F.; Duić, N. Synergetic Effect of Sewage Sludge and Biomass Co-Pyrolysis: A Combined Study in Thermogravimetric Analyzer and a Fixed Bed Reactor. *Energy Convers Manag.* **2016**, *118*, 399–405. [CrossRef]
41. Wang, L.; Xie, L.; Ma, H.; Zhou, J. Co-Pyrolysis of Pine Sawdust with Nickel Formate for Understanding Interaction Mechanisms and Enhancing Resistance Toward Biochar Deposition. *J. Mater. Res. Technol.* **2022**, *18*, 3751–3763. [CrossRef]
42. Gendek, A.; Aniszewska, M.; Owoc, D.; Tamelová, B.; Malaťák, J.; Velebil, J.; Krilek, J. Physico-Mechanical and Energy Properties of Pellets Made from Ground Walnut Shells, Coniferous Tree Cones and Their Mixtures. *Renew Energy* **2023**, *211*, 248–258. [CrossRef]
43. Singh, E.; Mishra, R.; Kumar, A.; Shukla, S.K.; Lo, S.L.; Kumar, S. Circular Economy-Based Environmental Management Using Biochar: Driving Towards Sustainability. *Process Saf. Environ. Prot.* **2022**, *163*, 585–600. [CrossRef]
44. Meyer, S.; Glaser, B.; Quicker, P. Technical, Economical, and Climate-Related Aspects of Biochar Production Technologies: A Literature Review. *Environ. Sci. Technol.* **2011**, *45*, 9473–9483. [CrossRef]
45. Vuppaladadiyam, A.K.; Vuppaladadiyam, S.S.V.; Sahoo, A.; Murugavelh, S.; Anthony, E.; Bhashkar, T.; Zheng, Y.; Zhao, M.; Duan, H.; Zhao, Y.; et al. Bio-Oil and Biochar from the Pyrolytic Conversion of Biomass: A Current and Future Perspective on the Trade-Off Between Economic, Environmental, and Technical Indicators. *Sci. Total Environ.* **2023**, *857*, 159155. [CrossRef] [PubMed]
46. Majumder, S.; Neogi, S.; Dutta, T.; Powel, M.A.; Banik, P. The Impact of Biochar on Soil Carbon Sequestration: Meta-Analytical Approach to Evaluating Environmental and Economic Advantages. *J. Environ. Manag.* **2019**, *250*, 109466. [CrossRef]
47. Lisowski, A.; Matkowski, P.; Dąbrowska, M.; Piątek, M.; Świętochowski, A.; Klonowski, J.; Mieszkalski, L.; Reshetiuk, V. Particle Size Distribution and Physicochemical Properties of Pellets Made of Straw, Hay, and Their Blends. *Waste Biomass Valorization* **2020**, *11*, 63–75. [CrossRef]
48. Tripathi, M.; Sahu, J.N.; Ganesan, P. Effect of Process Parameters on Production of Biochar from Biomass Waste Through Pyrolysis: A Review. *Renew. Sustain. Energy Rev.* **2016**, *55*, 467–481.
49. Wilk, M.; Magdziarz, A.; Kalemba, I.; Gara, P. Carbonisation of Wood Residue into Charcoal During Low Temperature Process. *Renew Energy* **2016**, *85*, 507–513. [CrossRef]
50. Aniszewska, M.; Gendek, A.; Hýsek, Š.; Malat'ák, J.; Velebil, J.; Tamelová, B. Changes in the Composition and Surface Properties of Torrefied Conifer Cones. *Materials* **2020**, *13*, 5660. [CrossRef]
51. *The European Parliament and the Council of the European Union Directive (EU) 2023/2413 of the European Parliament and of the Council of 18 October 2023 Amending Directive (EU) 2018/2001, Regulation (EU) 2018/1999 and Directive 98/70/EC as Regards the Promotion of Energy from Renewable Sources, and Repealing Council Directive (EU) 2015/652*; European Parliament: Bruxelles, Belgium, 2023.
52. Jeníček, L.; Tunklová, B.; Malaťák, J.; Neškudla, M.; Velebil, J. Use of Spent Coffee Ground as an Alternative Fuel and Possible Soil Amendment. *Materials* **2022**, *15*, 6722. [CrossRef]
53. Jenicek, L.; Neskudla, M.; Malatak, J.; Velebil, J.; Passian, L. Spruce and Barley Elemental and Stochiometric Analysis Affected by the Impact of Pellet Production and Torrefaction. *Acta Technol. Agric.* **2021**, *24*, 166–172. [CrossRef]

54. Tunklová, B.; Jeníček, L.; Malaťák, J.; Neškudla, M.; Velebil, J.; Hnilička, F. Properties of Biochar Derived from Tea Waste as an Alternative Fuel and Its Effect on Phytotoxicity of Seed Germination for Soil Applications. *Materials* **2022**, *15*, 8709. [CrossRef]
55. Jeníček, L.; Tunklová, B.; Malaťák, J.; Velebil, J.; Malaťáková, J.; Neškudla, M.; Hnilička, F. The Impact of Nutshell Biochar on the Environment as an Alternative Fuel or as a Soil Amendment. *Materials* **2023**, *16*, 2074. [CrossRef]
56. Chen, W.H.; Kuo, P.C. Torrefaction and Co-Torrefaction Characterization of Hemicellulose, Cellulose and Lignin as Well as Torrefaction of Some Basic Constituents in Biomass. *Energy* **2011**, *36*, 803–811. [CrossRef]
57. Chen, W.H.; Lin, B.J.; Lin, Y.Y.; Chu, Y.S.; Ubando, A.T.; Show, P.L.; Ong, H.C.; Chang, J.S.; Ho, S.H.; Culaba, A.B.; et al. Progress in Biomass Torrefaction: Principles, Applications and Challenges. *Prog. Energy Combust. Sci.* **2021**, *82*. [CrossRef]
58. Lesy ČR O Dřevě | Lesy České Republiky, s. p. Available online: https://lesycr.cz/drevo/ (accessed on 30 September 2022).
59. International Coffee Organization—What's New. Available online: https://www.ico.org/ (accessed on 26 August 2022).
60. Rajesh Banu, J.; Yukesh Kannah, R.; Dinesh Kumar, M.; Preethi; Kavitha, S.; Gunasekaran, M.; Zhen, G.; Awasthi, M.K.; Kumar, G. Spent Coffee Grounds Based Circular Bioeconomy: Technoeconomic and Commercialization Aspects. *Renew. Sustain. Energy Rev.* **2021**, *152*, 111721. [CrossRef]
61. ICO International Coffee Organization—Trade Statistics Tables. Available online: https://www.icocoffee.org/documents/cy2024-25/coffee-development-report-2022-23.pdf (accessed on 26 August 2022).
62. ČSÚ Spotřeba Potravin—2020 | ČSÚ. Available online: https://www.czso.cz/csu/czso/spotreba-potravin (accessed on 1 October 2022).
63. Campos-Vega, R.; Loarca-Piña, G.; Vergara-Castañeda, H.A.; Oomah, B.D. Spent Coffee Grounds: A Review on Current Research and Future Prospects. *Trends Food Sci. Technol.* **2015**, *45*, 24–36. [CrossRef]
64. Tokimoto, T.; Kawasaki, N.; Nakamura, T.; Akutagawa, J.; Tanada, S. Removal of Lead Ions in Drinking Water by Coffee Grounds as Vegetable Biomass. *J. Colloid. Interface Sci.* **2005**, *281*, 56–61. [CrossRef] [PubMed]
65. Mussatto, S.I.; Machado, E.M.S.; Martins, S.; Teixeira, J.A. Production, Composition, and Application of Coffee and Its Industrial Residues. *Food Bioprocess Technol.* **2011**, *4*, 661–672. [CrossRef]
66. Vivek, V. Global Market Report: Tea | International Institute for Sustainable Development. Available online: https://www.iisd.org/publications/report/global-market-report-tea (accessed on 1 October 2022).
67. Doublet, G.; Jungbluth, N. Life Cycle Assessement of Drinking Darjeeling Tea. In *Conventional and Organic Darjeeling Tea*; ESU-services Ltd.: Uster, Switzerland, 2010.
68. Cabrera, C.; Artacho, R.; Giménez, R. Beneficial Effects of Green Tea—A Review. *J. Am. Coll. Nutr.* **2006**, *25*, 79–99. [CrossRef]
69. Akbayrak, S.; Özçifçi, Z.; Tabak, A. Activated Carbon Derived from Tea Waste: A Promising Supporting Material for Metal Nanoparticles Used as Catalysts in Hydrolysis of Ammonia Borane. *Biomass Bioenergy* **2020**, *138*, 105589. [CrossRef]
70. Taşar, Ş. Thermal Conversion Behavior of Cellulose and Hemicellulose Fractions Isolated from Tea Leaf Brewing Waste: Kinetic and Thermodynamic Evaluation. *Biomass Convers Biorefin.* **2022**, *12*, 2935–2947. [CrossRef]
71. Xu, J.; Wang, M.; Zhao, J.; Wang, Y.H.; Tang, Q.; Khan, I.A. Yellow Tea (*Camellia Sinensis* L.), a Promising Chinese Tea: Processing, Chemical Constituents and Health Benefits. *Food Res. Int.* **2018**, *107*, 567–577. [CrossRef]
72. Mizuno, S.; Ida, T.; Fuchihata, M.; Namba, K. Effect of Specimen Size on Ultimate Compressive Strength of Bio-Coke Produced from Green Tea Grounds. *Mech. Eng. J.* **2016**, *3*, 15–00441. [CrossRef]
73. Pua, F.L.; Subari, M.S.; Ean, L.W.; Krishnan, S.G. Characterization of Biomass Fuel Pellets Made from Malaysia Tea Waste and Oil Palm Empty Fruit Bunch. *Mater. Today Proc.* **2020**, *31*, 187–190. [CrossRef]
74. Intagun, W.; Kanoksilapatham, W.; Maden, A.; Nobaew, B. Effect of Natural Additive on Pellets Physical Properties and Energy Cost. In Proceedings of the 2019 IEEE 2nd International Conference on Renewable Energy and Power Engineering, REPE 2019, Toronto, ON, Canada, 2–4 November 2019; pp. 130–134. [CrossRef]
75. Zhang, J.; Guo, Y. Physical Properties of Solid Fuel Briquettes Made from *Caragana korshinskii* Kom. *Powder Technol.* **2014**, *256*, 293–299. [CrossRef]
76. Zhang, L.; Xu, C.C.; Lei, H.; Wang, H.L.; Ning, T.T.; Hao, W.; Hu, X.D. Effects of Addition of Various Ingredients During Pelletizing on Physical Characteristics of Green Tea Residue Pellets. *Appl. Eng. Agric.* **2014**, *30*, 49–53. [CrossRef]
77. McCaffrey, Z.; Torres, L.; Sen Chiou, B.; Ferreira, S.R.; Silva, L.E.; Wood, D.F.; Orts, W.J. Torrefaction of Almond and Walnut Byproducts. *Front. Energy Res.* **2021**, *9*, 91. [CrossRef]
78. Almond Board of California Crop Reports | Almond Almanac Report. Available online: https://www.almonds.com/tools-and-resources/crop-reports/almond-almanac (accessed on 1 September 2022).
79. 2021 California Agricultural Statistics Review. Available Online | California Avocado Commission. Available online: https://www.californiaavocadogrowers.com/articles/2021-california-agricultural-statistics-review-available-online (accessed on 1 September 2022).
80. Espina, R.U.; Barroca, R.B.; Abundo, M.L.S. Proximate Analysis of the Torrefied Coconut Shells. *Int. J. Renew. Energy Res. IJRER* **2022**, *12*, 489–494. [CrossRef]

81. Silva, M.P.; Nieva Lobos, M.L.; Piloni, R.V.; Dusso, D.; González Quijón, M.E.; Scopel, A.L.; Moyano, E.L. Pyrolytic Biochars from Sunflower Seed Shells, Peanut Shells and Spirulina Algae: Their Potential as Soil Amendment and Natural Growth Regulators. *SN Appl. Sci.* **2020**, *2*, 1926. [CrossRef]
82. Ahmad, M.; Lee, S.S.; Dou, X.; Mohan, D.; Sung, J.K.; Yang, J.E.; Ok, Y.S. Effects of Pyrolysis Temperature on Soybean Stover-and Peanut Shell-Derived Biochar Properties and TCE Adsorption in Water. *Bioresour. Technol.* **2012**, *118*, 36–54. [CrossRef]
83. Apaydin-Varol, E.; Pütün, E.; Pütün, A.E. Slow Pyrolysis of Pistachio Shell. *Fuel* **2007**, *86*, 1892–1899. [CrossRef]
84. Yuan, H.R.; Liu, R.H. Study on Pyrolysis Kinetics of Walnut Shell. *J. Therm. Anal. Calorim.* **2007**, *89*, 983–986. [CrossRef]
85. Açıkalın, K. Thermogravimetric Analysis of Walnut Shell as Pyrolysis Feedstock. *J. Therm. Anal. Calorim.* **2011**, *105*, 145–150. [CrossRef]
86. Wang, Q.; Sarkar, J. Pyrolysis Behaviors of Waste Coconut Shell and Husk Biomasses. *Int. J. Energy Prod. Manag.* **2018**, *3*, 34–43. [CrossRef]
87. Putri, R.W.; Rahmatullah; Haryati, S.; Santoso, B.; Hadi, A.A. The Residence Time and Slow Pyrolysis Temperature Effect on Chemical Composition Pyrolysis Gas Product of Durian (*Durio zibethinus* Murr) Skin. *Chem. Eng. Trans.* **2022**, *97*, 247–252. [CrossRef]
88. Trubetskaya, A.; Grams, J.; Leahy, J.J.; Johnson, R.; Gallagher, P.; Monaghan, R.F.D.; Kwapinska, M. The Effect of Particle Size, Temperature and Residence Time on the Yields and Reactivity of Olive Stones from Torrefaction. *Renew Energy* **2020**, *160*, 998–1011. [CrossRef]
89. *ISO 18125:2017*; Solid Biofuels—Determination of Calorific Value. BSI: London, UK, 2017. Available online: https://www.iso.org/standard/61517.html (accessed on 16 January 2025).
90. *ISO 18122:2022*; Solid Biofuels—Determination of Ash Content. ISO: Geneva, Switzerland, 2022. Available online: https://www.iso.org/standard/83190.html (accessed on 16 January 2025).
91. *ISO 18134-2:2024*; Solid Biofuels—Determination of Moisture Content—Part 2: Simplified Method. ISO: Geneva, Switzerland, 2024. Available online: https://www.iso.org/standard/86024.html (accessed on 16 January 2025).
92. Recent Developments in Biomass Pelletization—A Review: BioResources. Available online: https://bioresources.cnr.ncsu.edu/resources/recent-developments-in-biomass-pelletization-a-review/ (accessed on 8 June 2024).
93. Malaťák, J.; Jankovský, M.; Malaťáková, J.; Velebil, J.; Gendek, A.; Aniszewska, M. Substituting Solid Fossil Fuels with Torrefied Timber Products. *Materials* **2023**, *16*, 7569. [CrossRef]
94. Arias, B.; Pevida, C.; Fermoso, J.; Plaza, M.G.; Rubiera, F.; Pis, J.J. Influence of Torrefaction on the Grindability and Reactivity of Woody Biomass. *Fuel Process. Technol.* **2008**, *89*, 169–175. [CrossRef]
95. Fermoso, J.; Mašek, O. Thermochemical Decomposition of Coffee Ground Residues by TG-MS: A Kinetic Study. *J. Anal. Appl. Pyrolysis.* **2018**, *130*, 249–255. [CrossRef]
96. Mašek, O.; Brownsort, P.; Cross, A.; Sohi, S. Influence of Production Conditions on the Yield and Environmental Stability of Biochar. *Fuel* **2013**, *103*, 151–155. [CrossRef]
97. Mayson, S.; Williams, I.D. Applying a Circular Economy Approach to Valorize Spent Coffee Grounds. *Resour. Conserv. Recycl.* **2021**, *172*, 105659. [CrossRef]
98. Silva, M.A.; Nebra, S.A.; Machado Silva, M.J.; Sanchez, C.G. The Use of Biomass Residues in the Brazilian Soluble Coffee Industry. *Biomass Bioenergy* **1998**, *14*, 457–467. [CrossRef]
99. Colantoni, A.; Paris, E.; Bianchini, L.; Ferri, S.; Marcantonio, V.; Carnevale, M.; Palma, A.; Civitarese, V.; Gallucci, F. Spent Coffee Ground Characterization, Pelletization Test and Emissions Assessment in the Combustion Process. *Sci. Rep.* **2021**, *11*, 5119. [CrossRef]
100. Sermyagina, E.; Saari, J.; Kaikko, J.; Vakkilainen, E. Integration of Torrefaction and CHP Plant: Operational and Economic Analysis. *Appl. Energy* **2016**, *183*, 88–99. [CrossRef]
101. Vassilev, S.V.; Baxter, D.; Andersen, L.K.; Vassileva, C.G. An Overview of the Chemical Composition of Biomass. *Fuel* **2010**, *89*, 913–933. [CrossRef]
102. Gendek, A.; Piętka, J.; Aniszewska, M.; Malaťák, J.; Velebil, J.; Tamelová, B.; Krilek, J.; Moskalik, T. Energy Value of Silver Fir (*Abies alba*) and Norway Spruce (*Picea abies*) Wood Depending on the Degree of Its Decomposition by Selected Fungal Species. *Renew Energy* **2023**, *215*, 118948. [CrossRef]
103. Kovalcik, A.; Obruca, S.; Marova, I. Valorization of Spent Coffee Grounds: A Review. *Food Bioprod. Process.* **2018**, *110*, 104–119.
104. McNutt, J.; He, Q. (Sophia) Spent Coffee Grounds: A Review on Current Utilization. *J. Ind. Eng. Chem.* **2019**, *71*, 78–88.
105. Nepal, R.; Kim, H.J.; Poudel, J.; Oh, S.C. A Study on Torrefaction of Spent Coffee Ground to Improve Its Fuel Properties. *Fuel* **2022**, *318*, 123643. [CrossRef]
106. Bejenari, V.; Marcu, A.; Ipate, A.M.; Rusu, D.; Tudorachi, N.; Anghel, I.; Şofran, I.E.; Lisa, G. Physicochemical Characterization and Energy Recovery of Spent Coffee Grounds. *J. Mater. Res. Technol.* **2021**, *15*, 4437–4451. [CrossRef]
107. Ballesteros, L.F.; Teixeira, J.A.; Mussatto, S.I. Chemical, Functional, and Structural Properties of Spent Coffee Grounds and Coffee Silverskin. *Food Bioproc. Technol.* **2014**, *7*, 3493–3503. [CrossRef]

108. Sermyagina, E.; Mendoza Martinez, C.L.; Nikku, M.; Vakkilainen, E. Spent Coffee Grounds and Tea Leaf Residues: Characterization, Evaluation of Thermal Reactivity and Recovery of High-Value Compounds. *Biomass Bioenergy* **2021**, *150*, 106141. [CrossRef]
109. Liu, Z.; Jiang, Z.; Cai, Z.; Fei, B.; Yu, Y.; Liu, X. Effects of Carbonization Conditions on Properties of Bamboo Pellets. *Renew Energy* **2013**, *51*, 1–6. [CrossRef]
110. Skanderová, K.; Malaťák, J.; Bradna, J. Energy Use of Compost Pellets for Small Combustion Plants. *Agron. Res.* **2015**, *13*, 413–419.
111. Weber, K.; Quicker, P. Properties of Biochar. *Fuel* **2018**, *217*, 240–261. [CrossRef]
112. van Loo, S.; Koppejan, J. The Handbook of Biomass Combustion and Co-Firing. In *The Handbook of Biomass Combustion and Co-Firing*; Routledge: London, UK, 2012; pp. 1–442. [CrossRef]
113. Biomasa pro Výrobu Energie—Jan Malaťák, Petr Vaculík, Brožovaná Vazba, Český Jazyk | Knihy Na Martinus.Cz. Available online: https://www.martinus.cz/?uItem=73504 (accessed on 30 September 2022).
114. Zhang, W.; Jiang, S.; Wang, K.; Wang, L.; Xu, Y.; Wu, Z.; Shao, H.; Wang, Y.; Miao, M. Thermogravimetric Dynamics and FTIR Analysis on Oxidation Properties of Low-Rank Coal at Low and Moderate Temperatures. *Int. J. Coal Prep. Util.* **2015**, *35*, 39–50. [CrossRef]
115. Haykiri-Acma, H.; Yaman, S. Synergy in Devolatilization Characteristics of Lignite and Hazelnut Shell During Co-Pyrolysis. *Fuel* **2007**, *86*, 373–380. [CrossRef]

materials

MDPI

Article

Comparative Analysis of Treatment Effects of Different Materials on Thin Oil Films

Xiuli Wu [1,*], Bo Zheng [2,*], Haiping Dai [3], Yongwen Ke [3] and Cheng Cai [3]

1 School of Environmental Science and Engineering, Tianjin University, Tianjin 300350, China
2 College of Urban and Rural Construction, Hebei Agricultural University, Baoding 071001, China
3 Shandong Jinwei Hairun Special Separation Equipment Co., Ltd., Weifang 261000, China; daihaiping2004@126.com (H.D.); keyongwen@mtshj.com (Y.K.); caicheng@mtshj.com (C.C.)
* Correspondence: wunan@mtshj.com (X.W.); zhengbo@hebau.edu.cn (B.Z.); Tel./Fax: +86-022-83595299 (X.W.); +86-021-67791327 (B.Z.)

Abstract: With the continuous and rapid development of global industries, issues such as offshore oil spills, leakage of organic chemicals, and the direct discharge of industrial oily sewage have caused serious damage to the ecological environment and water resources. Efficient oil–water separation is widely recognized as the solution. However, there is an urgent need to address the difficulties in treating thin oil films on the water surface and the low separation efficiency of existing oil–water separation materials. In view of this, this study aims to investigate high-efficiency oil–water separation materials for thin oil films. Four types of oil–water separation materials with different materials are designed to treat thin oil films on the water surface. The effects of factors such as oil film thickness, pressure, and temperature on the oil–water separation performance of these materials are studied. The viscosities of kerosene and diesel oil are tested, and the adsorption and separation effects of the oil–water separation materials on different oil products and oily organic solvents are examined. In addition, the long-term stability of the movable and portable oil–water separation components is verified. The results show that the oil-absorbing sponge-based oil–water separation membrane has an excellent microporous structure and surface roughness, endowing the membrane surface with excellent hydrophobicity and lipophilicity, and exhibiting good oil–water separation performance. The filtration flux of oil increases with the increase in pressure and temperature. It has good adsorption and separation performance for different oil products and oily organic solvents. Moreover, it maintains stable operation performance during the 12-month long-term oil–water separation process for kerosene and diesel oil.

Keywords: thin oil film; super lipophobic and hydrophobic; hollow fiber membrane; separation of oil and water; regeneration

1. Introduction

In recent years, with the rapid development of the national industrial manufacturing industry, oil pollution in water bodies has intensified. In addition to oil spills in rivers and the ocean, an increasing amount of oily industrial wastewater generated during the industrial manufacturing process is being discharged into the environment [1]. Polluted oil can form a thin oil film on the water surface that is insoluble in water [2], which not only causes damage to the ecological environment but also has a significant impact on aquatic organisms.

Oily industrial wastewater mainly comes from industrial processes such as petroleum and natural gas, mining, metal processing, chemical engineering, textile printing and dyeing, etc., [3]. Among them, the polluted oil in oily industrial wastewater mainly includes light petroleum hydrocarbons [4] (such as gasoline, kerosene, diesel oil, etc.), heavy petroleum hydrocarbon aromatic hydrocarbons [5], halogenated hydrocarbons [6] (such as benzene, toluene, chloroform, etc.) and other oily organic solvents that are insoluble or hardly soluble in water. The forms of oil in oily wastewater [7] include floating oil (particle size > 100 μm), dispersed oil (particle size 10–100 μm), emulsified oil (particle size < 10 μm), and dissolved oil (particle size < 0.1 μm). Floating oil [8] will form a thin oil film or oil layer on the surface of the water body.

The presence of the thin oil film will seriously affect the gas exchange process of the water body, leading to oxygen deficiency in the water body and causing the death of a large number of aquatic plants and animals. Moreover, the thin oil film will also affect the sunlight transmittance of the water body, weakening the photosynthesis of aquatic plants. In addition, most polluted oils are flammable or toxic substances [9]. They are not only flammable [10] but also can affect human health through biological enrichment [11], causing immeasurable long-term harm. Therefore, the oily wastewater generated in oil spill accidents or industrial processes must be properly treated. Thus, it is very necessary to find a method with high treatment efficiency and remarkable treatment effect.

Traditional treatment methods for thin oil films include biological methods, chemical methods, and physical methods. Biological methods mainly include the microbial metabolism method [11], which uses the metabolic action of oil-degrading microorganisms to convert oil pollution into the organic components of microorganisms or to proliferate into new microorganisms for purification purposes, the activated sludge method [12] that realizes purification by using suspended microbial flocs, and the biofilm treatment method [13] based on aerobic biological treatment. Chemical methods mainly refer to the coagulation/flocculation method, and the added agents as coagulating agents [14], which separates and removes the thin oil film by forming flocs. Compared with biological methods, chemical methods are simple to operate and have low operating costs. However, the oil removal effect is largely restricted by factors such as the type and dosage of the flocculant, the initial concentration of the oil, the temperature of the oily wastewater, and the pH value [15–18].

Physical methods mainly include the gravity separation method [19] that separates based on the density difference between oil and water, the adsorption method [20] that realizes separation by adsorbing the thin oil film with oil-absorbing materials, and the membrane separation method [21] that selectively separates oily wastewater by virtue of the selective permeability of the membrane. Among them, the gravity separation method and the adsorption method are slow and inefficient [22]. The membrane separation method, with its technical advantages such as energy conservation, no phase change, simple operation, and no secondary pollution, is expected to achieve efficient recovery of thin oil films [23–25].

However, membrane materials still face many challenges in practical applications. Polymer membranes have disadvantages such as weak chemical corrosion resistance, poor mechanical properties, and easy contamination [26,27]. While the brittleness, high manufacturing cost, and easy contamination of ceramic membranes are their main challenges [28,29]. In addition, the short membrane life and the trade-off between the permeation flux and the rejection rate are inevitable problems for both polymer membranes and ceramic membranes.

This study aims to develop high-efficiency oil–water separation materials for thin oil films. Four types of materials are designed, and the effects of factors like oil film thickness, pressure, and temperature on their separation performance are investigated. The viscosities of kerosene and diesel, adsorption and separation effects on different oils, and long-term stability of the separation components are also studied.

2. Experimental Section

2.1. Materials

Pure Kerosene and Pure Diesel: Provided by Sinopec (Beijing, China). Kerosene is a light petroleum hydrocarbon mainly composed of hydrocarbons with carbon numbers typically in the range of C9–C16. It has a relatively low viscosity (1.6 mPa·s at 20 °C), which is beneficial for its flow through separation materials. Diesel, also from Sinopec, is a heavier petroleum hydrocarbon mixture with carbon numbers usually in the range of C10–C22. Its higher viscosity (3.7 mPa·s at 20 °C) compared to kerosene affects its separation performance.

Waste Kerosene and Waste Diesel: Supplied by Tianjin Yahuan Recycling Resources Co., Ltd., (Tianjin, China). Their chemical compositions are similar to pure kerosene and diesel but may contain impurities from industrial use, which could impact the separation process and the performance evaluation of the separation materials.

Red Marker Sudan III: Analytical-grade, provided by Tianjin Comio Chemical Reagent Co., Ltd., (Tianjin, China). It is used to dye kerosene and diesel in the experiments for better visualization of the oil–water separation process. Chemically, Sudan III is a diazo dye.

Distilled Water: Self-made in the laboratory. It serves as the water phase in the oil–water mixture experiments, and its purity is crucial as impurities could interfere with the separation process.

2.2. Preparation Process

2.2.1. Preparation of Oil–Water Separation Materials

The experimental materials are four types of oil–water separation materials, which are, respectively: Oil-absorbent cotton (provided by Handan Hengyong Protective and Clean Products Co., Ltd., OAC, Shijiazhuang, China), oil-absorbent cotton-based oil–water separation membrane (prepared in the laboratory, Oil-absorbent cotton impregnated with graphene, OAC-G), PVC-modified oil-absorbent cotton (prepared in the laboratory, Oil absorbent cotton impregnated with PVC and graphene mixture, OAC-P/G), and double-layer oil-absorbent cotton with sponge (provided by Handan Hengyong Protective and Clean Products Co., Ltd., Double oil absorbent cotton with sponge, DOAC-S).

Among them, the oil-absorbent cotton-based oil–water separation membrane is prepared by uniformly coating PVDF nanofibers on the surface of the oil-absorbent cotton using the synchronous electrospinning–electrospraying method. The PVC-modified oil-absorbent cotton is prepared by impregnating the oil-absorbent cotton with a mixture of PVC and graphene using the impregnation method. The specific preparation steps and conditions are described in detail in our previous research [30].

The above four types of oil–water separation materials are, respectively, made into corresponding oil–water separation components, and the specifications of the components are listed in Table 1.

Table 1. Specifications of four oil–water separation components.

Code	Specification
Component length (m)	0.08
Component diameter (m)	0.014
Component area (m^2)	0.0035

2.2.2. Experimental Facility

The continuous oil–water separation test device is shown in Figure 1. At room temperature, kerosene and diesel oil dyed with Sudan III are, respectively, mixed with a certain volume of distilled water and placed in a water tank to prepare kerosene (diesel oil)/water mixtures. Different oil-absorbing membrane components are horizontally placed in this water tank so that the components float on the oil–water interface. The interfaces of the components are, respectively, connected to a pressure gauge, a valve, an oil storage tank, and a vacuum pump. The vacuum pump is started, enabling the oil-absorbing membrane components to continuously separate the oil–water mixtures, and the separated oil products are stored in the oil storage tank.

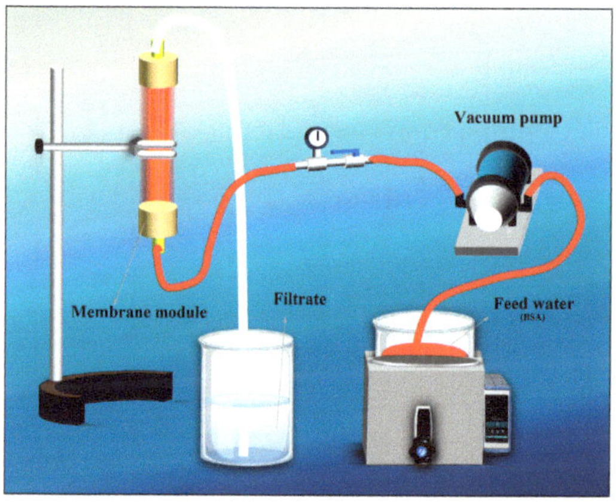

Figure 1. Schematic diagram of oil–water separation test device.

2.3. Characterization

2.3.1. Experiment on Flux Test of Oil–Water Separation Materials with Different Materials in Pure Kerosene and Pure Diesel

Under the conditions of a pressure of 0.08 MPa, a temperature of 20 °C, and a filtration time of 2 min, the fluxes of the OAC, OAC-G, OAC-P/G, and DOAC-S oil–water separation materials in pure kerosene and pure diesel oil were tested.

2.3.2. Experiments on the Influence of Thin Oil Film Thickness, Pressure, and Temperature on the Filtration Effects of Oil–Water Separation Materials with Different Materials

Thin Oil Film Thickness: Under the conditions of a pressure of 0.08 MPa and a temperature of 20 °C, the testing device is a water tank with a length × width of 0.29 m × 0.29 m. The OAC, OAC-G, OAC-P/G, and DOAC-S oil–water separation components are, respectively, immersed in the water tank. A certain amount of water is added to the water tank, and then 420.5 mL of kerosene/diesel oil is added. The oil floats on the water to form a thin oil film with a thickness of 5 mm. Every time 84.1 mL of kerosene/diesel oil is filtered out (i.e., the oil amount of a 1 mm thick thin oil film), the time is recorded. Then, the filtration times of the corresponding kerosene/diesel oil when the thicknesses of the thin oil films are 5 mm, 4 mm, 3 mm, 2 mm, and 1 mm are obtained, respectively, and the fluxes at the corresponding thin oil film thicknesses are calculated.

Pressure: An appropriate amount of kerosene (diesel oil)/water mixture is added to the water tank, and the OAC, OAC-G, OAC-P/G, and DOAC-S oil–water separation components are, respectively, immersed in the water tank. At a temperature of 20 °C and a thin oil film thickness of 5 mm, a vacuum pump is used to adjust the filtration pressure. The amounts of oil filtered out within 2 min under the conditions of pressures of 0.05 MPa,

0.06 MPa, 0.07 MPa, 0.08 MPa, and 0.09 MPa are, respectively, recorded, and the membrane filtration fluxes of the four oil–water separation materials for kerosene and diesel oil under different pressures are obtained.

Temperature: An appropriate amount of kerosene (diesel oil)/water mixture is added to the water tank, and the OAC, OAC-G, OAC-P/G, and DOAC-S oil–water separation components are, respectively, immersed in the water tank. Under the conditions of a pressure of 0.08 MPa, an oil film thickness of 5 mm, and a filtration time of 2 min, a water bath pot is used to heat and control the temperature of the oil–water mixture. The volumes of the filtered oil under the conditions of temperatures of 15 °C, 20 °C, 25 °C, 30 °C, 35 °C, and 40 °C are, respectively, measured, and the membrane filtration fluxes of the four oil–water separation materials for kerosene and diesel oil under different temperatures are obtained.

2.3.3. Static Adsorption and Dynamic Separation of Different Oil Products and Oily Organic Solvents

The viscosity of the oil products is tested by a kinematic viscosity tester for petroleum products (Model SYP1003-6B, produced by Shanghai Fuli Instrument Equipment Co., Ltd., Shanghai, China). Select OAC-G as a representative to test the static adsorption and dynamic separation effects of OAC-G on different oil products and oily organic solvents.

The different oil products and oily organic solvents for static adsorption include lubricating oil, hydraulic oil, soybean oil, palm oil, triethyl phosphate, N,N-dimethylacetamide, dimethylformamide, diesel oil, kerosene, and toluene. First, under the conditions of a temperature of 20 °C and an oil film thickness of 1 mm, OAC-G is, respectively, immersed in the above different oil products and oily organic solvents for 120 s. After taking it out, it is placed on a stainless steel filter screen. After 60 s, OAC-G is weighed, and the adsorption capacity is calculated according to the following formula:

$$Q = \frac{W_b - W_a}{W_a} \times 100\%$$

In the formula: W_a and W_b, respectively, represent the mass (g) of OAC-G before and after adsorption, and Q is the adsorption capacity (g/g) of OAC-G.

The different oil products and oily organic solvents for dynamic separation include palm oil, xylene, kerosene, diesel oil, butyl acrylate, methyl methacrylate, benzene, toluene, and nonanol. The different oil products and oily organic solvents for dynamic separation testing are all oil–water mixtures. Under the conditions of an oil film thickness of 1 mm, a temperature of 20 °C, a pressure of 0.05 MPa, and a filtration time of 2 min, the fluxes of the OAC-G oil–water separation component for the above different oil products and oily organic solvents are, respectively, tested.

2.3.4. Morphological Characterization

A field emission scanning electron microscope (Model S-4800, Hitachi, Tokyo, Japan) is selected to observe the morphology of the surface and cross-section of the materials. A sample with a length of 6 mm is cut, and the sample is pasted onto the test sample stage with conductive adhesive. After drying and sputtering with gold, the test sample is obtained. At the same time, a sample with a size of 2 mm × 2 mm is pasted onto the sample stage of an atomic force microscope (Model XE-100, Niddatal, Germany), and the test is carried out in the tapping mode. Each sample is measured three times during the test.

2.3.5. XRD Characterization

The surface chemical structure and composition of the materials are characterized by X-ray photoelectron spectroscopy. A uniform sample with a size of 5 mm × 5 mm is cut and fixed on the test sample stage of the energy spectrometer (Thermo Kalpha, Tokyo, Japan). During the test, an Al Kα light source of 1486.6 eV is selected, the binding energy internal standard correction (C1s 284.6 eV) is carried out, and a full spectrum test is performed on the sample. Each type of sample is tested three times.

2.3.6. Specific Surface Area Characterization

The specific surface area is tested by a Brunauer–Emmett–Teller (BET) specific surface area tester (BET-600, Tokyo, Japan). A sample with a size of 1 cm × 1 cm is pasted into the sample introduction unit of the specific surface area tester (SZB-9), and the test is carried out in the automatic mode. Each sample is measured five times during the test.

2.3.7. Static Water Contact Angle Characterization

The surface hydrophilicity of the samples is tested by a semi-automatic contact angle measuring instrument (Model DSA-100, Wiernsheim, Germany). The sample is fixed on a glass slide, and a program is set to drop 1 mL of distilled water onto the surface of the membrane sample. At the same time, the device automatically records the morphological changes in the water droplets and the corresponding contact angle values. Each type of sample is tested five times.

3. Results and Discussion

3.1. The Viscosities of Kerosene and Diesel Oil

Under the condition of 20 °C, the viscosity test results of kerosene and diesel oil are shown in Table 2. As can be seen from Table 2, the viscosities of kerosene and diesel oil are 1.6 mPa·s and 3.7 mPa·s, respectively. According to relevant reports, the greater the viscosity of the oil product, the greater the transmembrane pressure difference required when it passes through the oil–water separation material. Therefore, under the same pressure, the oil product with a lower viscosity has a higher membrane flux.

Table 2. The viscosity of different oils.

Oil Type	Viscosity (mPa·s)
Kerosene	1.6
Diesel	3.7

3.2. Surface Morpography Analysis

The surface morphologies and surface roughness of the four oil–water separation materials are shown in Figures 2 and 3, respectively. As can be seen from Figure 2, the surfaces of all four oil–water separation materials exhibit a microporous structure. Among them, the surface microporous structures of the two oil–water separation materials, OAC and DOAC-S, have better penetration and higher porosity. However, after the coating was performed with graphene and PVC/G, the porosity of the surface microporous structures of the oil-absorbent cotton-based oil–water separation material OAC-G and the PVC-modified oil-absorbent cotton oil–water separation material OAC-P/G decreases, and an obvious dense layer appears on the surface, but they still maintain good microporous permeability.

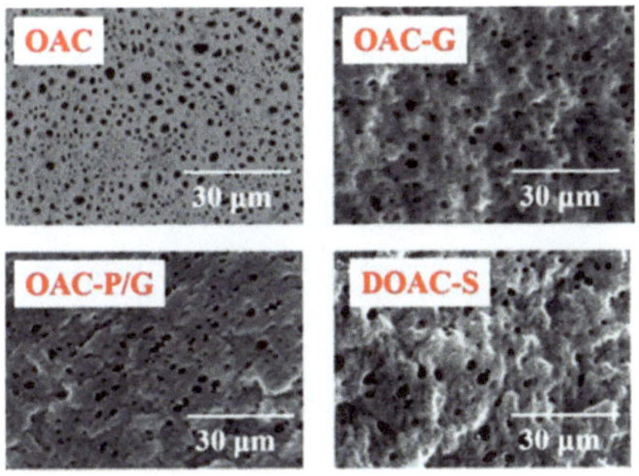

Figure 2. Surface morphologies of oil–water separation materials.

Figure 3. Surface roughness of oil–water separation material.

As can be seen from Figure 3, the surface roughness of the two oil–water separation materials, OAC-G and OAC-P/G, is significantly reduced, being 1.22 nm and 2.78 nm, respectively. This is mainly because after the sponge is impregnated with graphene, graphene has good spreadability on the sponge surface, resulting in a better overall flatness of the membrane surface. The surface roughness of the two oil-water separation materials, OAC and DOAC-S, is relatively high, being 8.07 nm and 6.24 nm, respectively.

3.3. Surface Structure and Composition Analysis

The XPS analysis results of the surfaces of the four oil–water separation materials are shown in Figure 4. As can be seen from Figure 4, characteristic signal peaks appear at the binding energies of 286 eV and 535 eV on the surfaces of all four oil–water separation materials, corresponding to the characteristic peaks of C 1s and O 1s in the structure of the oil-absorbing cotton OAC, respectively.

Compared with OAC, a characteristic signal peak appears at the binding energy of 687.3 eV for OAC-G, corresponding to the characteristic peak of F 1s in the PVDF structure, and the content of the C element on the surface is significantly reduced. The results confirm that the PVDF nanofibers are successfully coated on the surface of the oil-absorbing cotton by the synchronous electrospinning–electrospraying method.

In addition, a characteristic signal peak appears at the binding energy of 205.9 eV for OAC-P/G, corresponding to the characteristic peak of Cl 2p in the PVC structure. The results confirm that PVC and graphene are successfully coated on the surface of the oil-absorbing cotton by the impregnation method.

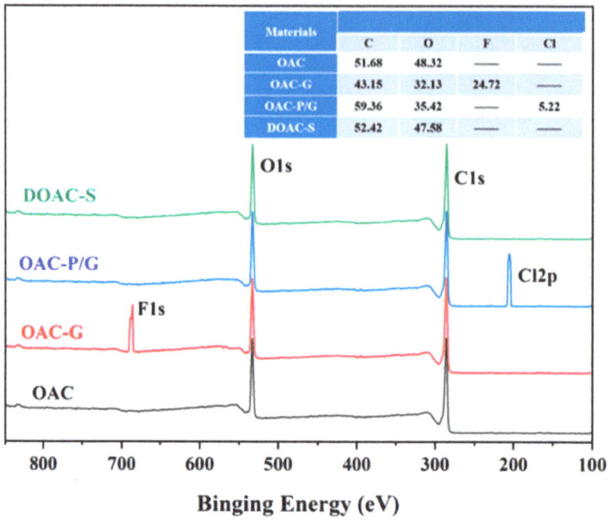

Figure 4. XPS spectrum and element composition of oil–water separation materials.

3.4. Membranes of Kerosene, Diesel Flux

The membrane flux of four oil–water separation materials in kerosene and diesel under a vacuum condition of 0.08 MPa is shown in Figure 5. As indicated in Figure 5, during the diesel test, OAC-G demonstrated the highest oil–water separation flux of 9096 L/(m^2·h), followed by DOAC-S, OAC, and OAC-P/G with fluxes of 8976 L/(m^2·h), 6827 L/(m^2·h), and 5989 L/(m^2·h), respectively. In the kerosene test, OAC-G still showed the highest flux of 17,498.00 L/(m^2·h), followed by OAC, OAC-P/G, and DOAC-S with fluxes of 14,875 L/(m^2·h), 14,487 L/(m^2·h), and 12,010 L/(m^2·h), respectively. It is evident that the highest oil–water separation flux for diesel is significantly lower than that for kerosene when using the same oil–water separation material, OAC-G. For OAC-G, the flux in the diesel test is 9096 L/(m^2·h), while in the kerosene test, it reaches 17,498.00 L/(m^2·h). This large disparity can likely be attributed to the differences in the physical properties of diesel and kerosene. Kerosene generally has a lower viscosity and surface tension compared to diesel. These properties affect the flowability of the fluids through the oil–water separation materials. With a lower viscosity and lower surface tension, kerosene can more easily pass through the pores of the material, resulting in a higher flux. In contrast, the relatively higher viscosity and surface tension of diesel may impede its flow through the material, leading to a lower separation flux. Additionally, the chemical composition of diesel and kerosene might also have an impact on their interaction with the oil–water separation materials, potentially influencing the separation efficiency and flux. This indicates that when selecting oil–water separation materials for different oil types, the physical and chemical properties of the oils need to be carefully considered to optimize the separation performance. Comparison shows that OAC-G exhibited superior separation performance for both kerosene and diesel compared to other materials. This may be attributed to the modified oil-absorbing cotton-based material having a smoother surface and larger specific surface area (Figure 6, OAC-G specific surface area is 2733 m^2/g, while OAC-P/G, DOAC-S, and OAC show values of 1815 m^2/g, 1763 m^2/g, and 1554 m^2/g, respectively), leading to better wettability towards kerosene and diesel and thus faster filtration rates.

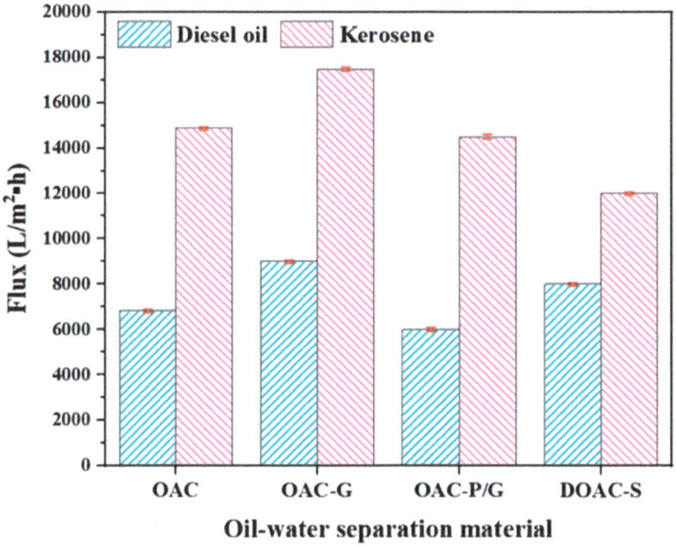

Figure 5. Kerosene, diesel flux of oil–water separation material.

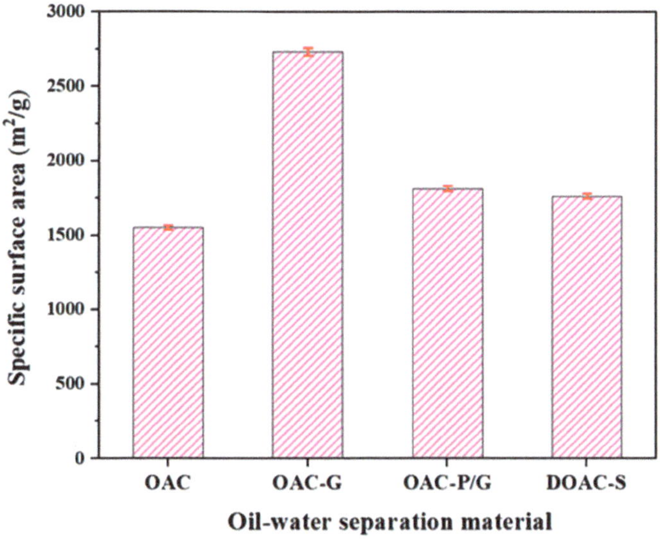

Figure 6. Specific surface area of oil–water separation materials.

3.5. Influence of Different Thin Oil Film Thicknesses on the Separation Performance of Four Oil–Water Separation Materials

The influence of the thin oil film thicknesses of kerosene and diesel on the separation performance of the four oil–water separation materials is shown in Figures 7 and 8, respectively. As can be seen from the figures, the separation ability of the four oil–water separation materials for kerosene is higher than that for diesel. When the thin oil film thickness is less than 2 mm, the fluxes of both kerosene and diesel are relatively small. Especially when the thin oil film thickness is less than or equal to 1 mm and during the final process of recovering the oil slick, the fluxes decrease significantly. This may be because when the thin oil film is too thin, the effective area of the oil–water separation material immersed in the thin oil film becomes smaller, resulting in a decrease in the flux.

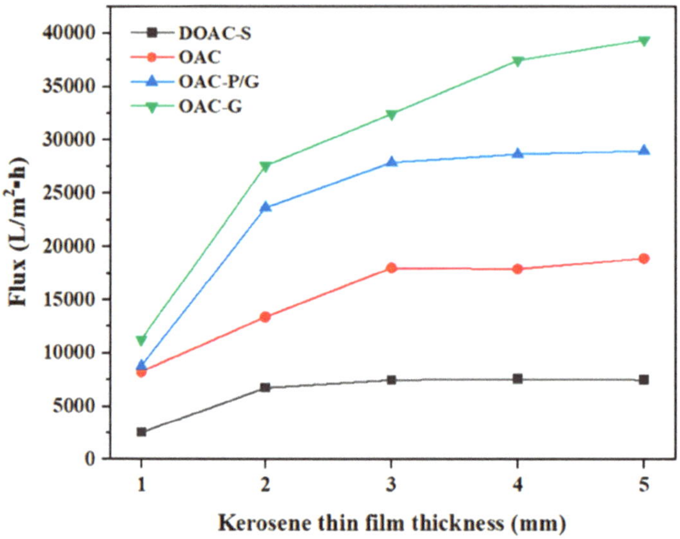

Figure 7. Flux of oil–water separation materials under different kerosene thin film thicknesses.

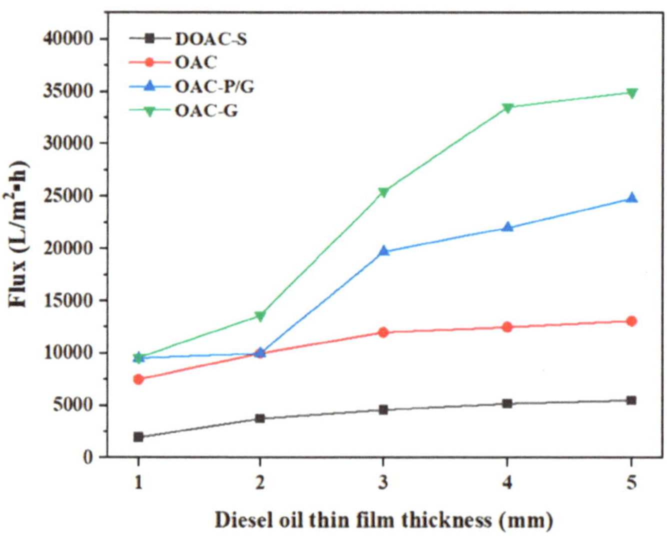

Figure 8. Flux of oil–water separation materials under different thin film thicknesses of diesel oil.

When the thin oil film thickness is between 3 and 5 mm, the membrane fluxes of the four oil–water separation materials gradually increase. OAC-G has the highest flux, with membrane fluxes for kerosene and diesel of 39,431 L/(m^2·h) and 3375 L/(m^2·h), respectively. This may be because after the oil-absorbing cotton is impregnated with graphene, its hydrophobicity is improved (Figure 9 shows that the water contact angle of OAC-G is 133°, while the water contact angles of the other three materials, OAC-P/G, DOAC-S, and OAC, are 125°, 100°, and 98°, respectively), and its lipophilicity is enhanced. Therefore, its flux is higher than that of the oil–water separation materials made of other materials.

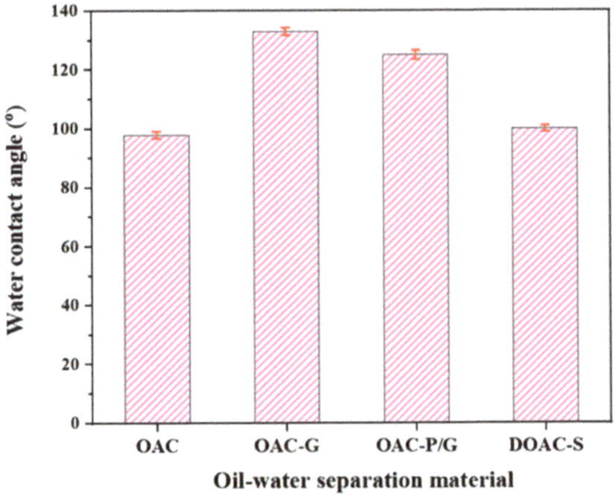

Figure 9. Water contact angle of different materials.

When comparing the oil–water separation fluxes of diesel and kerosene for DOAC-S and OAC, despite the fact that their fluxes are relatively small overall, there are still notable differences between the two oils. For these two materials, the flux for kerosene is higher than that for diesel. This can be related to the inherent properties of diesel and kerosene. As mentioned earlier, kerosene typically has lower viscosity and surface tension compared to diesel. Even for materials like DOAC-S and OAC which are described as thicker and more compact, the more fluid nature of kerosene enables it to pass through the membranes more easily, thus resulting in a higher flux. On the other hand, the higher viscosity of diesel makes it more resistant to flow through the relatively restricted pores of these thicker and more compact materials, leading to a lower separation flux. This further emphasizes that the physical characteristics of the oils play a crucial role in determining the oil–water separation flux, and for materials with certain structural properties (such as being thick and compact), the differences in oil properties have a more pronounced impact on the separation performance.

In the case of OAC-G, although it has relatively high fluxes for both oils, the difference in flux between kerosene and diesel still exists, following the same trend of kerosene having a higher flux. This suggests that regardless of the overall performance level of the material, the nature of the oil being separated significantly influences the efficiency of the oil–water separation process. It also implies that when applying these materials in practical scenarios, choosing the appropriate material based on the specific oil type is essential for achieving optimal separation results.

3.6. Influence of Different Pressures on the Filtration Performance of the Oil-Absorbent Cotton-Based Oil–Water Separation Material

In Figure 10, it could be seen that "The membrane flux of OAC-G for kerosene increases from 32,695 L/(m^2·h) to 35,520 L/(m^2·h), and the membrane flux for diesel increases from 35,415 L/(m^2·h) to 39,988 L/(m^2·h)". However, according to the bar chart (Figure 10), the initial flux (at 0.05 MPa) for kerosene is approximately 35,500 L/(m^2·h), and for diesel it is approximately 32,500 L/(m^2·h). So, the initial flux values in the text are reversed for kerosene and diesel compared to the figure. The text correctly states that "as the pressure increases, the membrane fluxes of OAC-G for both kerosene and diesel gradually increase". This is consistent with the bar chart in Figure 10, where we can see that as the pressure increases from 0.05 MPa to 0.09 MPa, the height of both the pink (kerosene) and blue (diesel) bars increases, indicating an increase in flux. In summary, the text has incorrect initial flux values for kerosene and diesel when compared to Figure 10, but the description of the flux-increasing trend with increasing pressure is consistent with the figure.

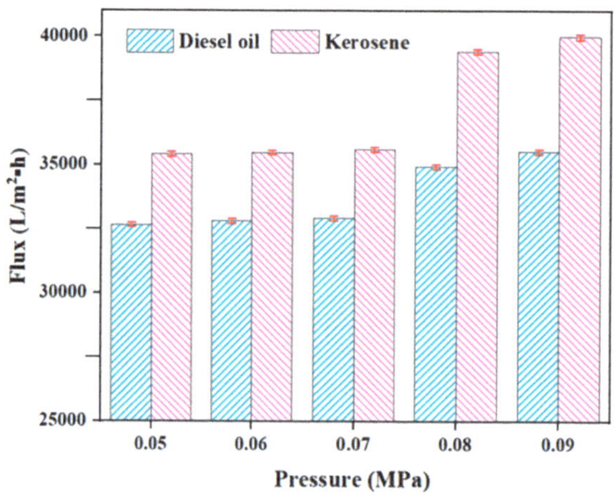

Figure 10. OAC-G fluxes of kerosene and diesel at different pressures.

3.7. Influence of Different Temperatures on the Filtration Performance of the Oil-Absorbent Cotton-Based Oil–Water Separation Material

Figure 11 shows the filtration flux results of OAC-G for filtering kerosene and diesel under the conditions of 15 °C, 20 °C, 25 °C, 30 °C, 35 °C, and 40 °C. As can be seen from Figure 11, with the gradual increase in temperature, the membrane fluxes of OAC-G for both kerosene and diesel gradually increase. The membrane flux of OAC-G for kerosene increases from 37,815 L/(m^2·h) to 40,950 L/(m^2·h), and the membrane flux for diesel increases from 34,275 L/(m^2·h) to 35,350 L/(m^2·h). The reason for this phenomenon is that as the temperature rises, the viscosity of the oil products decreases, the membrane mass transfer resistance decreases, and it is easier for the oil products to pass through the pores of OAC-G.

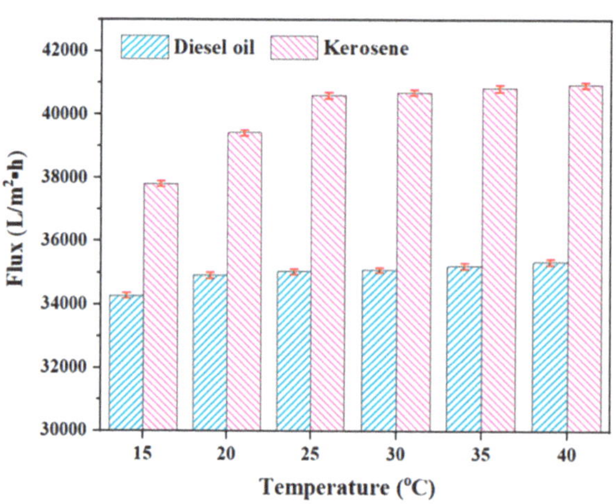

Figure 11. OAC-G fluxes of kerosene and diesel at different temperatures.

3.8. Analysis of the Recovery Tests of Different Oil Products and Oily Organic Solvents

The leakage of different oil products and oily organic solvents can have a significant impact on both the ecological environment and the human environment. Considering that the oil–water separation membrane has an excellent recovery effect on kerosene and diesel, the performance of the OAC-G material in adsorbing and separating different oil products and oily organic solvents on the water surface is verified and evaluated. All the different oil products and oily organic solvents used in the experiment are lightweight and can float on the water surface. The results of the static adsorption experiment are shown in Figure 12.

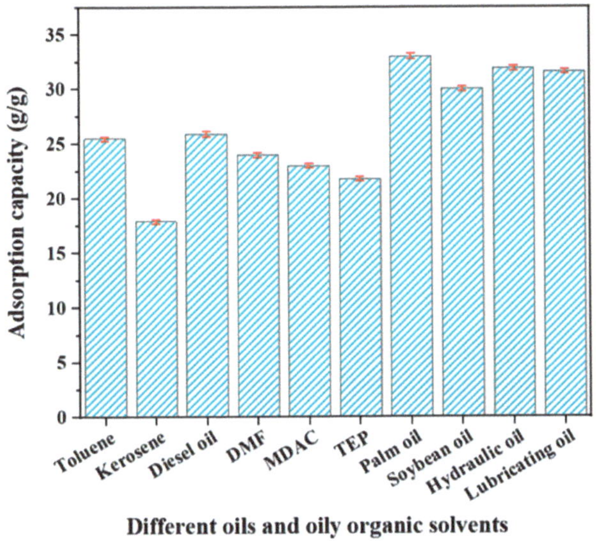

Figure 12. Static adsorption capacity of OAC-G for different oils and organic solvents.

As can be seen from Figure 12, OAC-G has a very good adsorption effect on diesel, kerosene, and other different oil products and oily organic solvents. Among them, the adsorption capacity for palm oil is the highest, which is 33 g/g. It also has a good adsorption effect on diesel and kerosene, which are 25.9 g/g and 17.9 g/g, respectively. This may be because OAC-G has a large number of microporous structures. After being modified by graphene, its hydrophobicity is enhanced and its lipophilicity is greatly improved, which is consistent with the previous analysis results of the membrane surface morphology and hydrophilic performance.

After the static adsorption test is completed, the OAC-G component is connected to the power system for the dynamic separation experiment, and the results are shown in Figure 13. As can be seen from Figure 13, the OAC-G component also has a very good dynamic separation effect on different oil products and oily organic solvents. The one with the highest membrane flux is dimethylformamide, and its membrane flux is 1530 L/(m^2·h). Followed by dimethylacetamide, kerosene, toluene, soybean oil, diesel, lubricating oil, palm oil, triethyl phosphate, and hydraulic oil. The one with the lowest flux is palm oil. In addition to the structure of the component itself, the main factors affecting the filtration flux may also include the molecular structure, polarity, and viscosity of the oil products and oily organic solvents.

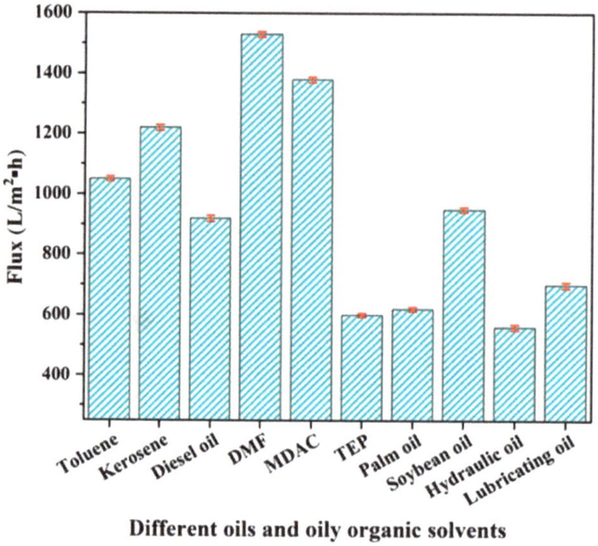

Figure 13. Dynamic membrane flux of OAC-G for different oils and oily organic solvents.

3.9. Analysis of the Long-Term Operation Effect of OAC-G Material

Based on the above analysis, considering the long-term performance of the membrane, under the condition of setting a constant flux of 35,000 L/(m^2·h) for the membrane unit, continuous and stable operation of separating kerosene and diesel using OAC-G was carried out for 12 months, respectively. The cleaning cycle was 4 months, and the cleaning process was a combined cleaning scheme of first soaking and cleaning with 1000 mg/L NaOH and then soaking and cleaning with 1000 mg/L HCl. The transmembrane pressure difference data of the two membrane units for separating kerosene and diesel were monitored and recorded. The results are shown in Figure 14. The peaks in the curve represent the pressure difference state of the membrane after being contaminated, and the troughs represent the state of the membrane pollution recovery and the decrease in the pressure difference after cleaning.

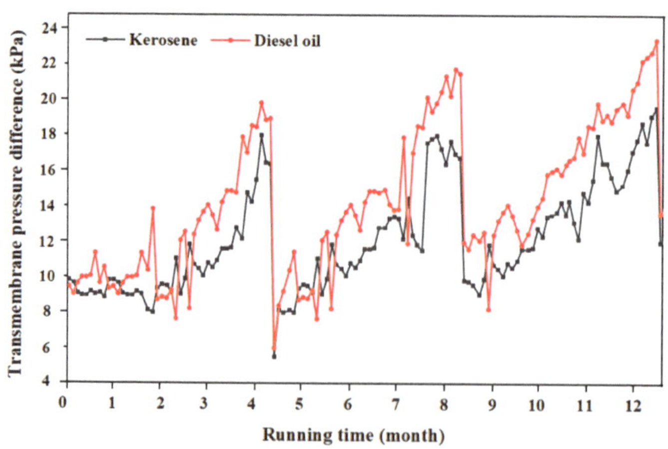

Figure 14. The OAC-G material changes the transmembrane pressure difference in kerosene and diesel oil for 12 consecutive months.

As can be seen from Figure 14, when the flux remains unchanged, the membrane oil filtration resistance continuously increases, and the transmembrane pressure difference also continuously increases. When it increases to 18–22 kPa at around the 4th month, it indicates that the membrane has been contaminated. Then, the first off-line chemical cleaning was carried out on the two groups of membranes, respectively. After cleaning, the transmembrane pressure difference in the kerosene oil–water separation system decreased from 17.9 kPa to 5.2 kPa, and the transmembrane pressure difference in the diesel oil–water separation system decreased from 19.6 kPa to 5.8 kPa. Subsequently, two off-line chemical cleanings were carried out, respectively, after the 8th month and the 12th month. From the changes in the transmembrane pressure difference curve data before and after cleaning, it can be seen that the SC membrane achieved a good flux recovery effect after cleaning. At the same time, it can also be seen from the figure that the overall transmembrane pressure difference before and after cleaning shows a slightly increasing trend.

4. Conclusions

In this study, four types of oil–water separation materials were used to treat thin oil films on the water surface. The effects of factors such as oil film thickness, pressure, and temperature on the filtration performance of the oil–water separation materials were investigated. By testing the viscosities of kerosene and diesel, as well as the adsorption and separation effects of the oil–water separation materials on different oil products and oily organic solvents, the long-term stability of the movable and portable components was verified.

The results show that the oil-absorbent cotton-based oil–water separation material has an excellent microporous structure and membrane surface roughness, endowing the membrane surface with excellent hydrophobicity and lipophilicity. It has good oil–water separation performance. The filtration flux of oil increases with the increase in pressure and temperature, and it has good adsorption and separation performance for different oil products and oily organic solvents. At the same time, it maintains stable operation performance during the 12-month long-term oil–water separation process for kerosene and diesel.

The research results can effectively address the existing problems of the difficulty in treating thin oil films on the water surface and the low separation efficiency of current oil–water separation materials, and achieve efficient oil–water separation of thin oil films.

Author Contributions: X.W.: Conceptualization, Methodology, Investigation, Instrumentation, Software, Writing—Original draft preparation. B.Z.: Conceptualization, Methodology, Investigation, Instrumentation, Software. H.D.: Project administration, Conceptualization, Formal analysis, Funding acquisition, Writing—review and editing, Supervision. Y.K.: Investigation, Formal analysis. C.C.: Formal analysis. All authors have read and agreed to the published version of the manuscript.

Funding: This study was supported by the Talent Introduction Project of Hebei Agricultural University (YJ2022019) and the Basic Scientific Research Business Expenses Program of Hebei Provincial Colleges and Universities (KY2022022).

Institutional Review Board Statement: Not applicable.

Informed Consent Statement: Not applicable.

Data Availability Statement: The original contributions presented in this study are included in the article. Further inquiries can be directed to the corresponding authors.

Conflicts of Interest: Authors Haiping Dai, Yongwen Ke and Cheng Cai are employed by the Shandong Jinwei Hairun Special Separation Equipment Co., Ltd. The remaining authors declare that the research was conducted in the absence of any commercial or financial relationships that could be construed as a potential conflict of interest.

References

1. Zhang, N.; Qi, Y.; Zhang, Y.; Luo, J.; Cui, P.; Jiang, W. A Review on Oil/Water Mixture Separation Material. *Ind. Eng. Chem. Res.* **2020**, *59*, 14546–14568.
2. Su, Y.; Fan, T.; Cui, W.; Li, Y.; Ramakrishna, S.; Long, Y. Advanced electrospun nanofibrous materials for efficient oil/water separation. *Adv. Fiber Mater.* **2022**, *4*, 938–958.
3. Baig, U.; Faizan, M.; Waheed, A. A review on super-wettable porous membranes and materials based on bio-polymeric chitosan for oil-water separation. *Adv. Colloid Interface Sci.* **2022**, *303*, 102635. [PubMed]
4. José, M.H.; Canejo, J.P.; Godinho, M.H. Oil/water mixtures and emulsions separation methods-an overview. *Materials* **2023**, *16*, 2503. [CrossRef]
5. Han, P.; Yu, B.; Chen, Y.; Nie, G.; Wang, K.; Sun, X.; Li, X.; Shao, W.; Liu, F.; He, J. Electrospun Polyacrylonitrile-Fluorinated Polyurethane/Polysulfone Nanofiber Membranes for Oil–Water Separation. *ACS Appl. Nano Mater.* **2024**, *7*, 4336–4348.
6. Abu Elella, M.H.; Abu-Thabit, N.Y.; Uwaezuoke, O.J.; Azad, A.K. Superwetting cotton textiles for separation of oil/water mixtures. *Cellulose* **2023**, *30*, 7427–7462. [CrossRef]
7. Xiao, F.; Zhang, H.; Wu, T.; Liu, J.; Liu, J.; Zhang, J.; Liu, W.; Liang, T.; Hu, J. Superhydrophobic/superlipophilic interface layer for oil-water separation. *Process Saf. Environ. Prot.* **2022**, *161*, 13–21.
8. Shan, X.; Huang, P.; Yang, L.; Feng, R.; Wang, Z. Robust polypropylene/ethylene-propylene-diene terpolymer thermoplastic vulcanizates film for green oil–water separation. *J. Polym. Res.* **2022**, *29*, 94.
9. Dong, B.; Lv, Y.; Zhi, Z.; Yang, J.; Hu, Q.; Wang, L.; Abadikhah, H.; Xu, X.; Wang, G.; Min, Z.; et al. Robust superhydrophobic ceramic fiber braid for oil water separation. *Ceram. Int.* **2023**, *49*, 11725–11729.
10. Wang, D.D.; Shen, X.Y.; Song, Y.Y.; Wang, S.T. Application Progress of Emerging Janus Particles for Oil-Water Separation. *Acta Chim. Sin.* **2023**, *81*, 1187.
11. Zhang, J.; Fang, W.; Zhang, F.; Gao, S.; Guo, Y.; Li, J.; Zhu, Y.; Zhang, Y.; Jin, J. Ultrathin microporous membrane with high oil intrusion pressure for effective oil/water separation. *J. Membr. Sci.* **2020**, *608*, 118201. [CrossRef]
12. Sun, X.D.; Yang, H.; Liang, Y.; Yan, K.; Liu, L.; Gao, D.; Ma, J. Light-Propelled Super-Hydrophobic Sponge Motor and its Application in Oil-Water Separation. *ACS Appl. Mater. Interfaces* **2023**, *15*, 43205–43215. [CrossRef] [PubMed]
13. Zhong, X.; Guo, Z. Simple and Ultrahigh Efficient Superhydrophilic Polydopamine-coated TiO_2 Cotton for Oil-water Separation. *J. Bionic Eng.* **2023**, *20*, 900–909. [CrossRef]
14. Lu, J.; Zhu, X.; Miao, X.; Wang, B.; Song, Y.; Ren, G.; Li, X. An unusual superhydrophilic/superoleophobic sponge for oil-water separation. *Front. Mater. Sci.* **2020**, *14*, 341–350. [CrossRef]
15. Chen, M.; Wang, Z.; Weng, S.; Yue, J.; Wang, Z.; Zhang, H.; Huang, Z.; You, X.; Li, R.; Fu, Q. Submicron-ultrathin PE/PDVB composite membrane for efficient oil/water separation. *J. Membr. Sci.* **2024**, *701*, 122701. [CrossRef]
16. Meng, H.; Xu, T.; Li, C.; Bai, J. Electrospinning janus membranes for on-demand oil/water collection and emulsion separation. *J. Polym. Sci.* **2024**, *62*, 3047–3057. [CrossRef]
17. Li, B.F.; Qi, B.; Guo, Z.Y.; Wang, D.; Jiao, T. Recent developments in the application of membrane separation technology and its challenges in oil-water separation: A review. *Chemosphere* **2023**, *327*, 138528. [CrossRef]
18. Wu, S.Q.; Yan, D.F.; Chen, Y.; Song, J. Self-driven oil/water separator with super-high separation rate. *Nano Energy* **2024**, *119*, 109066. [CrossRef]
19. Wang, J.Y.; Xiong, Z.M.; Guo, L.; Zhang, Y.-F.; Zhang, F.; Du, F.-P. Enhanced superhydrophobicity and durability of modified cotton cloth for efficient oil-water separation. *Sep. Purif. Technol.* **2025**, *354*, 128716. [CrossRef]
20. Abuhatab, F.; Hasan, S.W. Green fabrication methods and performance evaluation of eco-friendly membranes for enhanced oil-water separation. *J. Environ. Chem. Eng.* **2024**, *12*, 113993. [CrossRef]
21. Zhang, N.; Yang, X.W.; Wang, Y.L.; Qi, Y.; Zhang, Y.; Luo, J.; Cui, P.; Jiang, W. A review on oil/water emulsion separation membrane material. *J. Environ. Chem. Eng.* **2022**, *10*, 107257. [CrossRef]
22. Maguire-Boyle, S.J.; Barron, A.R. A new functionalization strategy for oil/water separation membranes. *J. Membr. Sci.* **2011**, *382*, 107–115. [CrossRef]
23. Zhu, H.Y.; Guo, Z.Y.; Yu, W.; Yuan, S.; Shen, L.; Zhao, D.L.; Lin, H. Photocatalytic and photothermal membranes for sustainable oil-water separation. *Water Res.* **2025**, *272*, 122919. [CrossRef] [PubMed]
24. Padaki, M.; Surya RAbdullah, M.S.; Abdullah, M.S.; Misdan, N.; Moslehyani, A.; Kassim, M.A.; Hilal, N.; Ismail, A.F. Membrane technology enhancement in oil-water separation. A review. *Desalination* **2015**, *357*, 197–207. [CrossRef]
25. Dmitrieva, E.S.; Anokhina, T.S.; Novitsky, E.G.; Volkov, V.V.; Borisov, I.L.; Volkov, A.V. Polymeric membranes for oil-water separation: A review. *Polymers* **2022**, *14*, 980. [CrossRef]

26. Mousa, H.M.; Fahmy, H.S.; Ali, G.A.M.; Abdelhamid, H.N.; Ateia, M. Membranes for oil/water separation: A review. *Adv. Mater. Interfaces* **2022**, *27*, 2200557. [CrossRef]
27. Li, B.F.; Qian, X.W.; Ran, L.F.; Han, J.; Yang, C.; Jiao, T. Applications and challenges of superwetted oil-water separation membranes in air under liquids and in specific environments. *Prog. Org. Coat.* **2024**, *195*, 108673. [CrossRef]
28. Liu, W.L.; Liu, Q.Y.; Liu, Z.; Liu, Z.; Hu, B.; Ding, R.; Deng, H.; Zheng, Y.; Yang, Z.; Zhang, R.; et al. In-situ construction of nanocomposite coating by electrostatic enhanced surface segregation toward antifouling oil-water separation membrane. *J. Membr. Sci.* **2025**, *717*, 123663. [CrossRef]
29. He, Y.X.; Guo, Z.G. Natural polymers-based separation membrane for high-efficient separation of oil water mixture. *Nanotoday* **2024**, *57*, 102367. [CrossRef]
30. Zhang, T.; Xiao, C.; Zhao, J.; Liu, X.; Ji, D.; Zhang, H. One-step facile fabrication of PVDF/graphene composite nanofibrous membrane with enhanced oil affinity for highly efficient gravity-driven emulsified oil/water separation and selective oil absorption. *Sep. Purif. Technol.* **2021**, *254*, 117576. [CrossRef]

materials

MDPI

Article

Properties and Behavior of Sandy Soils by a New Interpretation of MICP

Masaharu Fukue [1], Zbigniew Lechowicz [2], Catherine N. Mulligan [3,*], Seiichi Takeuchi [4], Yuichi Fujimori [5] and Kentaro Emori [6]

1 Japanese Geotechnical Association for Housing Disaster Prevention, 1622, Oshikiri, Shimizu-ku, Shizuoka 424-0008, Japan; fukue@scc.u-tokai.ac.jp
2 Department of Geotechnical Engineering, Institute of Civil Engineering, Warsaw University of Life Sciences, Nowoursynowska 159, 02-776 Warsaw, Poland; zbigniew_lechowicz@sggw.edu.pl
3 Department of Building, Civil and Environment Engineering, Concordia University, 1455 de Maisonneuve Blvd. W., Montreal, QC H3G 1M8, Canada
4 Fudo Tetra Co., 7-2, Koami-Cho, Nihonbashi, Chuo-ku, Tokyo 103-0016, Japan; seiichi.takeuchi@fudotetra.co.jp
5 Chubu Sokuchi Research Institute Co., 801-1 Konami, Suwa City 392-0131, Japan; fujimori-yuichi@chubusokuchi-lab.co.jp
6 Sanko Kaihatsu Co., Ltd., 1320 Gokanjima, Fuji City 416-0946, Japan; k.emori@sankoukaihathu.co.jp
* Correspondence: mulligan@civil.concordia.ca

Abstract: Research on MICP technology for ground improvement began in the early 2000s, and since then, it has been considered as innovative research. The field of applications is showing signs of expanding from sandy soil stabilization to remediation. However, the research has not always progressed, because it is extremely difficult to evaluate the ability (viability rate) related to microorganisms and how to handle them quantitatively. In fact, this problem hinders the consensus of research results in terms of quantitative evaluation of microorganisms and the cross-comparison (evaluation) and use of MICP technology research. The crucial disadvantage of using bacteria is that their properties are not constant due to changes over time and in the surrounding environment. Therefore, for engineering purposes, we used the carbonate formation rate (CPR), instead of urease activity, as a function of the microbial mass (OD) with viable bacteria. Thus, the standard OD−CPR relationship was defined experimentally, and the estimation method of viability was established. The required amount of microorganisms for testing was given by OD*, and the relationship "OD = Rcv OD*" was defined to convert from OD* to OD. Rcv was defined as the viable bacterial rate. It was found that the Ca^{2+}/OD ratio controls the inhibition behavior in MICP. At a Ca^{2+}/OD ratio of >8.46 M, then inhibition occurs, while at Ca^{2+}/OD = 8.46 M, CPR = 8.46 OD and the CPR is proportional to the viable OD, Rcv, and OD*. We show that it is possible to perform an experiment using OD* with aged bacteria, obtain Rcv from the standard OD−CPR and OD*−CPR relationships, convert OD* to OD and to perform a unified evaluation without actually determining the viability rate.

Keywords: MICP process; optical density (OD); cell viability; carbonate formation rate (CPR); OD−CPR relationship; sandy soils

1. Introduction

Microbial-induced carbonate precipitation (MICP) has been studied as one of the next-generation technological developments, because it is an eco-friendly natural technology learned from nature. The technology uses biominerals as binding materials, which has been introduced in the fields, such as sand stabilization [1], loess surface erosion control [2], reduction of liquefaction potential [3], scour protection around monopile foundation [4], increase in pull-out response of concrete pile [5], erosion protection of clayey soils [6], wind erosion control of desert soil [7], development of biogrout injection techniques [8], liquefaction resistance [9], stabilization of foreshore slopes [10], and ground improvement by biogrout [11]. In remediation technology, the application of MICP is also becoming increasingly popular in the fields of soil and water purification [12], solidification and remediation of lead–zinc tailings [13], strontium incorporation [14], remediation of uranium (VI) [15], remediation of heavy metals [16–20], biomineralization [21–23], remediation in mines [24–27], etc. In these areas, there are a number of researchers and many journal submissions published and in progress.

However, these achievements are still insufficient because these developments are relatively new since 2000. In addition, since MICP research is highly interdisciplinary, the collaboration of many researchers with different expertise may be slowing progress.

At present, there are essentially three problems for applying MICP in geoenvironmental engineering. Bacteria are living organisms, but they can be damaged over time due to age, temperature, pH, etc. Thus, their properties and behavior in the MICP process can change [28]. This means that the experimental results obtained depend on the viability of cells. This results in the decrease in optical density of cells which will cause inhibition or retardation in MICP [29].

Until now, the evaluation of microorganisms in MICP has been performed by determining the optical density (OD) of microorganisms in the culture medium. For example, the urease activity also has been evaluated by obtaining the optical density of bacteria, OD_{600nm}, which has been assumed as the density of the viable bacteria solution [29–31]. However, the optical density determination cannot distinguish between live and dead bacteria. In addition, the density of viable bacteria changes with aging, cold and heat shock, and various other elements. Thus, it is noted that caution should be exercised in the use of optical density.

In reality, despite these shortcomings, the use of OD is common. A possible shortcoming is that the inhibition and retardation effects of Ca^{2+} and OD on MICP are not well understood [29]. According to Fukue et al. [28], when the Ca^{2+}/OD value is greater than a certain value, then inhibition due to Ca^{2+} occurs. Therefore, when the OD value decreases with aging or a change in a factor, inhibition will occur under the same concentration of Ca^{2+}. The critical Ca^{2+}/OD value can be confirmed using only OD which is viable optical density, as was achieved in Fukue et al. [28]. If the descriptions above are correct, the comparison and identification of the data made in the past based on the conventional OD values are uncertain.

To use unified OD values in engineering purposes, Fukue et al. [28] found a method to convert from the apparent OD* to viable OD, in which the viable OD is defined as OD = Rcv OD*. The viable bacteria rate (Rcv) plays a role in cell viability. In addition, both OD and Rcv OD* are related to the carbonate precipitation rate (CPR) which is defined as the experimentally obtained standard OD–CPR relationship. The standard OD–CPR relationship is also used to judge if inhibition or retardation occurs. The OD*–CPR relationships have been obtained to investigate the capacity of bacteria for carbonate precipitation [29].

At present, the concentration of bacteria is not considered in the chemical reaction of the MICP process. In other words, concepts of mixture design as used in construction materials have not been applied. To solve this problem, the blending design method was established [28,29], using a unique (standard) carbonate precipitation (CPR)–OD relationship. At the same time, the conversion was enabled from an experimentally tentative optical density (OD*) to OD. Using the OD conversion and blending design of the biocement solution (BCS), the inhibition and retardation behavior in MICP were interpreted well, because experimental results obtained using different OD* were able to be combined and compared to the OD conversion [29]. It is highlighted that the optimum blending design of the BCS becomes possible. In other words, CPR has been able to be controlled with bacterial concentration in the BCS. By making full use of these methods, it is likely that significant advances and developments in MICP technology will be obtained. In this study, the objective is to show a new approach and to interpret the mechanical properties and behavior of soils in MICP processes, which have been unknown. While this study seems to be only from a macroscopic point of view, it is based on the microscopic point of view, both theoretically and experimentally for more than 15 years.

2. Materials and Methods

In this study, two typical granular materials, natural limestone [30] and MICP-treated fine sand [11], were examined in terms of dry density (ρ_d) and unconfined compressive strength (UCS), and the strength properties on the elemental tests and the behavior of the biomineralization processes were compared.

The ρ_d−UCS fitting curves for Ryukyu limestone were obtained by Kogure et al. [30]. The Ryukyu limestone consists of solidified sediments of coral sand and gravels, with some binding between particles. The formation has not been well understood. However, the lysis-reprecipitation of coral sand and gravel particles and/or foraminifer and coccolith have possibly solidified from the granular to solid materials, similar to general marine sediments [31–33]. The increase in dry density of the limestone may be due to the binding materials (biominerals).

As a result, the UCS of limestone was expressed with dry density, ρ_d, as:

$$UCS = 2.2 \times 10^{-2}\ {\rho_d}^{8.8} \qquad \text{(MPa)} \tag{1}$$

The square sectional area ($d \times d$) and 2d-height specimens were used for the UCS measurement, and the effects of d were investigated using various d values. The d changed from 10 to 100 mm with 5 stages, i.e., 10, 25, 50, 75, and 100 mm. The numbers of specimens increased with decreasing d, corresponding to 2, 5, 8, 10, and 20, respectively. As a result, the coefficient of determination R^2 obtained was 0.78 [30]. However, in Equation (1), the role of carbonate content is uncertain.

On the other hand, artificial MICP was applied to a large-sized fine sand embankment by van Paassen [11]. The ρ_d−UCS fitting curve was obtained by:

$$UCS = 4 \times 10^{-5} \exp(6\ \rho_d) \qquad \text{(MPa)} \tag{2}$$

where the coefficient of determination R^2 is 0.84. In this test, the biogrout solution was infiltrated laterally, and it was found that carbonate content apparently increased the UCS of fine sand and that the dry density increased with increasing carbonate contents. The increased dry density was assumed to be equal to the increased biogrout (biocement) in this analysis. It is important to note that the UCS can only be developed by bonding particles for granular materials. There is no doubt that the binding materials were precipitated carbonates. It was noted that the R^2 of the C−UCS relationship obtained was lower than that in Equation (2) [11].

Next, the similarity in solidification properties and behavior of the two types of granular materials under different solidification mechanisms was examined to understand essential binding mechanisms due to carbonation for granular materials.

The comparison between the ρ_d–UCS relationships for the limestone and fine sand is presented in Figure 1. It is noted that the curves lower than 1.7 t/m^3 were not based on actual data. Nevertheless, the extension curves by Equations (1) and (2) were used to discuss both materials with a relatively low UCS. It is because these low ranges in the UCS are very important for material improvement.

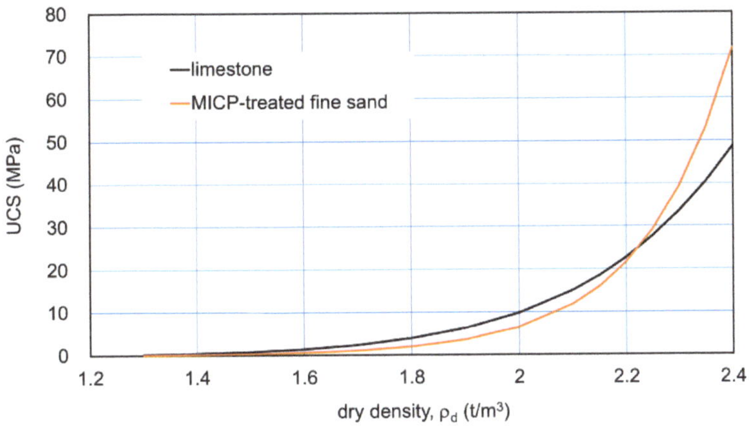

Figure 1. Dry density–UCS fitting curves for Ryukyu limestone [30] and for MICP-treated fine sand [11].

In Figure 1, the effects of C on UCS were hidden, because no C axis is presented. The R^2 values were 0.78 and 0.84 for the limestone and fine sand, respectively. The initial states for the two curves were assumed as presented in Table 1. The assumed values as the typical examples of sediments were not extreme for the comparison. In general, the factors affecting the minimum and maximum densities are the geometrical factors of particles, such as grain size distribution and grain shape, as well as the density of particles. The maximum void ratio of calcareous sediments is usually higher than that of ordinary sand (Table 1), in which it is not necessary to consider the density of particles.

Because of the differences in the textures of particles, formation mechanisms to rocks, physical and chemical properties, and behavior between natural limestone and MICP sandstone, it seems that the comparison may be meaningless. However, it is possible that both materials have become solidified as a result of carbonate precipitation, regardless of the mechanism [32,34]. Further comparison showing a similarity of the two curves in Figure 1 was examined in terms of the newly defined relative density.

In geotechnical engineering, the unification in properties among different types of granular soils are often examined using relative density. In this study, the definition of "modified relative density Dr*" was changed from the conventional definition of granular soils, because the transformation from granular state to rock (solid) materials was regarded as continuous. Therefore, the minimum density e_{min} was assumed as zero.

Accordingly, the modified relative density of materials consisting of granular to rock states, Dr*, was defined by:

$$Dr^* = \frac{e_{max} - e}{e_{max}} \quad (3)$$

The comparison with modified relative density–UCS relationships converted from those in Figure 1 is shown in Figure 2. Figure 2b indicates that the two curves for a wide range of dry density were close together and that limestone showed a higher UCS than that of MICP-treated fine sand. The details are discussed later in further data for sandy soils.

Table 1. Assumed constants of limestone and fine sand for comparison by a modified relative density.

	Specimen	Assumed			
	Shape	min. ρ_d (t/m^3)	ρ_s (t/m^3)	e_{max}	e_{min}
Limestone	Square Column	1.2	2.8	1.333	0
Biogrout and fine sand	Cylinder	1.4	2.65	0.893	0

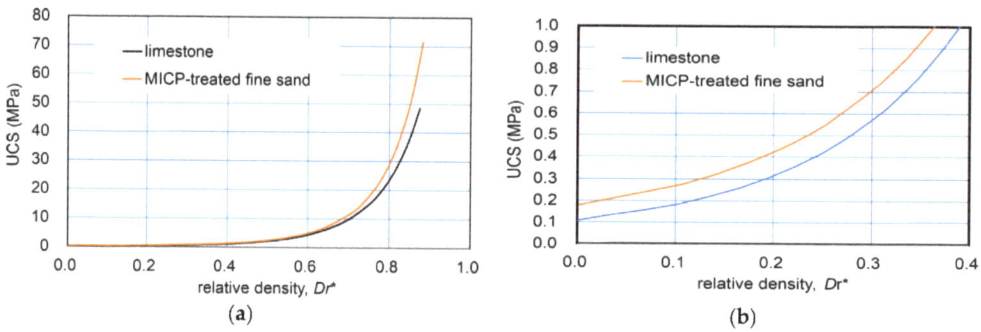

Figure 2. Comparison of Dr*−UCS relationships for limestone and MICP-treated fine sand: (**a**) wide range of Dr* (0–0.85); and (**b**) a narrow range of Dr* (0–0.4).

Figures 1 and 2 indicate that the natural diagenesis of limestone and MICP processes on fine sand may show similar properties and behavior despite their different natures. In other words, their varied nature and solidification process can be primarily described in terms of strength hardening due to both the increases in carbonate and dry density. This will be discussed in the experimental results for sandy soils later.

2.1. Sandy Soils Used in MICP

To investigate the C−UCS relationship in MICP for various granular soils, six different types of sandy soils (obtained from Tokai Sand Co., Omaezaki City, Japan. http://tokaisand.co.jp/archives/145 accessed on 5 January 2025) were used, which are as follows:

- Vietnam sand, crushed sand with high-electrostatic charges;
- Coarse sand with a low uniformity coefficient;
- Fine sand a with high liquefaction potential;
- Medium sand;
- Fine sand b with high liquefaction potential;
- Fine sand c (densified fine sand b);
- River sand with a high uniformity coefficient.

The numerical data are presented later using the six types of sandy soils (seven different specimens).

2.2. Problems in UCS Measurement for Loose Sandy Soils

The block samples of MICP-treated sandy soils were used for the mechanical tests in this study. The test method will be described in Section 2.3.1. Two types of specimens were used, depending on the test types. Unconfined (uniaxial) compression (using equipment from KS-12210, Tesco Ltd., Arakawa-ku, Japan), and triaxial compression tests (with equipment from KS-12230, Tesco Ltd., Arakawa-ku, Japan) were performed on cylindrical specimens trimmed from the block samples by Japanese standard test methods (JIS A 1216 and JGS 0521, respectively). The preparation method of the specimens was the same as used in geotechnical engineering, except for the handling difficulty. On the other hand, a pocket penetrometer test was performed directly on the block samples.

Since the binding carbonates built crystalline skeleton structures, the failure strains in limestone and MICP under uniaxial compression was very small, which was described as the failure of brittle materials.

If the specimens of very loose sandy soils were prepared in MICP, the UCS would be extremely low in comparison to the unconfined compressive strength estimated from the triaxial compression test, as shown in Figure 3. The unconfined compressive strength was expressed by qu, while UCSp was measured by the pocket penetrometer obtained from CL-700, SOIL TEST INC. CHICAGO USA and used according to ASTM WK27337. The difference between the two values was apparent. Note that the Mohr' s circle under a low σ_3 stress seemed to be the appropriate size.

On the other hand, UCSp, qu_{est}, and Mohr's circles 1, 2, and 3 showed reasonable Mohr's circles with the Mohr—Coulomb failure criterion, as shown in Figure 3. Thus, only qu was extremely low in comparison to the other experimental data under the triaxial failure criterion.

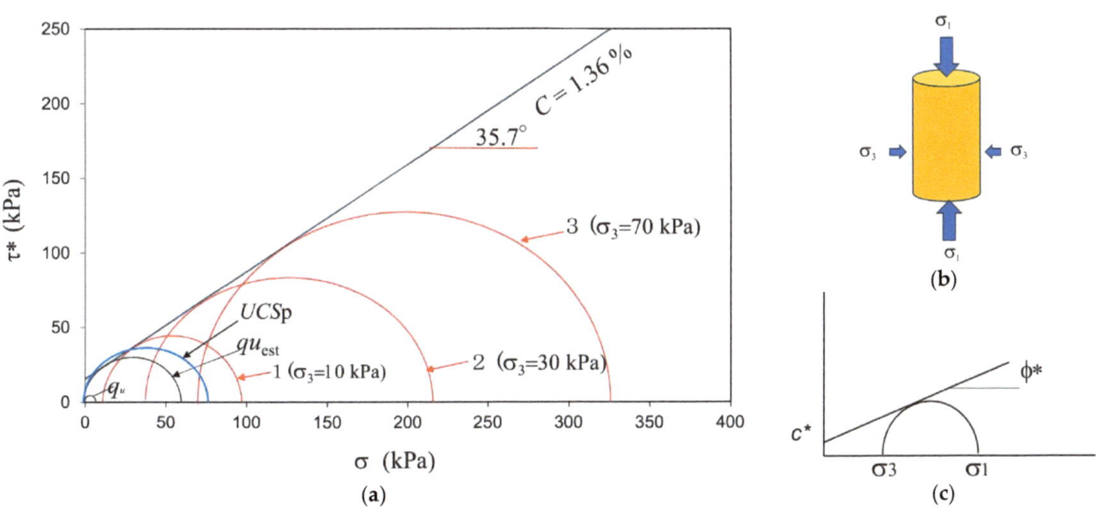

Figure 3. (**a**) Various Mohr's circles in triaxial compression; (**b**) specimen in triaxial compression; (**c**) the Mohr—Coulomb failure criterion. * Denotes the values for MICP treated soils.

It was considered that the specimens in the uniaxial compression tests failed in tensile forces, micro-failures, as shown in Figure 4. The failure patterns were dissimilar from general shear failure in the ordinal soil specimen in the triaxial compression test. This resulted from the brittle properties of MICP-treated loose sands.

The failure types depend on the porosity of soil specimen which may concern the inhomogeneity of structure consisting of particle arrangement. The looser the sand, the more complicated the properties and behavior became, as shown in Figure 4. The failure patterns could be expressed in terms of micro-shear or progressive tensile failures, as shown in Figure 4a,b, or tensile failure, which was more explainable theoretically, as shown in Figure 4c. In addition, it is possible that UCS (solidification) may affect the failure pattern. At present, it is difficult to predict failure patterns of MICP-treated loose sand in uniaxial compression as it is different from the ordinary sand and rock.

The results shown in Figure 4 indicate an example of brittle failures of MICP-treated soils of low density that can be avoided by applying a low lateral confinement stress.

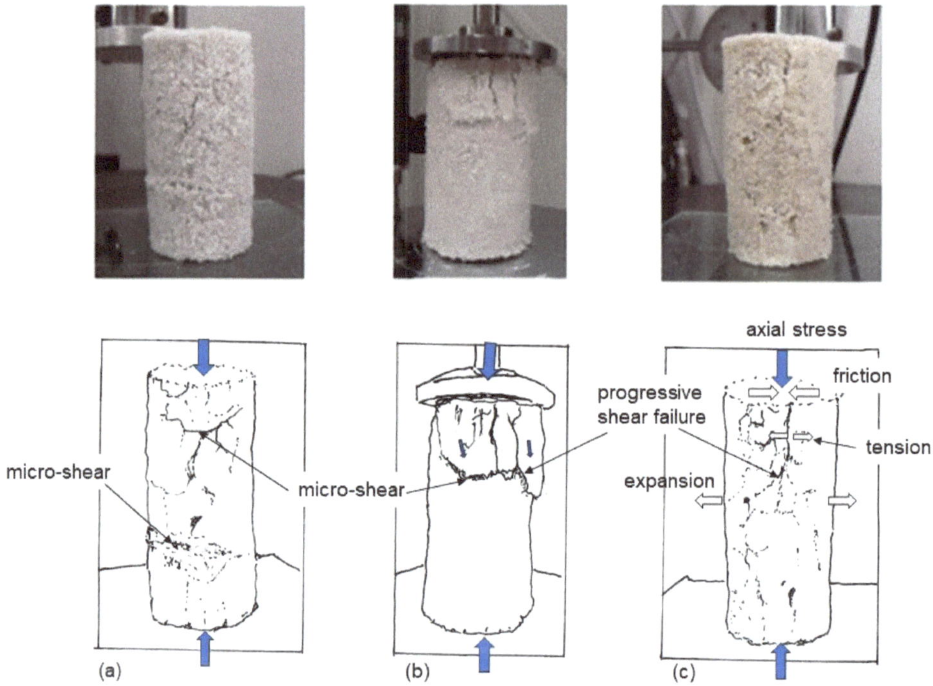

Figure 4. Micro-failure patterns in tension for loose MICP-treated sandy soils in uniaxial compression tests. (**a–c**) indicate three MICP-treated specimens.

The pocket penetrometer used is shown in Figure 5. The test method is simple. Some discussion on the test method and applications have been published [35–37]. The disadvantage in the use of the penetrometer is a limited maximum UCS, i.e., 460 kPa, but it is sufficient for the MICP study on low-density sandy soils with liquefaction potential.

In summary, the pocket penetrometer is useful to measure UCS for low-density MICP-treated sandy soils from the viewpoints of time and cost saving. The results obtained may be a little influenced by confined lateral pressure during penetration of the tip into soil. However, it seems that the developed confined pressure, due to the penetrating effect by the penetrometer's tip, was neglected or at most less than 10 kPa. It is noted that liquefaction potential is of the most concern for this type of sand.

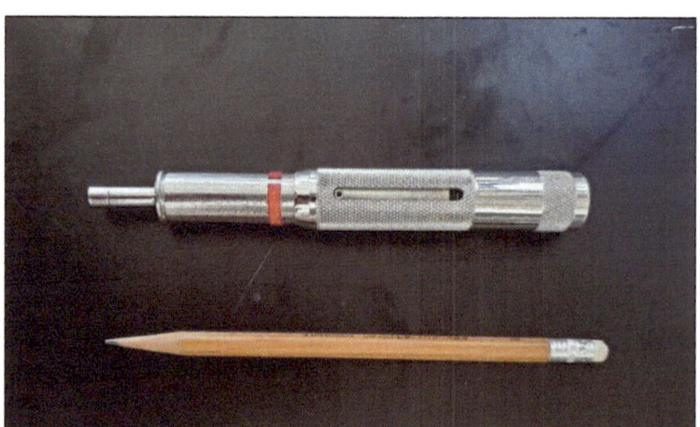

Figure 5. Pocket penetrometer (ASTM WK27337).

2.3. Experiments

2.3.1. Preparation of Specimens of Seven Types of Sandy Soils

To investigate the carbonate (C)−UCS relationships, the six types of sandy soils were loosely packed in the three boxes made of styrofoam. To examine the effects of densification, fine sand B was also densified into three styrofoam boxes. In the BCS, 0.3 M Ca^{2+} (obtained from Tokuyama Corporation,1-1, Shunan City, Japan), 0.3 M urea (supplied by Kanto Chemical Corporation Inc., Chuo-ku, Japan), and NO-A10 strains (obtained from Life Engineering Co., Shimizu-ku, Japan. https://life-engineering.jimdoweb.com accessed on 5 January 2025) were contained. The OD value of NO-A10 strains was unknown, and the amounts of bacteria were based on the experiment. The initial density of soils was calculated from mass, volume, and water content of air-dried soils. The surfaces of the packed soils were covered by unlaid cloth and netting. The drainage was allowed from the bottom side with a tube. The injection volume of the BCS was approximately 1.2 -fold that of the pore volume. The injection of the BCS was conducted by spraying without considering the homogeneity of the ultimate C value, because the study required the deviation of C values through the specimens. Since the thickness of specimens was relatively thin, the injection was conducted after the production of white suspensions and crystalline carbonate began in the BCS.

The three boxes for one soil sample were used to vary the number of injections. The number of injection cycles was varied from 1 to 3. After the injections, the specimens were rinsed with sufficient tap water and left for at least for one month. The curing setup is shown in Figure 6.

2.3.2. Carbonate Content Determination

The carbonate contents (with equipment from Life Engineering Co., Shizuoka, Japan. https://life-engineering.jimdoweb.com accessed on 5 January 2025) were determined by measuring the CO_2 gaseous pressure induced in a pressure cell obtained from DMC-203N11, No.92AF 015, Okano Works Ltd., Neyagawa City, Japan. https://okanoworks2.com/products/pressure accessed on 5 January 2025), with 3 N HCl (obtained from Wako Pure Chemical Industries, Ltd., Chuo-ku, Japan). The $CaCO_3$ contained in the soil samples was converted to the mass of $CaCO_3$ using a linear $CaCO_3$ gaseous pressure calibration curve [38]. The method is simple and easy to measure using several grams of the dry MICP-treated sample. Generally, it only takes a few minutes of operation per single sample, except for the preparation of samples, i.e., drying of samples in the oven and weighing.

The carbonate content measured on MICP-treated soil was expressed by:

$$C_t = CaCO_3/\text{dry soil sample} \times 100 \qquad (\%) \tag{4}$$

where C_t is the total of natural carbonate after MICP treatment.

In fact, a natural soil sample usually contains a small amount of $CaCO_3$ as soil particles [31,33], which does not contribute to the strength. Therefore, to eliminate the natural $CaCO_3$, a similar test should be performed using the original natural soil sample. To present the carbonate content alone by MICP, the naturally contained $CaCO_3$ content must be deducted from C_t in Equation (4).

Figure 6. Curing setup of MICP experiments.

3. Test Results

3.1. Carbonate Content and UCS

The UCSs of six soil samples, i.e., seven types of specimens, were measured with the pocket penetrometer. The UCS increased with carbonate content. As a result, for the respective specimens, the higher the carbonate content, the greater the UCS. The results showed that most of the UCS values increased linearly with increasing carbonate content (Figure 7). This type of C−UCS relationship has been presented and discussed by many studies. However, there is no consensus based on only the C−UCS relationships, as they are dependent on the soil type.

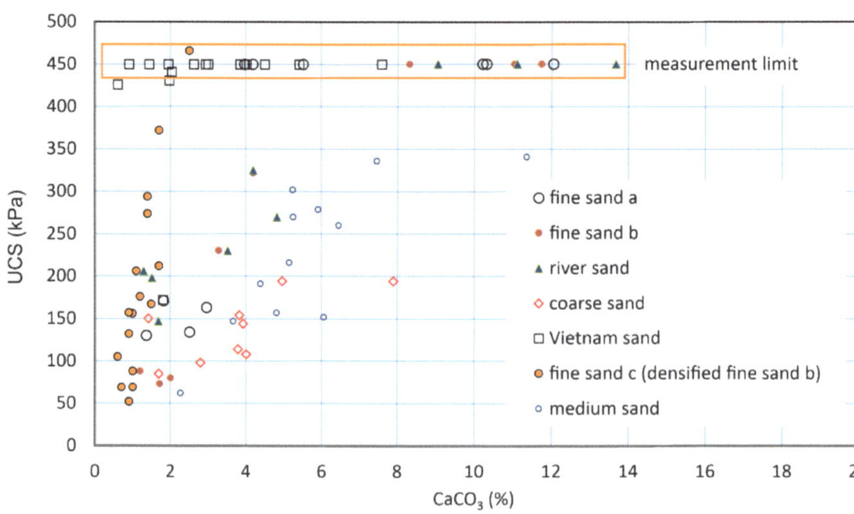

Figure 7. C−UCS relationships for various sandy soils.

The soil properties and a-values obtained are presented in Table 2. Note that the a-values were estimated as the slopes of the fitting C−UCS curves by:

$$UCS = a\,C \quad \text{(kPa)} \tag{5}$$

However, the a-values might be underestimated, because some of the measurements were possibly influenced by the disturbance of samples. For example, for Vietnam sand, a-values seemed to between about 180 to 690 kPa/%. This may result from the very high electrostatic charges of the Vietnam sand particles (see Figure 7).

Therefore, the a-value that can be specific for a soil type is significant. In this study, the a-value was defined to be a distinct mechanical behavior in MICP. In addition, the dependency of various dry densities on a-value will be discussed later.

The maximum and minimum dry densities of soil samples were measured by the method in Japanese Industrial Standard, JIS A 1224. Theoretically, the initial dry density must accord with the minimum dry density. However, the initial dry density of soil specimens was sometimes lower than that of the minimum dry density because the specimens contained macropores (Table 2). D_{max}, D_{60}, D_{30}, and D_{10} represent the maximum particle diameter and the particle size at which 60%, 30%, and 10% of the particles are smaller, respectively.

Table 2. Soil properties used and a-values.

Type of Sand	Particle Density ρ_s (t/m^3)	D_{max} (mm)	D_{60} (mm)	D_{30} (mm)	D_{10} (mm)	Uniformity U_c	ρ_{max} (t/m^3)	ρ_{min} (t/m^3)	Initial C (%)	Initial ρ_d (t/m^3)	a (kPa/%)
Vietnam sand	2.662	2	0.61	0.48	0.34	1.79	1.695	1.416	0.22	1.4	186–690
Coarse sand	2.65	2	1.35	1.05	0.87	1.55	1.568	1.336	0.2	1.28	34.5
Fine sand a	2.693	2	0.303	0.205	0.135	2.24	1.719	1.41	0.3	1.4	64.4
Medium sand	2.664	4.75	0.33	0.28	0.2	1.65	1.619	1.308	0.47	1.26	44
Fine sand b	2.679	4.75	0.295	0.195	0.13	2.27	1.683	1.377	0.2	1.38	53.9
Fine sand c	2.679	4.75	0.295	0.195	0.13	2.27	1.683	1.377	0.2	1.58	120
River sand	2.65	4.75	0.5	0.301	0.155	3.23	1.752	1.404	0.38	1.56	68.7

3.2. *UCS as a Function of Carbonate Content and Dry Density*

One of the features of MICP is that there is no change in soil volume. This is due to the low-pressure infiltration of BCS. On the other hand, MICP must increase the dry density due to the precipitation of solid carbonates.

In marine sediments, the dead bodies of foraminifera and coccolith are always present, which contribute to the binding of the sediment particles. This is also applicable to natural limestone formation from coral sand and gravel [35,36]. In any case, the build-up of a skeleton structure in granular soils including coral sediments requires particle-to-particle bonding, which is due to the MICP process and dissolution and reprecipitation under the change in micro-circumstances, such as partial pressure of CO_2, redox potential, pH, and Ostwald' ripening and/or sintering under appropriate conditions [33,39–45].

Assuming that the carbonates induced by MICP and metamorphisms increased the dry density of sediments, carbonate content C and dry density ρ_d relationship was given by:

$$\rho_d* = \rho_d\ (1 + \frac{C}{100})\quad (t/m^3) \tag{6}$$

where ρ_d* is the dry density after MICP or carbonate diagenesis, ρ_d is the initial dry density, and C is the carbonate content increased by MICP or carbonate diagenesis.

Figure 8 shows the ρ_d*−UCS relationship obtained using Equation (6). In Figure 8 as described earlier, the linear C−UCS relationship obtained was clearly positioned in the ρ_d*−UCS relationship, as well as in Equation (6).

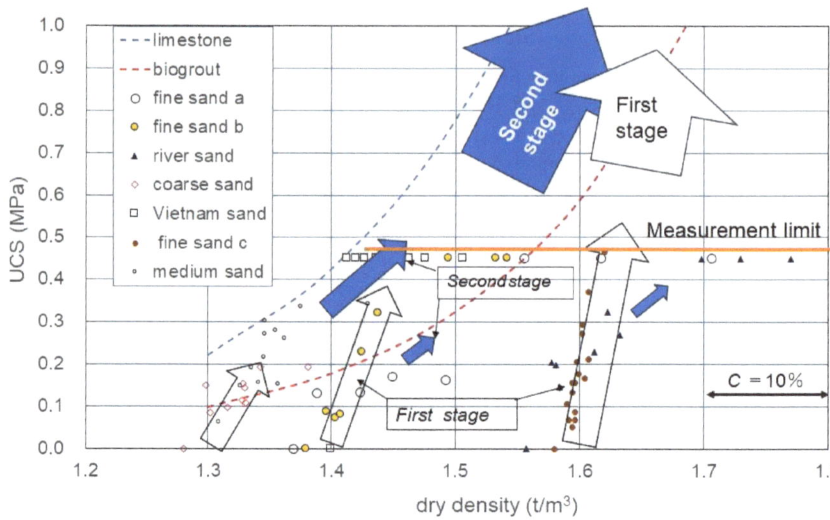

Figure 8. UCS as a function of C and ρ_d, assuming that dry density was increased with increased carbonate precipitation.

Though the UCS measured was plotted on the ρ_d axis, the carbonate content was converted into the incremental dry density.

3.3. Definition of the First and Second Stages in the Carbonation Process

Table 2 and Figure 8 show that the UCS in MICP-treated soils varied with soil type including grain size, dry density, etc. Furthermore, it was shown that for the incremental rate of UCS with carbonate content C, the a-value depended on the initial dry density (Figure 8). This fact can easily be explained with the correlation between the number of contacts between particles and dry density, because the a-value increased proportionally with the increased contact points between particles due to the increase in dry density.

The initial effects in the MICP process were described as the rapid bonding at the contact between particles because the binding started with zero strength. The denser the skeleton structure, the higher the UCS. It was experimentally found that the rapid increase in UCS was proportional to the C value. In this study, this initial process was defined as the first stage in MICP, as expressed by the empty arrows in Figure 8.

After the skeleton structure was developed to some extent, newly induced carbonates covered the surface of soil particles. However, they lacked the strengthening power of the skeletal structure. The slow MICP process after the first stage was defined as the second stage in MICP, as presented by the filled arrows in Figure 8. It is noted that there might be an infinite number of processes with different soil types in MICP. For example, with respect to the initial dry density of a soil, an infinite MICP process was considered.

3.4. Initial Dry Density and a-Value

It was indicated that the a-value can be one of the mechanical factors in various MICP-treated soils. From Equation (4), the a-value was given by:

$$a = UCS/C \qquad (kPa/\%) \tag{7}$$

The a-values of the seven types in the test series are presented in Table 2. The initial dry density and a-value for the various sandy soils are presented in Figure 9. It seems that the plots deviate widely. This indicates that the soil samples used had properties previously mentioned. In other words, Figure 9 indicates the different properties expressed by the a-value at the zero potential energy state (like at the initial dry density). The initial void ratio or initial dry density of soils affected the e−log p relationships for the soil, which was explained by the concept of the equation of state in soils [43].

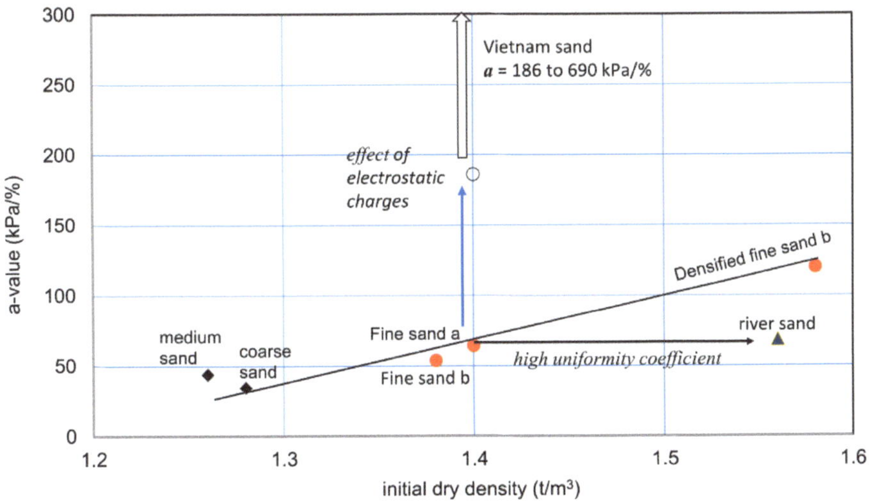

Figure 9. Initial dry density and a-value relationships for various sandy soils. The vertical arrow indicates the increase in a value due to the special action due to particle's surface forces, under a con-stant distribution of particle sizes. On the other hand, the horizontal arrow shows a constant a-value for particles from ordinary or averaged particle distribution at a given dry density.

The extremely high a-value in Vietnam sand, i.e., 186 to 690 kPa/%, may result from the effects of high electrostatic charges of particles. The relatively low a-value for river sand may be attributed to its high uniformity coefficient (Table 2), in which relatively greater particles caused a relatively high density without increases in the a-value.

The effect of compaction was examined by comparing the original (fine sand b) and compacted specimens (fine sand c). The result suggests that there may be a unique relationship between dry densities and a-values for similar granular soils with a similar range in uniformity coefficients. Therefore, unless soils are special like Vietnam sand and river sand, the initial dry density−a-value relationship for ordinary granular soils may not differ widely. This needs to be investigated further.

4. Discussion

4.1. The First and Second Stages in the MICP Process

In engineering purposes, if the initial dry density is known, the design (target) strength UCS can be expressed by a target C-value in Figure 10. If the target C or UCS is in the first stage, both values are expressed by UCS = a C. On the other hand, if the target C or UCS in the second stage, the second stage must be evaluated or estimated. In the case, a continuous increase in the C value results in a non-linear C−UCS relationship, as illustrated in Figure 10. The non-linear behavior results from the transformation from the first to second stages in MICP. Note that there must be an inflection in the slope for the ρ_d−UCS curve, as illustrated in Figure 10.

It is unknown if the two second stages started from different first stages for similar soil types. The skeleton structures of the two first and second states are apparently different because the respective skeletons are built at the initial dry density. That is why the a-values for the two different first stages are different. In other words, the second stages will keep their respective skeleton structures from the first stage, and the pore spaces will continuously be filled with carbonates by precipitation.

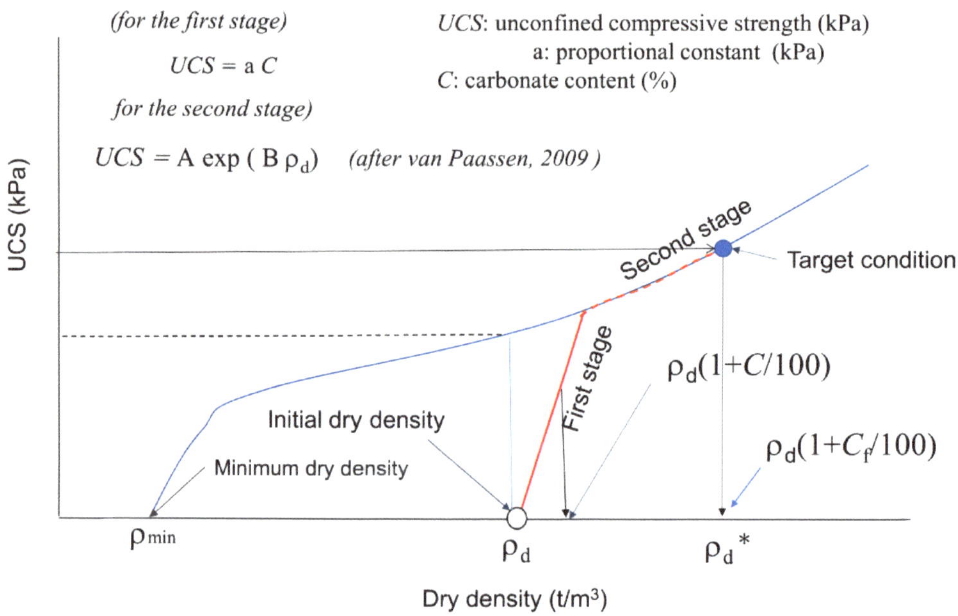

Figure 10. Definition of the first and second stages in the MICP process [11].

In Figure 10, the second stage is illustrated based on Equation (2) [11], which can be recognized as an example for the general expression of the second stage of MICP in fine sand. In summary, when the effects of carbonate content on UCS or other strength factors are investigated, the dry density should be incorporated. In the design of MICP, C and ρ_d should be considered as factors of strength. Therefore, to avoid confusion, it is important to distinguish the first and second stages in MICP. Note that in this study, the a-value was only defined in the first stage. At present, the estimation (fitting curve) by van Paassen [11] will provide an index for the second stage of fine sand.

4.2. Conversion Between C and the CPR

The CPR and C terms have different definitions. The CPR is precipitated carbonate (M) in the BCS, while the C value (%) is carbonate content defined as the mass ratio of $CaCO_3$ by soil particles. Both values are conveniently used for individual purposes. Thus, the CPR is used in the case of the BCS without the involved soil, and the C value is always used in the case of UCS when soil is involved.

In procedures of engineering design, carbonate content C (%) is given based on the design strength, e.g., UCS. If the a-value of the soil is known, the C value is predicted by UCS/a. Then, the C is theoretically converted to the carbonate precipitation rate, CPR, in the BCS [28].

When bacteria are mixed with the BCS components such as urea, the hydrolysis of urea is initiated. The hydrolysis occurs in the cell and produces $2NH_3$ and CO_2 extracellularly. Ca^{2+} reacts with $CO_3{}^{2-}$ from the CO_2 produced by the hydrolysis. In the optimizing case of the Ca^{2+}/NO-A10 ratio, the entire process from the mixing of bacteria with other components to the complete precipitation of carbonates requires from approximately at least 30 min to 48 h [28,29]. Therefore, various physical and chemical phenomena occur, such as dispersion of the components, development of a Ca^{2+} diffuse double layer, aggregation of bacteria with the Ca^{2+} diffuse double layer [44,45] and crystal growth, adsorption of the aggregates on soil particles, and settlement of the aggregates.

Figure 11 shows a segment of a video depicting the BCS in a glass tube, including the movement of bacterial aggregates with the Ca^{2+} diffuse double layer by convection. Aggregates adsorbed on the wall of the glass tube, and crystalline growth formed in aggregates. In this case, the video was shot using a digital microscope for approximately 40 min after the BCS was prepared. Accordingly, the bacteria with Ca^{2+} diffuse double layers and the aggregate shapes could not be recognized with the naked eye. The growth of carbonates started around the cells which were seen as cloudy suspensions. It took generally more than 30 min from the mixing of the BCS.

Thus, the organic reaction is not as rapid as the inorganic reaction. For engineering purposes to produce good-quality cementation effects, a slow reaction is required which may contribute to adsorption on aggregate-soil particles.

Bacteria with negative surface charges can be adsorbed on positively charged glass, while negatively charged soil surfaces can adsorb the bacteria with Ca^{2+}. Therefore, the beginning adsorption behavior of bacteria with or without Ca^{2+} may be different in a glass tube or soils. However, for MICP, Ca^{2+} is a good binder between soil particles and bacteria.

Figure 11. Digital image of aggregation, adsorption, and crystalline in the MICP process.

Adsorbed aggregates could grow by absorbing advected aggregates. Crystalline growth depended on the size of aggregates. The precursor of crystals was usually amorphous which then produced crystalline forms in the aggregates (Figure 12). Ultimate crystals were various types of calcites. If only Ca^{2+} was used as a metal element, single crystals were sometimes produced, but they were usually polycrystals (Figure 13).

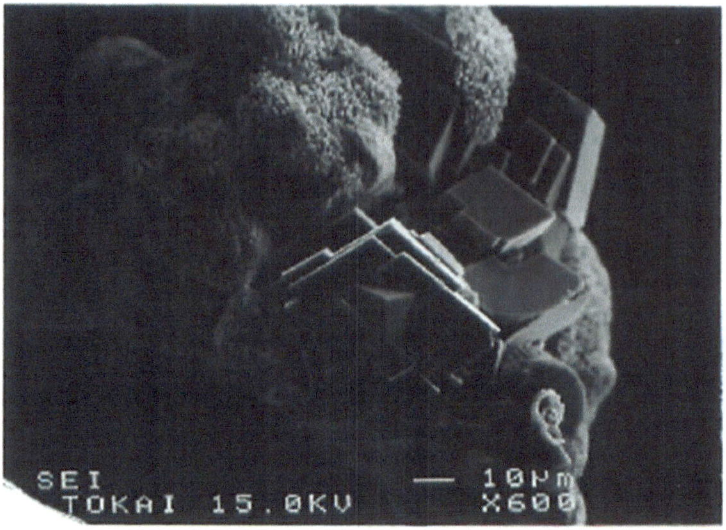

Figure 12. SEM images of calcite growth from amorphous aggregates formed in the original bacteria–Ca^{2+} system in the BCS.

(**a**)

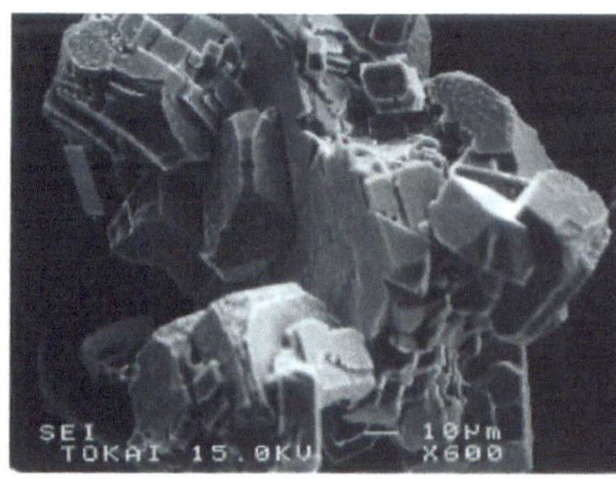

(**b**)

Figure 13. SEM images of ultimate single and polycrystals in MICP: (**a**) single crystals of calcite; (**b**) polycrystals of calcite.

According to the application of the adsorption theory [46,47], bacterial adsorption seems to be simply expressed by the Langmuir and Freundlich or modified theories. However, in MICP, the adsorption of bacteria on soil particles occurs as bacteria aggregates with Ca^{2+} and partial amorphous and crystalline $CaCO_3$. Under the ideal blending of the BCS [28], the formation of the complex aggregates may depend on at least the concentrations of bacteria and initial Ca^{2+}, flow velocity of the BCS, and soil dry density.

In this study, a concept based on the bacteria-dependent CPR was used [28] and was presented by Equation (8).

$$\text{CPR} = 8.46\ \text{OD} - 17.633\ (\text{OD})^2 \qquad (\text{M}) \tag{8}$$

If after an elapsed time the urease activity started is sufficient to produce the full amount of carbonate, only the first term of the right side in Equation (9) can be used as follows:

$$\text{CPR} = 8.46\ \text{OD} \qquad (\text{M}) \tag{9}$$

The CPR (M) is the carbonate precipitation rate in the BCS, while C (%) is the carbonate content in soil. The carbonate content (%) was defined in terms of the CPR by:

$$\text{C} = \frac{\text{m}_\text{c}}{\text{m}_\text{s}} \times 100 = \text{CPR}\frac{n}{(1-\text{n})}\frac{(0.1)}{\rho_\text{s}} \times 100 \quad (\%) \tag{10}$$

where n is the porosity (in decimals), ρ_s is the density of soil particles (t/m^3), and the units of the CPR are in M = mol/L. Equation (10) approximately gives C = 2.5%, when n = 0.4, CPR = 1.0 M, and ρ_s is 2.65 t/m^3. Thus, it should be remembered that in an ordinary granular loose sandy soil, 1.0 M Ca^{2+} can produce approximately 2.5% of carbonates (calcite).

Substituting Equation (8) into Equation (10), the relationship between C and OD at 24 h precipitation can be obtained as:

$$\text{C} = \frac{\text{m}_\text{c}}{\text{m}_\text{s}} \times 100 = \{8.46\text{OD} - 17.633(\text{OD})^2\}\frac{\text{n}}{(1-\text{n})}\frac{(0.1)}{\rho_\text{s}} \times 100 \quad (\%) \tag{11}$$

Assuming that the retardation effects with OD^2 were negligible, because of the sufficient passage of time, the relationship between C and OD was given by:

$$\text{C} = 8.46\text{W} \times \text{OD} \quad (\%) \tag{12}$$

and

$$\text{W} = \frac{\text{n}}{(1-\text{n})}\frac{(0.1)}{\rho_\text{s}} \times 100 \quad (\%) \tag{13}$$

Assuming that the density of particles was 2.65 t/m^3, W was expressed as a function of porosity n, as shown in Figure 14. Thus, C was expressed as a function of OD through the CPR.

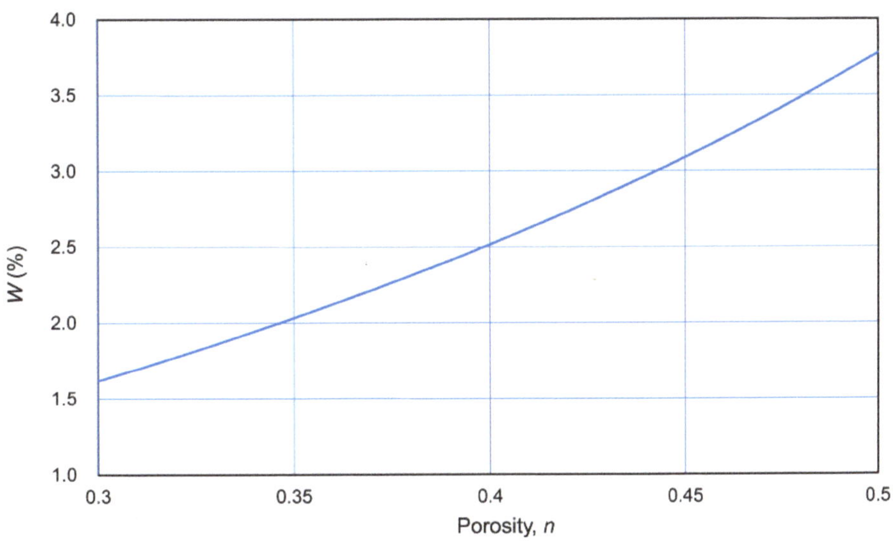

Figure 14. W−n relationship assuming ρ_s = 2.65 t/m^3.

From Equation (12), OD was expressed by:

$$OD = 0.118\ C/W \quad (14)$$

where OD is the viable OD which provides the standard OD−CPR relationship. Therefore, under the condition of bacteria-dependent MICP, which means no Ca^{2+} dependency, the distribution of bacteria represented by OD can be estimated from the C values measured with the gaseous pressure meter. Note that if OD* is needed, it is necessary to convert from OD to OD* by OD* = OD/Rcv. Thus, Equation (11) shows the relationship between OD (microscopic properties) and C (macroscopic properties), in which CPR, UCS, OD*, and Rcv values are all included. Therefore, the C value, which can be measured experimentally, can be one of the significant factors to anticipate the flow pattern of the BCS. The Ca^{2+} adsorbed by bacteria is adsorbed to the soil particles by Darcy's law. It is noted that whether or not the flow pattern obeys Darcy's flow is possibly related to the soil homogeneity.

5. Conclusions

One of the most important problems in the development of the MICP technique for engineering purposes is to introduce quantitative evaluation of urease-producing bacteria into chemical reactions. If this can be resolved, the test results can be comprehensively evaluated and consensus can be obtained. In this study, first, the method of quantitative evaluation of the bacteria was developed by establishing the standard OD−CPR relationship and the concept of experimental viability (Rcv). It was shown that the use of a bacterial evaluation method and Rcv can contribute to a comprehensive understanding of the properties and behavior of the MICP process. It was confirmed that the Ca^{2+}/OD ratio controls the inhibition behavior in MICP. For example, if $Ca^{2+}/OD > 8.46$ M in the BCS, then inhibition occurs, while at $Ca^{2+}/OD = 8.46$ M, CPR = 8.46 OD and the CPR is proportional to the viable OD, Rcv, and OD*. As the CPR in MICP is determined to be time-dependent, when the elapsed time is not sufficient, then retardation in MICP will occur. Although the Rcv test takes two days, this is acceptable for engineering applications.

The UCS process due to carbonate precipitation can be divided into an initial rapid process (the first stage) due to the bonding of particle contacts and a gradual increase in the process (the second stage) after the first stage. In addition, it was shown that the first stage depends on the type of soil, especially the particle size, its distribution, the initial dry density, and the adsorbed intensity among bacteria, Ca^{2+} and the surfaces of soil particles with electrostatic charges. The a-value defined in this study is a good factor for quantitatively evaluating the UCS in "the first stage" in the MICP process.

Experimental work showed that the properties and behavior of MICP-treated soils are not solely dependent on carbonate content but are influenced by other factors. The factors affecting UCS, except for carbonate content, are mainly dry density governed by grain size, grain shape, and grain size distribution. Furthermore, it was indicated that the electrical surface charge density contributed to a very strong UCS.

The dry density and UCS relationship can be divided into two types of behavior, i.e., the first and second stages of the MICP process. The confusion between these two types of behavior will result in a misunderstanding of the MICP process.

Author Contributions: Conceptualization, M.F. and Z.L.; methodology, M.F. and Z.L.; software, M.F. and K.E.; validation, Z.L., C.N.M., Y.F. and K.E.; formal analysis, M.F.; investigation, M.F., S.T., Y.F. and K.E.; resources, M.F. and S.T.; data curation, not applicable.; writing—original draft preparation, M.F.; writing—review and editing, M.F., Z.L. and C.N.M.; visualization, S.T.; supervision, M.F.; project administration, M.F. All authors have read and agreed to the published version of the manuscript.

Funding: This research received no external funding.

Institutional Review Board Statement: Not applicable.

Informed Consent Statement: Not applicable.

Data Availability Statement: The original contributions presented in this study are included in the article. Further inquiries can be directed to the corresponding author.

Acknowledgments: Bacteria NO-A10 and-N10 strains used for the experiment were supplied by Life Engineering, Co., Japan.

Conflicts of Interest: Author Seiichi Takeuchi was employed by the company Fudo Tetra Co. Author Yuichi Fujimori was employed by the company Chubu Sokuchi Research Institute Co. Author Kentaro Emori was employed by the company Sanko Kaihatsu Co., Ltd. The remaining authors declare that the research was conducted in the absence of any commercial or financial relationships that could be construed as a potential conflict of interest.

References

1. Chen, M.; Gowthaman, S.; Nakashima, K.; Komatsu, S.; Kawasaki, S. Experimental Study on Sand Stabilization Using Bio Cementation with Wastepaper Fiber Integration. *Materials* **2021**, *14*, 5164. [CrossRef] [PubMed]
2. Cheng, Y.; Tang, C.; Pan, X.; Liu, B.; Xie, Y.; Cheng, Q.; Shi, B. Application of microbial induced carbonate precipitation for loess surface erosion control. *Eng. Geol.* **2021**, *294*, 106387. [CrossRef]
3. DeJong, J.T.; Gomez, M.G.; San Pablo, A.C.; Graddy, C.M.; Nelson, D.C.; Lee, M.; Ziotopoulou, K.; El Kortbawi, M.; Montoya, B.; Kwon, T.H. State of the Art: MICP soil improvement and its application to liquefaction hazard mitigation. In Proceedings of the 20th International Conference on Soil Mechanics and Geotechnical Engineering, Sydney, Australia, 1–5 May 2002; pp. 405–508. ISBN 978-0-9946261-6-5.
4. Li, Y.; Guo, Z.; Wang, L.; Yang, H.; Li, Y.; Zhu, J. An Innovative Eco-friendly Method f or Scour Protection around Monopile Foundation. *Appl. Ocean Res.* **2022**, *123*, 103177. [CrossRef]
5. Lin, H.; Suleiman, M.; Jabbour, H.M.; Brown, D. Bio-grouting to enhance axial pull-out response of pervious concrete ground improvement piles. *Can. Geotechnol.* **2018**, *55*, 119–130. [CrossRef]
6. Liu, B.; Xie, Y.; Tang, C.; Pan, X.; Jiang, N.; Singh, D.; Cheng, Y.; Shi, B. Bio-mediated method for improving surface erosion resistance of clayey soils. *Eng. Geol.* **2021**, *293*, 106295. [CrossRef]
7. Meng, H.; Gao, Y.; He, J.; Qi, Y.; Hang, L. Microbially induced carbonate precipitation for wind erosion control of desert soil: Field-scale tests. *Geoderma* **2021**, *383*, 114723. [CrossRef]
8. Ramos, O.; Kwon, T.-H. Development of bio-grout injection strategy and design guide using reactive transport model for field-scale soil improvement based on microbially induced calcium carbonate precipitation (MICP). *Geomech. Energy Environ.* **2023**, *36*, 100509. [CrossRef]
9. Riveros, G.A.; Sadrekarimi, A. Liquefaction resistance of Fraser River sand improved by a microbially-induced cementation. *Soil Dyn. Earthq. Eng.* **2020**, *131*, 106034. [CrossRef]
10. Salifu, E.; MacLachlan, E.; Lyer, K.R.; Knapp, C.W.; Tarantino, A. Application of microbially induced calcite precipitation in erosion mitigation and stabilisation of sandy soil foreshore slopes: A preliminary investigation. *Eng. Geol.* **2016**, *201*, 96–105. [CrossRef]
11. van Paassen, L. Biogrout, Ground Improvement by Microbially Induced Carbonate Precipitation. Ph.D. Thesis, Delft University Technology, Delft, The Netherlands, 2009; 195p.
12. Achal, V.; Pan, X.; Zhang, D. Bioremediation of strontium (Sr) contaminated aquifer quartz sand based on carbonate precipitation induced by Sr resistant *Halomonas* sp. *Chemosphere* **2012**, *89*, 764–768. [CrossRef] [PubMed]
13. Dong, Y.; Gao, Z.; Di, J.; Wang, D.; Yang, Z.; Wang, Y.; Guo, X.; Li, K. Experimental study on solidification and remediation of lead–zinc tailings based on microbially induced calcium carbonate precipitation (MICP). *Constr. Build. Mater.* **2023**, *369*, 130611. [CrossRef]
14. Fujita, Y.; Redden, G.D.; Ingram, J.C.; Cortez, F.; Ferris, G.; Smith, R.W. Strontium incorporation into calcite generated by bacterial ureolysis. *Geochim. Cosmochim. Acta* **2004**, *68*, 3261–3270. [CrossRef]
15. Gerber, U.; Zirnsteinb, I.; Krawczyk-Bärscha, E.H.; Lünsdorf, H.; Arnolda, T.; Merround, M.L. Combined use of flow cytometry and microscopy to study the interactions between the gram-negative betaproteobacterium *Acidovorax facilis* and uranium (VI). *J. Hazard. Mater.* **2016**, *317*, 127–134. [CrossRef] [PubMed]
16. Kumar, A.; Song, H.-W.; Mishra, S.; Zhang, W.; Zhang, Y.-L.; Zhang, Q.-R.; Yu, Z.-G. Application of microbial-induced carbonate precipitation (MICP) techniques to remove heavy metal in the natural environment: A critical review. *Chemosphere* **2023**, *318*, 137894. [CrossRef] [PubMed]
17. Lin, H.; Zhou, M.; Li, B.; Dong, Y. Mechanisms, application advances and future perspectives of microbial-induced heavy metal precipitation: A review. *Int. Biodeter. Biodegr.* **2023**, *178*, 105544. [CrossRef]

18. Mark, J.; Jones, M.J.; Butchins, L.J.; Charnock, J.K.; Pattrick, R.A.D.; Small, J.S.; Vaughan, D.J.; Wincott, P.L.; Livens, F.R. Reactions of radium and barium with the surfaces of carbonate minerals. *Appl. Geochem.* **2011**, *26*, 1231–1238. [CrossRef]
19. Mitchell, A.C.; Ferris, G.F. The coprecipitation of Sr into calcite precipitates induced by bacterial ureolysis in artificial groundwater: Temperature and kinetic dependence. *Geochim. Cosmochim. Acta* **2005**, *69*, 4199–4210. [CrossRef]
20. Zhao, J.; Csetenyi, L.; Gadd, G.M. Fungal-induced $CaCO_3$ and $SrCO_3$ precipitation: A potential strategy for bioprotection of concrete. *Sci. Total Environ.* **2022**, *816*, 151501. [CrossRef] [PubMed]
21. Chen, M.; Li, Y.; Jiang, X.; Zhao, D.; Liu, X.; Zhou, J.; He, Z.; Zheng, C.; Pan, X. Study on soil physical structure after the bioremediation of Pb pollution using microbial-induced carbonate precipitation methodology. *J. Hazard. Mater.* **2021**, *411*, 125103. [CrossRef] [PubMed]
22. Han, L.; Li, J.; Xue, Q.; Chen, Z.; Zhou, Y.; Poon, C.S. Bacterial-induced mineralization (BIM) for soil solidification and heavy metal stabilization: A critical review. *Sci. Total Environ.* **2020**, *746*, 140967. [CrossRef]
23. Zeng, Y.; Chen, Z.; Lyu, Q.; Wang, X.; Du, Y.; Huan, C.; Liu, Y.; Yan, Z. Mechanism of microbiologically induced calcite precipitation for cadmium mineralization. *Sci. Total Environ.* **2022**, *852*, 158465. [CrossRef] [PubMed]
24. Bao, S.; Di, J.; Dong, Y.; Gao, Z.; Gu, Q.; Zhao, Y.; Zhai, H. Experimental study on the effect of cementation curing time on MICP bio-cemented tailings. *Constr. Build. Mater.* **2024**, *411*, 134263. [CrossRef]
25. He, Z.; Xu, Y.; Wang, W.; Yang, X.; Jin, Z.; Zhang, D.; Pan, X. Synergistic mechanism and application of microbially induced carbonate precipitation (MICP) and inorganic additives for passivation of heavy metals in copper-nickel tailings. *Chemosphere* **2023**, *311*, 136981. [CrossRef]
26. Zhou, G.; Xu, Y.; Wang, Y.; Zheng, L.; Zhang, Y.; Li, L.; Sun, B.; Li, S.; Zhu, Y. Study on MICP dust suppression technology in open pit coal mine: Preparation and mechanism of microbial dust suppression material. *J. Environ. Manag.* **2023**, *343*, 118181. [CrossRef]
27. Zúniga-Barra, H.; Toledo-Alarcon, J.; Torres-Aravena, A.; Jorquera, L.; Rivas, M.; Gutiérrez, L.; Jeison, D. Improving the sustainable management of mining tailings through microbially induced calcite precipitation: A review. *Miner. Eng.* **2022**, *189*, 107855. [CrossRef]
28. Fukue, M.; Lechowicz, Z.; Fujimori, Y.; Emori, K.; Mulligan, C.N. Incorporation of optical density into the blending design for a biocement solution. *Materials* **2022**, *15*, 1951. [CrossRef]
29. Fukue, M.; Lechowicz, Z.; Fujimori, Y.; Emori, K.; Mulligan, C.N. Inhibited and Retarded Behavior by Ca^{2+} and Ca^{2+}/OD Loading Rate on Ureolytic Bacteria in MICP process. *Materials* **2023**, *16*, 3357. [CrossRef] [PubMed]
30. Kogure, T.; Aoki, H.; Maekado, A.; Matsukura, Y. Effect of Specimen Size and Rock Properties on the Uniaxial Compressive Strength of Ryukyu Limestone. *J. Jpn. Soc. Eng. Geol.* **2005**, *46*, 2–8, (Text in Japanese with English Summary). [CrossRef]
31. Fukue, M.; Nakamura, T.; Kato, Y.; Naoe, K. Correlation among carbonate content, accumulation rate and topography of seabed. *Soils Found.* **1996**, *36*, 51–60. [CrossRef]
32. Fukue, M.; Nakamura, T.; Kato, Y. Cementation of soils due to calcium carbonate. *Soils Found.* **1999**, *39*, 55–64. [CrossRef] [PubMed]
33. Sanders, D. Syndepositional dissolution of calcium carbonate in neritic carbonate environments: Geological recognition, processes, potential significance. *J. Afr. Earth Sci.* **2003**, *36*, 99–134. [CrossRef]
34. Worden, R.H.; Burley, S.D. Sandstone diagenesis: The evolution of sand to stone. In *Sandstone Diagenesis: Recent and Ancient*; Burley, S.D., Worden, R.H., Eds.; International Association of Sedimentologists: Ghent, Belgium, 2003; ISBN 978-1-405-10897-3.
35. Budak, T.O.; Gurbuz, A.; Eksioglu, B. Practical transitions among undrained shear strengths of remolded samples from pocket penetrometer tests and other laboratory tests. *Catena* **2022**, *213*, 106148. [CrossRef]
36. Mousavi, F.; Abdi, E.; Ghalandarayeshi, S.; Page-Dumroese, D.S. Modeling unconfined compressive strength of fine-grained soils: Application of pocket penetrometer for predicting soil strength. *Catena* **2021**, *196*, 104890. [CrossRef]
37. O'Kelly, B.C. Discussion of "Practical transitions among undrained shear strengths of remolded samples from pocket penetrometer tests and other laboratory tests" by Budak T.O., Gurbuz A. and Eksioglu B. [Catena 213 (2022) 106148]. *Catena* **2023**, *228*, 107129. [CrossRef]
38. Fukue, M.; Ono, S.; Sato, Y. Cementation of sands due to Microbiologically induced Carbonate. *Soils Found.* **2011**, *51*, 83–93. [CrossRef]
39. Burdalski, R.; Ribeiro, B.G.O.; Gomez, M.G.; Gorman-Lewis, D. Mineralogy, morphology, and reaction kinetics of ureolytic bio-cementation in the presence of seawater ions and varying soil materials. *Sci. Rep.* **2022**, *12*, 17100. [CrossRef] [PubMed]
40. Garrels, R.M.; Christ, C.H. *Solution, Minerals, and Equilibria*; Croneis, C., Ed.; Happer's Geoscience Series; Happer & Row: New York, NY, USA, 1965.
41. Morais, M.M.; de Camargo, I.L.; Colombo, P.; Fortulan, C.A. Additive manufacturing of calcium carbonate parts through vat-photopolymerization and sintering in carbon dioxide atmosphere. *Open Ceram.* **2023**, *14*, 100348. [CrossRef]
42. Willibalda, C.; Dual, J.; Schneebeli, M.; Löwe, H. Size controls on the crossover from normal to self-inhibited sintering of ice spheres. *Acta Mater.* **2021**, *213*, 116926. [CrossRef]
43. Fukue, M.; Mulligan, C.N. Equations of state in soil compression based on statistical mechanics. *Soils Found.* **2009**, *49*, 99–114. [CrossRef]
44. Cruz, R.C.D.; Segadães, A.M.; Oberacker, R.; Hoffmann, M.J. Double layer electrical conductivity as a stability criterion for concentrated colloidal suspensions. *Colloids Surf. A Physicochem. Eng. Asp.* **2017**, *520*, 9–16. [CrossRef]

45. Poortinga, A.T.; Bos, R.; Norde, W.; Busscher, H.J. Electric double layer interactions in bacterial adhesion to surfaces. *Surf. Sci. Rep.* **2002**, *47*, 1–32. [CrossRef]
46. Baciocchia, R.; Bonib, M.R.B.; Lavecchia, R. Modeling of chlorophenols competitive adsorption on soils by means of the ideal adsorbed solution theory. *J. Hazard. Mater.* **2005**, *118*, 239–246. [CrossRef]
47. Zhao, Y.-Y.; Zhang, M.; Hu, X.-M.; Feng, Y.; Xue, D.; Wang, Q.-S.; Geng, Z.; Liu, Y.; Zhang, J.; Jia, X.-H. Study on adsorption and dust suppression mechanism of urease-producing bacteria on coal-soil mixed dust. *J. Environ. Chem. Eng.* **2023**, *11*, 111407. [CrossRef]

materials

MDPI

Article

An Assessment of the Environmental Impact of Construction Materials of Monocrystalline and Perovskite Photovoltaic Power Plants Toward Their Sustainable Development

Izabela Piasecka [1,*] and Zbigniew Kłos [2]

[1] Faculty of Mechanical Engineering, Bydgoszcz University of Science and Technology, al. Prof. S. Kaliskiego 7, 85-796 Bydgoszcz, Poland

[2] Faculty of Civil Engineering and Transport, Poznań University of Technology, 60-965 Poznań, Poland; zbigniew.klos@put.poznan.pl

* Correspondence: izapia000@pbs.edu.pl

Abstract: The interest in alternative energy sources, including the use of solar radiation energy, is growing year by year. Currently, the most frequently installed photovoltaic modules are made of single-crystalline silicon solar cells (sc-Si). However, one of the latest solutions are perovskite solar cells (PSC), which are considered the future of photovoltaics. Therefore, the main objective of this research was to assess the environmental impact of the construction materials of monocrystalline and perovskite photovoltaic power plants toward their sustainable development. The research object was the construction materials and components of two 1 MW photovoltaic power plants: one based on monocrystalline modules and the other on perovskite modules. The life cycle assessment (LCA) method was used for the analyses. The IMPACT World+, IPCC and CED models were used in it. The analyses were performed separately for five sets of elements: support structures, photovoltaic panels, inverter stations, electrical installations and transformers. Two post-consumer management scenarios were adopted: storage and recycling. The life cycle of a photovoltaic power plant based on photovoltaic modules made of perovskite cells is characterized by a smaller negative impact on the environment compared to traditional power plants with monocrystalline silicon modules. Perovskites, as a construction material of photovoltaic modules, fit better into the main assumptions of sustainable development compared to cells made of monocrystalline silicon. However, it is necessary to conduct further work which aims at reducing energy and material consumption in the life cycles of photovoltaic power plants.

Keywords: environmental impact; structural materials; renewable energy materials; sustainable development; photovoltaic power plant

1. Introduction

The photovoltaic industry is currently experiencing an intensive period, which continues to develop dynamically, bringing innovation, growth in installed capacities and strengthening the role of solar energy in the global energy mix. The photovoltaic sector is undergoing not only legislative changes, but also market and technological ones. Industry 4.0 tools are playing increasingly important roles in the photovoltaic industry—from big data to AI. It is predicted that the world will see further, significant development of photovoltaics in the coming years. By 2032, the capacities of the photovoltaic installations put into service each year are to be higher by an average of over 5% than the previous year. The reasons for such rapid development of the industry are seen in ambitious renewable energy goals, an increase in electrification, phasing out coal-fired power plants, concerns about energy security, expanding political support and falling costs of energy production from photovoltaics. According to forecasts, China is to remain the leader in photovoltaic investments for the coming years. This situation will not be changed by incentives for investment in the US, Europe and India. However, China's current huge advantage in photovoltaic investments is expected to decrease due to the larger scale of investment in the US. (Figure 1) [1–4].

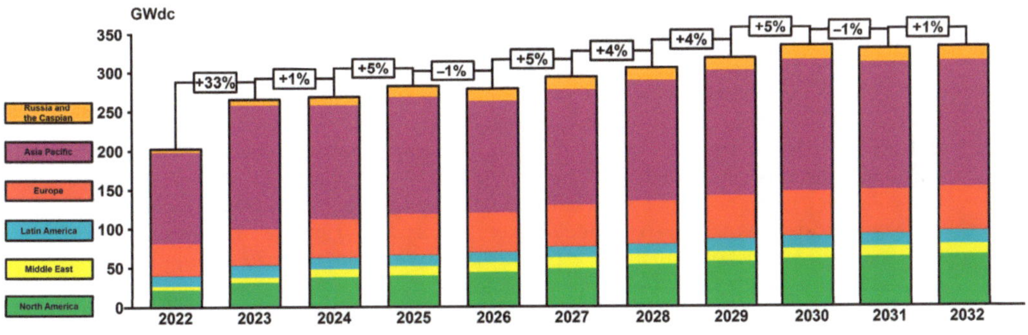

Figure 1. Forecast of growth in photovoltaic investments until 2032. Own work based on [4].

Solar energy is considered "green" and environmentally friendly. However, each source of energy, to a greater or lesser extent, affects the environment. Considering the life cycle of photovoltaic systems, they do not generate significant harmful effects on the environment during the exploitation stage. However, the stages of production and post-consumer management of their materials and components may potentially constitute sources of dangerous impacts, resulting in a decrease in the quality of the environment, an impact on human health and depletion of non-renewable raw materials [5–7].

There is a lack of research in the literature on life cycle analyses of photovoltaic power plants. Most often, one can find studies focusing on their key elements, like photovoltaic modules, and, in particular, on the various types of materials from which they are manufactured. Most research works contain analyses of the life cycles of cells and photovoltaic modules manufactured from silicon. Examples of such studies include Wang et al. [8], Kato et al. [9], Golroudbary et al. [10], Dones and Frischknecht [11], Heath et al. [12], Fthenakis and Kim [13], Reich et al. [14], Frankl [15] and Alsema [16], which are dedicated to single-crystalline silicon (sc-Si) modules. Other noteworthy research on these analyses includes studies conducted by Oliver and Jackson [17], Nomura [18], Kato [19], Ito [20,21], Dones and Frischknecht [11], Fthenakis and Kim [13], Fthenakis and Alsema [22] and Alsema [16], which analyze the life cycles of multi-crystalline silicon (mc-Si) modules. There have also been several studies on amorphous-silicon (a-Si) modules; for example, Kato [19], Ito [20] and Alsema [16], while other works, such as Bravi [23], discuss photovoltaic cells made of multi-junction thin-film silicon (μc-Si). There are also individual works devoted not only to silicon cells but also to those made of other materials; for example, studies by Ito [20], Fthenakis and Kim [13] and Fthenakis and Alsema [22] analyzed the life cycle of cadmium telluride (CdTe) modules. The topic of copper-indium-gallium-diselenide solar cells (CIGS) was taken up by Bravi [23], and the topic of dye-sensitized solar cells (DSSC) was taken up by Greijer [24].

However, there are also several studies on the life cycle of high-power photovoltaic systems in the literature. Examples include the analyses by Kato [9,25] covering 10, 30 and 100 MW power plants built from mc-Si, a-Si and CdTe cells, or the work by Schaefer and Hagedorn [26], which includes studies on 2.5 MW power plants built from sc-Si, mc-Si and a-Si cells.

Currently, the most commonly used type of photovoltaic module is made of monocrystalline silicon (sc-Si). However, research is constantly being conducted to find new solutions that are more efficient in terms of energy, economy and environment. One of them is perovskite modules (PSC), which, as the name suggests, contain a compound with a perovskite structure. Most often, it is a hybrid, organic-inorganic material based on lead or tin halide (the active layer that collects light). Perovskite materials are relatively cheap and easy to produce. The efficiency of this type of cell has increased from approx. 3.8% (2009) to approx. 25.5% (currently) in the case of systems with a single junction, and to approx. 28.8%, in the case of tandem cells based on silicon. For this reason, a decision was made to conduct a comparative analysis of the materials and structural components of two 1 MW photovoltaic power plants—the first one based on monocrystalline silicon modules and the second one—on perovskite modules [1,27–29].

Among the studies that assess the environmental impact of the life cycle of photovoltaic systems, there are no studies which conduct analyses using the relatively new IMPACT World+ method. Most of the studies conducted focus exclusively on the impact of global warming potential (GWP); for example, [19–26]. Usually, scientific studies do not consider a wider range of impacts that threaten human health, cause a decrease in the quality of ecosystems or deepen the depletion of raw materials. However, it was recognized that they required detailed analyses, especially in the context of the main assumptions of the sustainable development of the photovoltaic sector.

Therefore, the aim of the study was to assess the environmental impact of construction materials of monocrystalline and perovskite photovoltaic power plants toward their sustainable development.

2. Materials and Methods

2.1. Object and Plan of Analysis

The subjects of the study were the materials and structural elements of two 1 MW photovoltaic power plants. The first one was based on modules made of monocrystalline silicon, while the second one was based on perovskites (tandem type, based on silicon). The life cycle assessment (LCA) method was selected to assess the environmental impact of the structural materials of monocrystalline and perovskite photovoltaic power plants toward their sustainable development. Based on the ISO 14040 [30] (environmental management, life-cycle assessment, principles and framework) and the ISO 14044 [31] (environmental management, life-cycle assessment, requirements and guidelines) standards, it was decided that the life cycle analysis in this study would include four consecutive stages: determination of goals and scope, life-cycle inventory (LCI), life-cycle impact assessment (LCIA) and interpretation (Figure 2) [32–36].

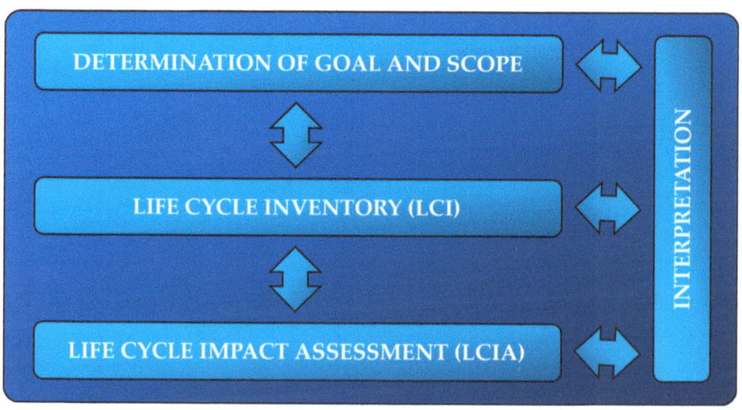

Figure 2. LCA stages of a photovoltaic power plant.

In the first step of the analyses, the aim and scope of the research were described (details are presented in Section 2.2). The analysis of the current state of knowledge and technology (Section 1) revealed that the literature lacks detailed studies on the assessment of the life cycle of photovoltaic power plants, including the impact of the construction materials of monocrystalline and perovskite power plants on the environment (considering not only CO_2 emissions, but also other impacts on the environment, human health and depletion of raw materials). When the aim and scope of the research was developed, it was very important to collect as much reliable data as possible on the technical objects under study. This was made possible via cooperation with companies producing materials and components for photovoltaic power plants and dealing with their operational and post-consumer management (details in Section 2.3). The next stage of the research included a detailed analysis of the life cycle of the power plants under consideration. It was performed using SimaPro 9.5 software (Ecoinvent 3.9.1), IMPACT World+, IPCC and CED models (details in Section 2.4). The obtained results and their interpretation are presented in Sections 3 and 4.

2.2. Determination of Goals and Scope

The first stage of the LCA analysis involves defining its purpose and scope. In this stage, decisions are made about the analytical models to be used, the boundaries of the system being studied are determined and the type of data that must be collected and analyzed is determined. In this phase, it is also necessary to analyze the relationships between the research object and the environment (considering all individual processes). Therefore, during this stage, it is crucial to define the product system (in this case, the photovoltaic power plant), the functional unit and the boundaries of the system [37–40].

The main objective of the conducted analyses was to compare the impact of construction materials occurring in the life cycle of a photovoltaic power plant based on monocrystalline silicon modules and perovskite modules on the environment (comparative analysis). The LCA analysis was conducted to identify potential differences in the size of the environmental impact of two photovoltaic power plants, the construction of which was based on two different technologies (sc-Si and PSC).

The systems of the studied power plants were built in such a way that they were comparable in terms of the breadth and depth of the analyses conducted. The companies with whom cooperation was established and which provided the data necessary for the research had a strong position on the European market, hence, the geographical horizon included the area of Europe. It was assumed that all materials, components and electrical energy necessary for the processes occurring in the life cycle were produced in this area. The time horizon for both facilities was also the same and amounted to 20 years. This time period corresponds to the recommendations of photovoltaic module manufacturers regarding the appropriate time of their operation and is consistent with the practical experience of photovoltaic companies in this area (reduction in energy conversion efficiency over time). Transport processes were not taken into account in the analyses, due to the possibility of large differences in this respect, depending on the potential location of the production plant and the area of assembly and subsequent operation of the test objects. In all analyses, the cut-off level was equal to 0.1%.

The conducted studies were classified as bottom-up analyses, which were used, on the one hand, to describe the existing reality (retrospective analysis), and, on the other hand, to model more sustainable construction solutions (prospective analysis). Due to the high level of advancement, the conducted studies can be classified as detailed analyses. The data used were obtained directly from the manufacturers and companies involved in the operational and post-consumer management of the studied photovoltaic power plants, or when this was not possible—from SimaPro software databases. The installed capacity of both technical facilities, 1 MW, was assumed as the functional unit. The environmental assessment included 27 impact categories of the IMPACT World+ model, 3 impact categories of the IPCC model and 6 impact categories of the CED model. The results were also grouped into two impact areas characterizing the impact on human health and the quality of ecosystems [41–45].

One of the basic research assumptions was to conduct the most detailed analyses possible, covering as many areas of impact on the environment as possible, both in terms of the impact on human health, environmental quality and the issue of resource depletion.

For this reason, it was decided to use the IMPACT World+ model (27 categories), which, alongside ReCiPe 2016 (22 categories), stands out with the largest number of available impact categories. An important aspect was also the possibility of assessing the potential impact on the environment in two time perspectives: short term and long term. Most life cycle analyses presented in the literature include the impact of photovoltaic modules on the global warming potential. In order to compare the results presented in this study with the results found in the literature, a decision was made to conduct additional analyses using the latest LCA model in this area—IPCC 2021. Nowadays, the issue of energy consumption is also an extremely important aspect, which may seem particularly interesting in relation to technical facilities that produce energy. For this reason, it was decided to conduct analyses with the CED model intended for this purpose.

2.3. Life Cycle Inventory (LCI)

The second stage of the assessment includes data collection and their preliminary analysis. Quantitative data are collected to determine the set of inputs and outputs of the examined facility. This is an activity necessary to achieving the purpose of the analysis and to creating a model of the life cycle of the photovoltaic power plants considered. During this phase, the collected input data (energy and material) and output streams (waste and emissions) are identified and quantified. The collected information must also be analyzed for its reliability, completeness and accuracy [46–48].

All processes that took place in the life cycle of the assessed photovoltaic power plants were interconnected by energy and matter flows. Data collected during the LCI were assigned to individual unit processes, to verify them later (based on the mass and energy balance). Models were systematically constructed and then filled with data, with the size of input data always corresponding to the size of output data. The input data included main and auxiliary materials and water demand. The output data included emissions and main products. The data on key processes were obtained by cooperating with the manufacturers and companies involved in operational and post-consumer management. A small amount of data on less important materials and processes (in terms of their impact on the environment) was downloaded from SimaPro 9.5 software (database: Ecoinvent 3.9.1). Due to the conclusion of a confidentiality agreement with the companies producing, operating and managing the photovoltaic power plants, this study does not disclose technological data or information on the structure of these facilities in detail [49–53].

Silicon is a material that is used in the production of both monocrystalline and perovskite cells. Monocrystalline silicon cells are made using the energy-intensive Czochralski method. Their efficiency is higher compared to other silicon cells, such as polycrystalline or amorphous (approx. 22%). A single monocrystalline photovoltaic cell consists of a silicon wafer. On the upper surface of the wafer, an electron-collecting electrode in the form of a grid is placed, and on the lower surface, a lower electrode in the form of a metallic layer is applied (Figure 3). The efficiency of perovskite-silicon tandem solar cells is higher than that of single-crystalline silicon solar cells (approx. 25%). In the analyzed case, the low-bandgap cell took the form of a silicon heterojunction (SHJ) cell, while the wide-bandgap cell took the form of a (Cs, FA) Pb $(I, Br)_3$ perovskite solar cell (Figure 4) [54,55].

Both analyzed power plants had an installed capacity of 1 MW and consisted of the same five groups of structural element assemblies: photovoltaic panels, support structure, inverter station, electrical installation and transformer. The photovoltaic panels were embedded in the ground using a double piling system (two bases). Both solutions used central inverter stations, which constitute a complete solution dedicated to photovoltaic power plants. It houses all the electrical devices necessary for the quick connection of the power plant to the medium-voltage power grid. It includes two central inverters, an optimized transformer, a medium-voltage switchboard, a monitoring system and constant voltage connections for photovoltaic panels. The monocrystalline power plant was built using 2632 PV modules with a total surface area of 4896 m^2. These were modules made of monocrystalline silicon (sc-Si) with a power of 380 W each, dimensions of 1769 × 1052 × 30 mm (1.86 m^2)

and a number of cells of 120 pieces per module. In the case of the perovskite power plant, there were 5556 PV modules with a total surface area of 4390 m^2. These were silicon-based tandem cells (PSC) with a power of 180 W each and dimensions of 1245 × 635 × 2.8 mm (0.79 m^2). Figure 5 shows a functional block diagram of the photovoltaic power plant (data obtained from the investor and manufacturers).

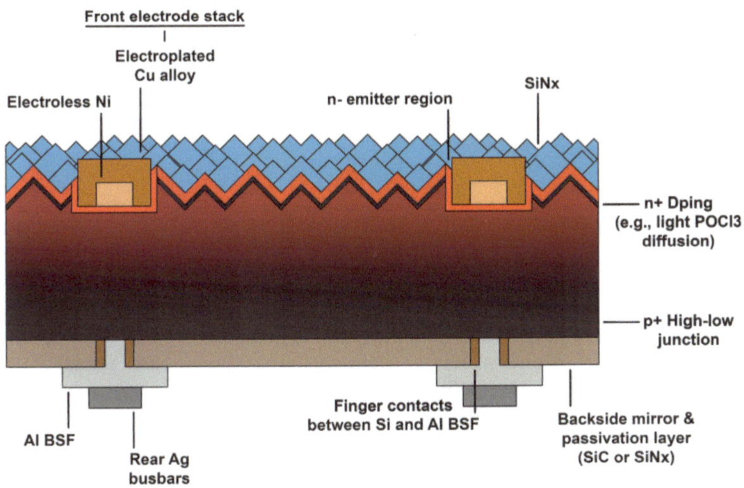

Figure 3. Structure of monocrystalline solar cell (sc-Si). Own work based on [54].

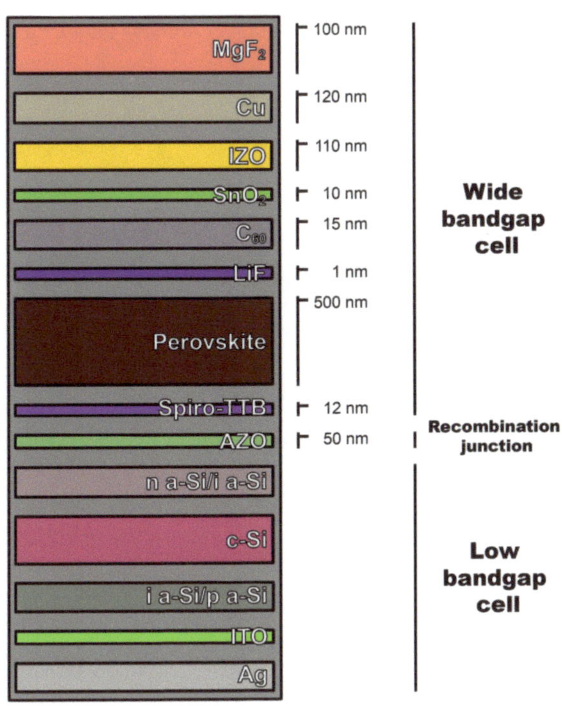

Figure 4. Schematic of perovskite-silicon tandem solar cell. Own work based on [55].

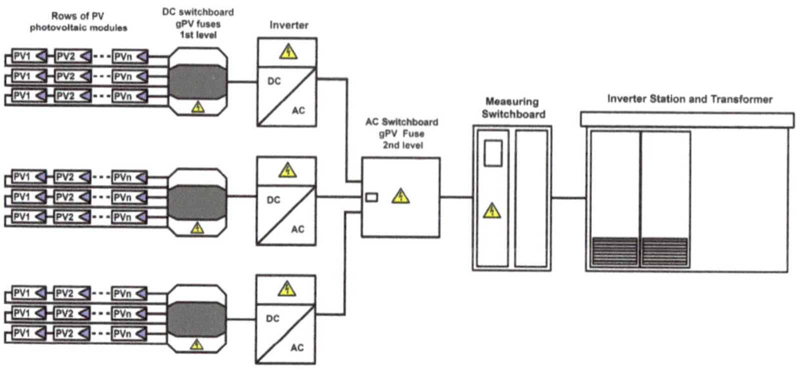

Figure 5. Functional block diagram of a photovoltaic power plant.

The total mass of the construction materials of the analyzed photovoltaic power plant based on monocrystalline modules (sc-Si) was about 200 tons; the total mass of the construction materials based on perovskite modules (PSC) was about 160 tons. The largest share in the mass of the sc-Si power plant was monocrystalline photovoltaic panels made of silicon—about 53% (of which about 47% was solar glass and about 45% was aluminum). In the case of the PSC power plant, there were also photovoltaic panels, comprising 48% of the share in mass. The mass percentage share of the remaining construction elements was similar for both power plants. In the case of the supporting structure, its share was about 17% for sc-Si and about 16% for PSC (mostly made of steel), for the inverter station it was about 13% for sc-Si and about 14% for PSC (elements made mainly of steel (about 42%) and aluminum (about 38%)), while for the electrical installation it was about 2% in both cases (it is mostly made of copper) (Figures 6 and 7) (data obtained from the investor and manufacturers).

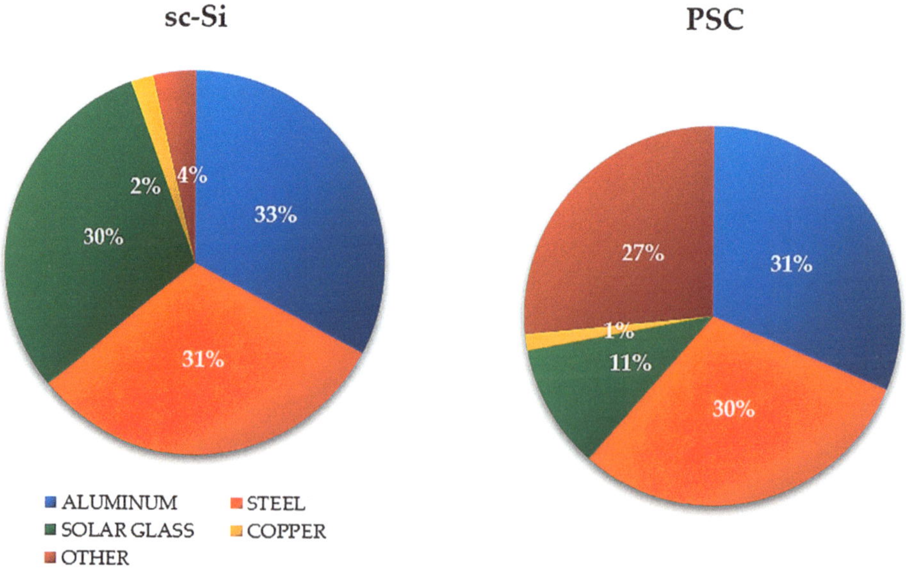

Figure 6. Mass percentage of the most important construction materials of a photovoltaic power plant based on monocrystalline (sc-Si) and perovskite (PSC) modules.

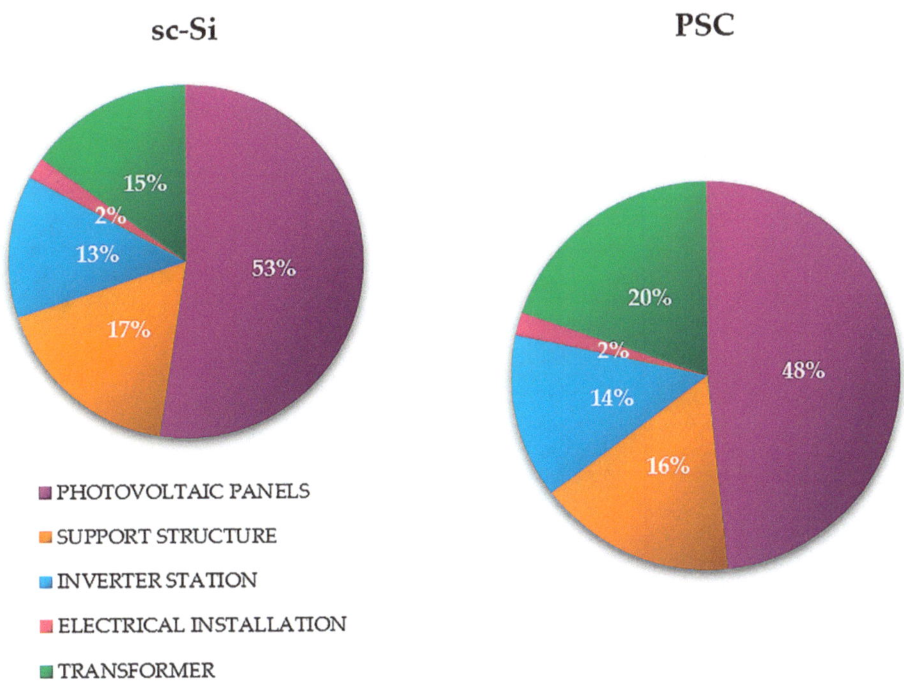

Figure 7. Mass percentage of the most important structural elements of a photovoltaic power plant based on monocrystalline (sc-Si) and perovskite (PSC) modules.

The energy sector is undergoing changes, and the gradual transition to renewable energy sources is more than obvious. Not every technical object that potentially seems sustainable remains so after its life cycle. This topic is one of the most common problems with photovoltaic panels. They are largely recyclable. Materials such as glass, aluminum and semiconductors can be recovered and then reused. So far, the most common recycling methods have been based on mechanical, thermal and chemical processes. Recycling photovoltaic panels involves dismantling the frame located on the PV modules to separate the aluminum and glass parts. Next, the cables and the junction box are removed. The prepared material is then cut and crushed. Almost all glass from PV modules (approx. 95%) can be reused. The same applies to the recycling of external metal parts. Photovoltaic cells are also recycled (approx. 80–90% can be reused). The remaining elements are subjected to high-temperature thermal treatment (up to 500 °C). In this process, the polymer materials surrounding the individual parts of the PV panels evaporate. In this way, the silicon cells become ready for further processing. Their recycling involves acid etching of the silicon particles and enriching the structure so that their original properties are restored. Therefore, the common belief that photovoltaic panels are not suitable for recycling is a myth. However, this is a process that requires time for wide implementation and further research to achieve the full potential of proper recycling of all PV panel components. For this reason, close cooperation between design and recycling units is necessary to ensure the ability to sustainably manage post-consumer waste through conscious eco-design. Although in most cases PV panels can be recycled and, thus, their negative impact on the environment can be minimized, one must be aware of the risk of improper post-consumer management. Landfilling waste poses a huge threat to the natural environment. Considering the heavy metals present in photovoltaic modules (e.g., lead and tin), this can cause serious pollution problems. In addition, they also contain valuable metals such as silver and copper, which further supports their recycling [54,55].

2.4. Life Cycle Impact Assessment (LCIA)

The third stage of LCA studies consists of determining the potential impact of the studied object on the environment. For this reason, the key element of LCIA is to characterize and then classify emissions resulting from the life cycle of the photovoltaic power plant, and then to link these emissions to the effects they may cause to the environment. The obtained results are presented in the form of category indicators (quantitative presentation of impact categories). The LCIA phase consists of mandatory and optional elements. The mandatory ones include the selection of impact categories, category indicators, characterization models, classification and characterization. The optional elements include normalization, grouping and weighting (Figure 8). All mandatory and optional elements were used in the assessment. The analyses were performed using SimaPro 9.5 (Ecoinvent 3.9.1) software, and the IMPACT World+, IPCC and CED models [56–59].

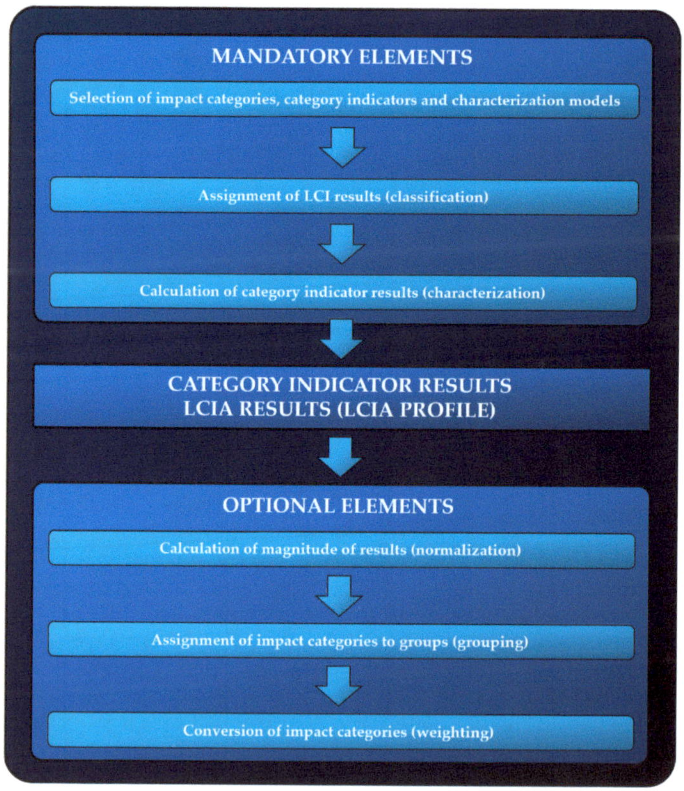

Figure 8. LCIA stages of a photovoltaic power plant.

Classification involves assigning LCI results to impact categories. Specialized analytical software is used for this purpose (in the case of this study, it was SimaPro 9.5) [60,61].

Characterization and conversion of LCI results into category indicator results are very complex processes. They consist of converting LCI results using specific characterization parameters and then presenting them in the form of relative shares in each category. The models used at this stage were IMPACT World+, IPCC and CED [62–64].

Normalization involves determining the size of the impact category indicator results in relation to reference information. It allows for determining the relative weight of the indicator in each geographical horizon (for example Europe) or in relation to a person (for example an average European resident) in an adopted time horizon. As part of the conducted assessment, the normalization procedure was used to prepare the LCIA results for the next step—grouping and weighting [65,66].

Grouping and weighting is a process in which a weighting factor is determined and assigned to impact categories, and then multiplied by normalized index values. In the conducted research, weighting was performed only on complete, internationally recognized sets of coefficients, developed for all considered categories. The implementation of this stage allowed results to be obtained in units, which could be compared (both within the impact category, as well as the impact areas and the final environmental indicator). Analyses of the characterization, normalization, grouping and weighting were performed using SimaPro 9.5 software [67–69].

The IMPACT World+ method is the latest update of the IMPACT 2002+, LUCAS and EDIP methods. It allows for the determination of environmental impacts within 27 impact categories, which can be grouped into two impact areas (impacts on ecosystem quality and human health) and an environmental final indicator. IMPACT World+ includes the broadest set of midpoint impact categories compared to other methods used in LCA analyses. Impact categories are further divided into those covering short-term damages (up to 100 years after emission) and long-term damages (occurring even after 100 years after emission) (Figure 9) [70–72].

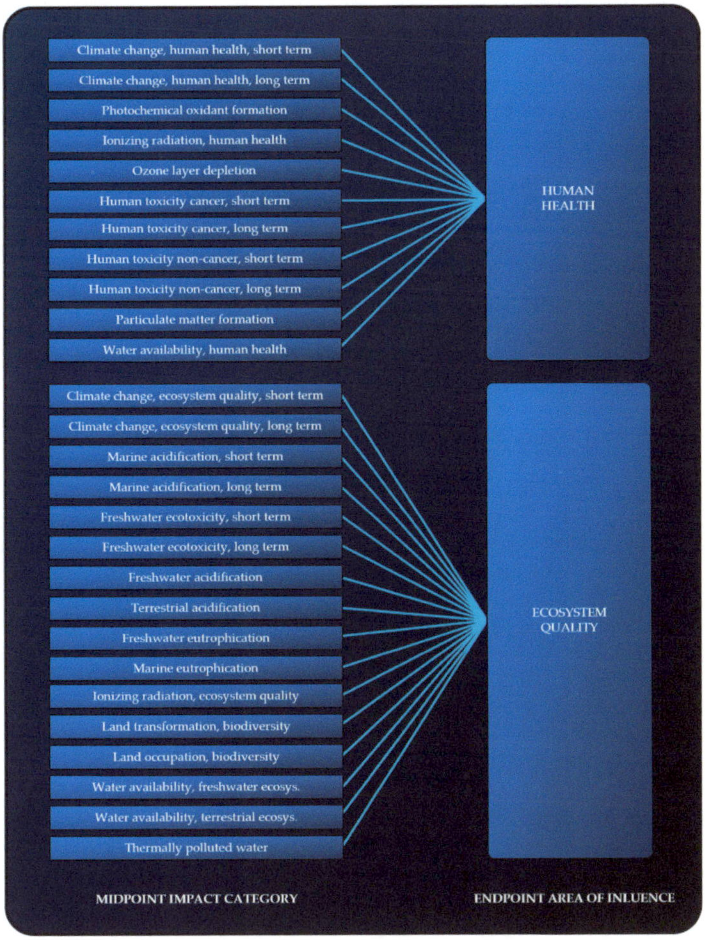

Figure 9. Impact categories and impact areas analyzed within the IMPACT World+ method.

The IPCC 2021 GWP model was also used in the analyses. It allowed the greenhouse potential (GWP) to be determined. It was developed by the Intergovernmental Panel on Climate Change. Carbon dioxide IV (CO_2) is the compound in this model to which the greenhouse potential is referred. For this reason, the results are presented in kg CO_2 eq. The study assumed a 100-year period of GHG impact on the greenhouse effect. The analysis of the impact on the environment included three impact categories: GWP100—fossil, biogenic and land transformation [73–76].

The last method used for the analyses was cumulative energy demand (CED). It allows for determining how much primary energy was used in the life cycle of the research object. The results are presented in MJ or MWh. As part of the assessment, the results were analyzed for all available impact categories—three covering non-renewable sources (fossil, nuclear and biomass) and three covering renewable sources (biomass, wind, solar, geothermal and water) [77–79].

2.5. Interpretation

Interpretation is not only the last stage of LCA; it is present in each of the earlier stages of the conducted procedure. Its aim is to analyze the results and verify them in terms of the previously formulated goal and scope of the research. The completeness of the analysis within the conducted research was checked with a positive result. The data necessary for the interpretation were completed. A compliance check was also performed. The methods used, the assumptions adopted, the detail and depth of the analysis and the accuracy of the data used were consistent with the goal and scope of the research. The obtained assessment results and their interpretation are presented in Sections 3 and 4 [80–82].

3. Results

The obtained results of the analyses were divided into three main sections covering the analytical models used: IMPACT World+ Endpoint version 1.02 (Section 3.1), IPCC 2021 GWP 100 version 1.01 (Section 3.2) and CED—Cumulative Energy Demand version 1.11 (Section 3.3). The Ecoinvent 3.9.1 database was used for the research. The data used were obtained directly from manufacturers and companies involved in the operational and post-consumer management of the studied photovoltaic power plants; when this was not possible, the data were obtained from SimaPro software databases. Within the IMPACT World+ model, the results were grouped into three areas: impact categories (covering 27 categories, Section 3.1.1), areas of influence (covering 2 areas, Section 3.1.2) and total impact (presenting the total impact on the environment, Section 3.1.3).

3.1. IMPACT World+

The presented research results using the IMPACT World+ model were expressed in three types of units: DALY, PDF × m^2 × yr, and EUR. DALY (disability-adjusted life years) is an indicator used to determine health status. It expresses the total number of years of life lost due to premature death or damage to health (because of injury or disease). It was developed by WHO. One DALY means the loss of one year of health, which may be caused by disability or death. The PDF × m^2 × yr unit describes the level of biodiversity. PDF means the potentially disappeared fraction of species, which is calculated on a given area (hence, m^2) or in each volume (hence, m^3), in each period (hence, the addition of years). EUR (euro, €) is the currency of the European Union.

3.1.1. Impact Categories

Among the impact categories characterized by the highest potential negative impact on human health in the life cycle of a photovoltaic power plant based on monocrystalline (sc-Si) modules, we can distinguish water availability, human health and climate change, long term. In both cases, photovoltaic panels were the elements of the power plant with the highest level of harmful impact; in the case of the categories first mentioned, it was 2.13×10^0 DALY (landfill) and 1.27×10^0 DALY (recycling), while for the second, it is was

-2.47×10^{-2} DALY (landfill) and 1.82×10^{-2} DALY (recycling). In the scope of impact categories related to ecosystem quality, the most potential harmful impacts were recorded for freshwater ecotoxicity, long term, climate change and ecosystem quality, long term. Also in these areas, photovoltaic panels were characterized by the highest values of destructive impacts. For the first category, they had the values of 1.58×10^{-2} PDF $\times$ m^2 $\times$ yr (landfill) and 6.78×10^{-4} PDF $\times$ m^2 $\times$ yr (recycling), while for the second, they had the values of 6.62×10^{-5} PDF $\times$ m^2 $\times$ yr (landfill) and 4.56×10^{-5} PDF $\times$ m^2 $\times$ yr (recycling). The main reason for such a high level of adverse impact on the environment is the very energy-intensive process of manufacturing photovoltaic cells from monocrystalline silicon, using the Czochralski method. It also produces a large amount of post-production waste during the cutting of the cells, from a round shape to one close to a rectangular shape. In the case of the remaining elements of the examined photovoltaic power plant (support structure, inverter station, electrical installation and transformer), the impact categories characterized by the maximum value of negative impact on human health included water availability, human health, human toxicity and non-cancer, long term. In the area of impact on ecosystem quality, these are the categories of freshwater ecotoxicity, long term, and climate change, ecosystem quality, long term (analogous to photovoltaic panels). The electrical installation was distinguished by the lowest total level of harmful impact on the environment (both on human health and ecosystem quality). This was due to the lowest mass of materials necessary for its production in comparison to the other groups of elements (approx. 2% of the total mass of the power plant—Figure 7). However, electrical wires (which constitute the largest share in its structure) are made primarily of copper (for which the processes of ore extraction and subsequent production are very energy-intensive) and polymer materials constituting insulation (recycling of which is a complex and technologically demanding process) (Table 1).

The second analyzed photovoltaic power plant was a facility based on perovskite modules (PSC). And, in this case, the highest level of adverse impact, both in terms of impact on human health and on the quality of the ecosystem, was characterized by the life cycle of the photovoltaic panels. The impact categories distinguished by the maximum value of potential harmful impact were again: water availability, human health (1.92×10^{0} DALY for landfill, and 1.15×10^{0} DALY for recycling), climate change, human health, long term (2.23×10^{-2} DALY for landfill, and 1.64×10^{-2} DALY for recycling), freshwater ecotoxicity, long term (1.43×10^{-3} PDF $\times$ m^2 $\times$ yr for landfill, and 6.14×10^{-4} PDF $\times$ m^2 $\times$ yr for recycling), and climate change, ecosystem quality, long term (5.99×10^{-5} PDF $\times$ m^2 $\times$ yr for landfill, and 4.12×10^{-5} PDF $\times$ m^2 $\times$ yr for recycling). Photovoltaic panels have the largest share in the total mass of the power plant (approx. 48%—Figure 7). However, their production is less energy- and material-intensive compared to monocrystalline silicon panels; hence, lower values of destructive impact were noted in the above-mentioned impact categories. In relation to other construction elements (support structures, inverter stations, electrical installations, and transformers), the impact categories with the highest degree of dangerous impact on health again include water availability, human health and human toxicity non-cancer, long-term. In turn, in terms of impact on the ecosystem, these are again the categories of freshwater ecotoxicity, long term, and climate change, ecosystem quality, long term (as in the case of sc-Si panels). Similarly, due to the lowest percentage share in the power plant mass (approx. 2%—Figure 7), the electrical installation elements were characterized by the lowest degree of adverse impact on the environment (both in terms of impact on human health and the quality of the ecosystem) (Table 2).

Table 1. Characterization of environmental consequences in the life cycle of a photovoltaic power plant based on monocrystalline (sc-Si) modules for considered impact categories and different post-consumer management scenarios (model IMPACT World+ Endpoint ver. 1.02).

Element of a Technical Object	Support Structure		Photovoltaic Panels		Inverter Station		Electrical Installation		Transformer		Unit
Form of Post-Use Management / Impact Category	Landfill	Recycling	Landfill	Recycling	Landfill	Recycling	Landfill	Recycling	Landfill	Recycling	
Climate change, human health, short term	3.11×10^{-5}	2.81×10^{-6}	4.70×10^{-3}	3.17×10^{-3}	6.29×10^{-5}	5.66×10^{-5}	2.67×10^{-5}	1.68×10^{-5}	4.53×10^{-5}	4.51×10^{-5}	DALY
Climate change, human health, long term	$\mathbf{9.42 \times 10^{-5}}$	$\mathbf{-7.62 \times 10^{-6}}$	$\mathbf{2.47 \times 10^{-2}}$	$\mathbf{1.82 \times 10^{-2}}$	$\mathbf{1.88 \times 10^{-4}}$	$\mathbf{1.34 \times 10^{-4}}$	$\mathbf{8.25 \times 10^{-5}}$	$\mathbf{4.07 \times 10^{-5}}$	$\mathbf{1.31 \times 10^{-4}}$	$\mathbf{1.09 \times 10^{-4}}$	**DALY**
Photochemical oxidant formation	4.15×10^{-7}	1.49×10^{-7}	1.35×10^{-5}	8.19×10^{-7}	6.95×10^{-7}	5.14×10^{-7}	3.96×10^{-7}	2.03×10^{-7}	4.93×10^{-7}	4.19×10^{-7}	DALY
Ionizing radiation, human health	1.66×10^{-7}	-1.12×10^{-7}	1.34×10^{-5}	9.33×10^{-6}	2.96×10^{-7}	2.41×10^{-7}	9.49×10^{-8}	4.72×10^{-8}	1.96×10^{-7}	1.29×10^{-6}	DALY
Ozone layer depletion	7.13×10^{-9}	-4.50×10^{-9}	9.47×10^{-7}	2.73×10^{-7}	1.41×10^{-8}	5.10×10^{-9}	5.38×10^{-9}	2.04×10^{-9}	1.14×10^{-8}	3.51×10^{-9}	DALY
Human toxicity cancer, short term	8.10×10^{-5}	9.38×10^{-6}	6.12×10^{-5}	3.93×10^{-4}	1.40×10^{-4}	7.37×10^{-5}	6.16×10^{-5}	7.35×10^{-6}	1.17×10^{-4}	8.58×10^{-5}	DALY
Human toxicity cancer, long term	3.49×10^{-6}	2.19×10^{-6}	2.22×10^{-5}	1.51×10^{-5}	9.99×10^{-6}	4.66×10^{-6}	1.68×10^{-6}	5.88×10^{-6}	4.03×10^{-6}	1.85×10^{-6}	DALY
Human toxicity non-cancer, short term	1.71×10^{-5}	4.00×10^{-5}	7.51×10^{-5}	4.24×10^{-5}	2.92×10^{-5}	4.49×10^{-5}	1.30×10^{-5}	3.41×10^{-5}	2.33×10^{-5}	3.56×10^{-5}	DALY
Human toxicity non-cancer, long term	2.01×10^{-4}	1.37×10^{-4}	7.06×10^{-4}	4.16×10^{-5}	3.43×10^{-4}	1.58×10^{-4}	1.47×10^{-4}	1.07×10^{-4}	2.76×10^{-4}	1.13×10^{-4}	DALY
Particulate matter formation	5.80×10^{-5}	1.60×10^{-5}	3.02×10^{-3}	2.08×10^{-3}	1.15×10^{-4}	6.09×10^{-5}	4.05×10^{-5}	1.73×10^{-5}	6.75×10^{-5}	4.03×10^{-5}	DALY
Water availability, human health	$\mathbf{1.89 \times 10^{-2}}$	$\mathbf{-1.72 \times 10^{-2}}$	$\mathbf{2.13 \times 10^{0}}$	$\mathbf{1.27 \times 10^{0}}$	$\mathbf{3.10 \times 10^{-2}}$	$\mathbf{1.03 \times 10^{-2}}$	$\mathbf{1.14 \times 10^{-2}}$	$\mathbf{2.16 \times 10^{-3}}$	$\mathbf{2.54 \times 10^{-2}}$	$\mathbf{1.01 \times 10^{-2}}$	**DALY**
Climate change, ecosystem quality, short term	7.70×10^{-8}	6.94×10^{-9}	2.16×10^{-5}	1.46×10^{-5}	1.36×10^{-7}	1.22×10^{-7}	5.77×10^{-8}	3.63×10^{-8}	9.80×10^{-8}	9.73×10^{-8}	PDF × m^2 × yr
Climate change, ecosystem quality, long term	$\mathbf{2.37 \times 10^{-7}}$	$\mathbf{-1.90 \times 10^{-8}}$	$\mathbf{6.62 \times 10^{-5}}$	$\mathbf{4.56 \times 10^{-5}}$	$\mathbf{4.14 \times 10^{-7}}$	$\mathbf{2.94 \times 10^{-7}}$	$\mathbf{1.81 \times 10^{-7}}$	$\mathbf{8.97 \times 10^{-8}}$	$\mathbf{2.89 \times 10^{-7}}$	$\mathbf{2.39 \times 10^{-7}}$	**PDF × m^2 × yr**
Marine acidification, short term	5.89×10^{-9}	7.53×10^{-10}	1.72×10^{-6}	1.19×10^{-6}	1.06×10^{-8}	7.66×10^{-9}	4.61×10^{-9}	2.29×10^{-9}	7.43×10^{-9}	6.14×10^{-9}	PDF × m^2 × yr
Marine acidification, long term	5.42×10^{-8}	6.94×10^{-9}	1.59×10^{-5}	1.09×10^{-5}	9.81×10^{-8}	7.06×10^{-8}	4.25×10^{-8}	2.11×10^{-8}	6.84×10^{-8}	5.66×10^{-8}	PDF × m^2 × yr
Freshwater ecotoxicity, short term	1.59×10^{-8}	1.33×10^{-8}	4.05×10^{-6}	2.05×10^{-6}	3.37×10^{-8}	1.73×10^{-8}	9.92×10^{-9}	3.62×10^{-9}	2.09×10^{-8}	1.47×10^{-8}	PDF × m^2 × yr
Freshwater ecotoxicity, long term	$\mathbf{5.12 \times 10^{-5}}$	$\mathbf{2.31 \times 10^{-5}}$	$\mathbf{1.58 \times 10^{-3}}$	$\mathbf{6.78 \times 10^{-4}}$	$\mathbf{1.45 \times 10^{-4}}$	$\mathbf{1.56 \times 10^{-5}}$	$\mathbf{4.17 \times 10^{-5}}$	$\mathbf{1.13 \times 10^{-6}}$	$\mathbf{6.01 \times 10^{-5}}$	$\mathbf{9.27 \times 10^{-6}}$	**PDF × m^2 × yr**
Freshwater acidification	1.34×10^{-8}	2.14×10^{-9}	8.80×10^{-7}	5.91×10^{-7}	3.97×10^{-8}	1.66×10^{-8}	1.05×10^{-8}	3.78×10^{-9}	1.68×10^{-8}	7.59×10^{-9}	PDF × m^2 × yr
Terrestrial acidification	8.64×10^{-8}	1.50×10^{-8}	5.74×10^{-6}	3.87×10^{-6}	2.46×10^{-7}	1.05×10^{-7}	7.78×10^{-8}	2.83×10^{-8}	1.05×10^{-7}	4.84×10^{-8}	PDF × m^2 × yr
Freshwater eutrophication	1.18×10^{-10}	-8.07×10^{-11}	6.19×10^{-9}	2.59×10^{-9}	2.29×10^{-10}	1.95×10^{-10}	7.55×10^{-11}	3.63×10^{-11}	1.57×10^{-10}	1.22×10^{-10}	PDF × m^2 × yr
Marine eutrophication	2.10×10^{-9}	1.48×10^{-9}	1.27×10^{-7}	8.17×10^{-8}	3.68×10^{-9}	2.59×10^{-9}	1.45×10^{-9}	1.37×10^{-9}	2.69×10^{-9}	1.93×10^{-9}	PDF × m^2 × yr
Ionizing radiation, ecosystem quality	1.80×10^{-16}	-9.66×10^{-17}	2.22×10^{-14}	1.45×10^{-14}	3.23×10^{-16}	1.48×10^{-16}	1.66×10^{-16}	3.82×10^{-17}	2.09×10^{-16}	7.89×10^{-17}	PDF × m^2 × yr
Land transformation, biodiversity	4.32×10^{-8}	3.38×10^{-9}	4.64×10^{-6}	2.77×10^{-6}	8.48×10^{-8}	5.84×10^{-8}	3.65×10^{-8}	2.28×10^{-8}	5.27×10^{-8}	3.39×10^{-8}	PDF × m^2 × yr
Land occupation, biodiversity	1.29×10^{-8}	7.19×10^{-9}	2.15×10^{-6}	1.49×10^{-6}	2.58×10^{-8}	1.84×10^{-8}	1.16×10^{-8}	7.18×10^{-9}	1.51×10^{-8}	1.10×10^{-8}	PDF × m^2 × yr
Water availability, freshwater ecosystem	1.95×10^{-10}	-1.58×10^{-10}	2.37×10^{-9}	1.09×10^{-9}	3.51×10^{-10}	6.55×10^{-12}	1.50×10^{-10}	9.90×10^{-12}	2.69×10^{-10}	8.04×10^{-12}	PDF × m^2 × yr
Water availability, terrestrial ecosystem	1.49×10^{-11}	6.07×10^{-12}	3.80×10^{-9}	2.32×10^{-9}	2.65×10^{-11}	1.79×10^{-11}	1.15×10^{-11}	5.80×10^{-12}	1.73×10^{-11}	1.26×10^{-11}	PDF × m^2 × yr
Thermally polluted water	1.13×10^{-12}	5.26×10^{-13}	5.19×10^{-10}	3.30×10^{-10}	2.60×10^{-12}	1.41×10^{-12}	1.07×10^{-12}	3.26×10^{-13}	2.19×10^{-12}	1.92×10^{-12}	PDF × m^2 × yr

Bold data—the highest levels of negative consequences.

Table 2. Characterization of environmental consequences in the life cycle of a perovskite photovoltaic power plant (PSC) for considered impact categories and different post-consumer management scenarios (model IMPACT World+ Endpoint ver. 1.02).

Element of a Technical Object	Support Structure		Photovoltaic Panels		Inverter Station		Electrical Installation		Transformer		Unit
Form of Post-Use Management	Landfill	Recycling	Landfill	Recycling	Landfill	Recycling	Landfill	Recycling	Landfill	Recycling	
Impact Category											
Climate change, human health, short term	2.82×10^{-5}	2.54×10^{-6}	4.26×10^{-3}	2.87×10^{-3}	5.69×10^{-5}	5.12×10^{-5}	2.41×10^{-5}	1.52×10^{-5}	4.10×10^{-5}	4.08×10^{-5}	DALY
Climate change, human health, long term	**8.52×10^{-5}**	**-6.90×10^{-6}**	**2.23×10^{-2}**	**1.64×10^{-2}**	**1.70×10^{-4}**	**1.21×10^{-4}**	**7.46×10^{-5}**	**3.68×10^{-5}**	**1.19×10^{-4}**	**9.85×10^{-5}**	**DALY**
Photochemical oxidant formation	3.76×10^{-7}	1.35×10^{-7}	1.22×10^{-5}	7.41×10^{-7}	6.29×10^{-7}	4.65×10^{-7}	3.58×10^{-7}	1.84×10^{-7}	4.46×10^{-7}	3.79×10^{-7}	DALY
Ionizing radiation, human health	1.50×10^{-7}	-1.02×10^{-7}	1.22×10^{-5}	8.44×10^{-6}	2.68×10^{-7}	2.18×10^{-7}	8.59×10^{-8}	4.27×10^{-8}	1.78×10^{-7}	1.17×10^{-7}	DALY
Ozone layer depletion	6.45×10^{-9}	-4.08×10^{-9}	8.57×10^{-7}	2.47×10^{-7}	1.27×10^{-8}	4.62×10^{-9}	4.87×10^{-9}	1.85×10^{-9}	1.03×10^{-8}	3.18×10^{-9}	DALY
Human toxicity cancer, short term	7.33×10^{-5}	8.48×10^{-6}	5.53×10^{-4}	3.56×10^{-4}	1.26×10^{-4}	6.67×10^{-5}	5.58×10^{-5}	6.65×10^{-6}	1.06×10^{-4}	7.77×10^{-5}	DALY
Human toxicity cancer, long term	3.16×10^{-6}	1.98×10^{-6}	2.01×10^{-5}	1.36×10^{-5}	9.04×10^{-6}	4.22×10^{-6}	1.52×10^{-6}	5.32×10^{-7}	3.65×10^{-6}	1.67×10^{-6}	DALY
Human toxicity non-cancer, short term	1.55×10^{-4}	3.62×10^{-5}	6.80×10^{-4}	3.84×10^{-4}	2.64×10^{-4}	4.06×10^{-5}	1.18×10^{-4}	3.08×10^{-5}	2.11×10^{-4}	3.22×10^{-5}	DALY
Human toxicity non-cancer, long term	1.81×10^{-4}	1.24×10^{-4}	6.39×10^{-4}	3.76×10^{-5}	3.11×10^{-4}	1.43×10^{-4}	1.33×10^{-4}	9.65×10^{-5}	2.50×10^{-4}	1.02×10^{-4}	DALY
Particulate matter formation	5.25×10^{-5}	1.45×10^{-5}	2.74×10^{-3}	1.88×10^{-3}	1.04×10^{-4}	5.51×10^{-5}	3.66×10^{-5}	1.57×10^{-5}	6.10×10^{-5}	3.65×10^{-5}	DALY
Water availability, human health	**1.71×10^{-2}**	**-1.55×10^{-2}**	**1.92×10^{0}**	**1.15×10^{0}**	**2.80×10^{-2}**	**9.36×10^{-3}**	**1.03×10^{-2}**	**1.95×10^{-3}**	**2.29×10^{-2}**	**9.10×10^{-3}**	**DALY**
Climate change, ecosystem quality, short term	6.97×10^{-8}	6.28×10^{-9}	1.95×10^{-5}	1.32×10^{-5}	1.23×10^{-7}	1.11×10^{-7}	5.22×10^{-8}	3.28×10^{-8}	8.87×10^{-8}	8.81×10^{-8}	PDF × m^2 × yr
Climate change, ecosystem quality, long term	**2.14×10^{-7}**	**-1.72×10^{-8}**	**5.99×10^{-5}**	**4.12×10^{-5}**	**3.74×10^{-7}**	**2.66×10^{-7}**	**1.64×10^{-7}**	**8.12×10^{-8}**	**2.61×10^{-7}**	**2.16×10^{-7}**	**PDF × m^2 × yr**
Marine acidification, short term	5.33×10^{-9}	6.81×10^{-10}	1.56×10^{-6}	1.07×10^{-6}	9.63×10^{-9}	6.93×10^{-9}	4.17×10^{-9}	2.07×10^{-9}	6.72×10^{-9}	5.55×10^{-9}	PDF × m^2 × yr
Marine acidification, long term	4.91×10^{-8}	6.28×10^{-9}	1.44×10^{-5}	9.90×10^{-6}	8.88×10^{-8}	6.39×10^{-8}	3.84×10^{-8}	1.91×10^{-8}	6.19×10^{-8}	5.12×10^{-8}	PDF × m^2 × yr
Freshwater ecotoxicity, short term	1.44×10^{-8}	1.20×10^{-8}	3.67×10^{-6}	1.85×10^{-6}	3.05×10^{-8}	1.56×10^{-8}	8.97×10^{-9}	3.27×10^{-9}	1.89×10^{-8}	1.33×10^{-8}	PDF × m^2 × yr
Freshwater ecotoxicity, long term	**4.63×10^{-5}**	**2.09×10^{-5}**	**1.43×10^{-3}**	**6.14×10^{-4}**	**1.32×10^{-4}**	**1.41×10^{-5}**	**3.78×10^{-5}**	**1.03×10^{-6}**	**5.44×10^{-5}**	**8.39×10^{-6}**	**PDF × m^2 × yr**
Freshwater acidification	1.21×10^{-8}	1.93×10^{-9}	7.96×10^{-7}	5.35×10^{-7}	3.59×10^{-8}	1.50×10^{-8}	9.53×10^{-9}	3.42×10^{-9}	1.52×10^{-8}	6.87×10^{-9}	PDF × m^2 × yr
Terrestrial acidification	7.82×10^{-8}	1.35×10^{-8}	5.19×10^{-6}	3.50×10^{-6}	2.23×10^{-7}	9.54×10^{-8}	7.04×10^{-8}	2.56×10^{-8}	9.48×10^{-8}	4.38×10^{-8}	PDF × m^2 × yr
Freshwater eutrophication	1.07×10^{-10}	-7.30×10^{-11}	5.60×10^{-9}	2.34×10^{-9}	2.08×10^{-10}	1.77×10^{-10}	6.83×10^{-11}	3.28×10^{-11}	1.42×10^{-10}	1.10×10^{-10}	PDF × m^2 × yr
Marine eutrophication	1.90×10^{-9}	1.34×10^{-9}	1.15×10^{-7}	7.39×10^{-8}	3.33×10^{-9}	2.35×10^{-9}	1.31×10^{-9}	1.24×10^{-9}	2.43×10^{-9}	1.75×10^{-9}	PDF × m^2 × yr
Ionizing radiation, ecosystem quality	1.63×10^{-16}	-8.74×10^{-17}	2.00×10^{-14}	1.31×10^{-14}	2.92×10^{-16}	1.34×10^{-16}	1.50×10^{-16}	3.45×10^{-17}	1.89×10^{-16}	7.14×10^{-17}	PDF × m^2 × yr
Land transformation, biodiversity	3.91×10^{-8}	3.06×10^{-9}	4.20×10^{-6}	2.51×10^{-6}	7.67×10^{-8}	5.28×10^{-8}	3.30×10^{-8}	2.06×10^{-8}	4.77×10^{-8}	3.07×10^{-8}	PDF × m^2 × yr
Land occupation, biodiversity	1.16×10^{-8}	6.51×10^{-9}	1.94×10^{-6}	1.35×10^{-6}	2.33×10^{-8}	1.66×10^{-8}	1.05×10^{-8}	6.50×10^{-9}	1.37×10^{-8}	9.99×10^{-9}	PDF × m^2 × yr
Water availability, freshwater ecosys.	1.76×10^{-10}	-1.43×10^{-10}	2.14×10^{-9}	9.82×10^{-10}	3.18×10^{-10}	5.93×10^{-12}	1.35×10^{-10}	8.96×10^{-12}	2.43×10^{-10}	7.28×10^{-12}	PDF × m^2 × yr
Water availability, terrestrial ecosys.	1.35×10^{-11}	5.49×10^{-12}	3.44×10^{-9}	2.10×10^{-9}	2.40×10^{-11}	1.62×10^{-11}	1.04×10^{-11}	5.25×10^{-12}	1.57×10^{-11}	1.14×10^{-11}	PDF × m^2 × yr
Thermally polluted water	1.02×10^{-12}	4.76×10^{-13}	4.70×10^{-10}	2.99×10^{-10}	2.35×10^{-12}	1.28×10^{-12}	9.72×10^{-13}	2.95×10^{-13}	1.98×10^{-12}	1.73×10^{-12}	PDF × m^2 × yr

Bold data—the highest levels of negative consequences.

According to the above, in the perspective of the life cycle of materials and elements of both analyzed photovoltaic power plants, the categories of impact with the highest level of harmful impact on human health include water availability, human health, climate change and human health, long term. In turn, in terms of impact on ecosystem quality, these are the categories of freshwater ecotoxicity, long term, and climate change, ecosystem quality, long term. For this reason, a decision was made to conduct a more detailed analysis of destructive impacts in these areas.

The first of the categories considered were water availability and human health. It illustrates how the life cycle of the tested materials and construction elements affects the availability of water, the deficiency of which can directly affect health (especially in areas with a lower level of economic development and those where access to water is very limited). The maximum value of the negative impact in this area characterizes the life cycle of monocrystalline photovoltaic panels (sc-Si), which would be stored in a landfill after the end of their use (2.13×10^0 DALY). The use of recycling processes would significantly reduce the amount of hazardous impact on health (less about 8.58×10^{-1} DALY). For all the assessed elements, the life cycle of a power plant based on perovskite cells (PSC) generates fewer adverse effects on water availability, compared to a power plant based on monocrystalline silicon cells (sc-Si). Among the processes related to the reduction in water availability in the life cycles of both technical facilities examined, we can distinguish primarily the consumption of water from various sources (lakes, rivers, wells, etc.) and its use for cooling turbines (during electricity generation). The issue of limited water availability is one of the key environmental, economic and health problems. People overuse available water resources and drain huge areas of land (e.g., for the purpose of building mines). Nowadays, its consumption is greater than the increase, which leads to the depletion of resources. Another problem is the reduction in water quality through its contamination caused by, for example, agriculture, industry, transport, low emissions, detergents or poorly secured or illegal landfills. The average water consumption in European countries is approx. 1200 m^3/yr, in USA—aprox. 2500 m^3/yr, and in Ethiopia—aprox. 360 m^3/yr (due to difficult accessibility). Almost 900 million people in the world do not have access to water considered suitable for drinking. In developing countries, lack of access to clean water is the cause of about 80% of diseases. Therefore, special attention should be paid to the issue of water demand in all industrial processes, including those in the life cycle of photovoltaic power plants (from the extraction of raw materials to the post-consumer management of materials and construction elements) (Figure 10) [83–85].

The second analyzed impact category is climate change, human health, long term. It includes emissions of chemical compounds throughout the entire life cycle of photovoltaic power plants, which contribute to the deepening of climate change and, consequently, are characterized by a harmful effect on health. The long-term perspective considers the potential impact of the assessed substances on the environment for a period exceeding 100 years from the moment of emission. In this case, the greatest degree of dangerous impact is also distinguished by the life cycle of monocrystalline photovoltaic panels, which would be placed in a landfill (2.47×10^{-2} DALY). However, recycling would, to some extent, make it possible to reduce the level of adverse impact on human health in the area under consideration (less about 6.54×10^{-3} DALY). The life cycle of the sc-Si power plant causes more harmful impacts in terms of climate change than the life cycle of the PSC power plant. Among the chemical compounds influencing climate change, in the life cycles of both studied power plants, the highest values of destructive impact were characterized by emissions: carbon dioxide, methane, dinitrogen monoxide, sulfur hexafluoride, tetrafluoromethane (CFC-14), hexafluoroethane (HFC-116), and trifluoromethane (HFC-23). Carbon dioxide is an important part of the carbon cycle in nature, because it is a product of combustion and respiration. Large amounts of it are emitted in the processes of manufacturing materials and components of the tested power plants, which are characterized by a high level of energy and material intensity. The increase in its concentration in the environment results in a deepening of the greenhouse effect and many other dangerous consequences for human health and the quality of the ecosystem. On the other hand, life on Earth would not be possible without CO_2. The problem is not the existence of carbon dioxide itself, but the large increase in its concentration, which is currently occurring at an increasingly rapid pace. (Figure 11) [86,87].

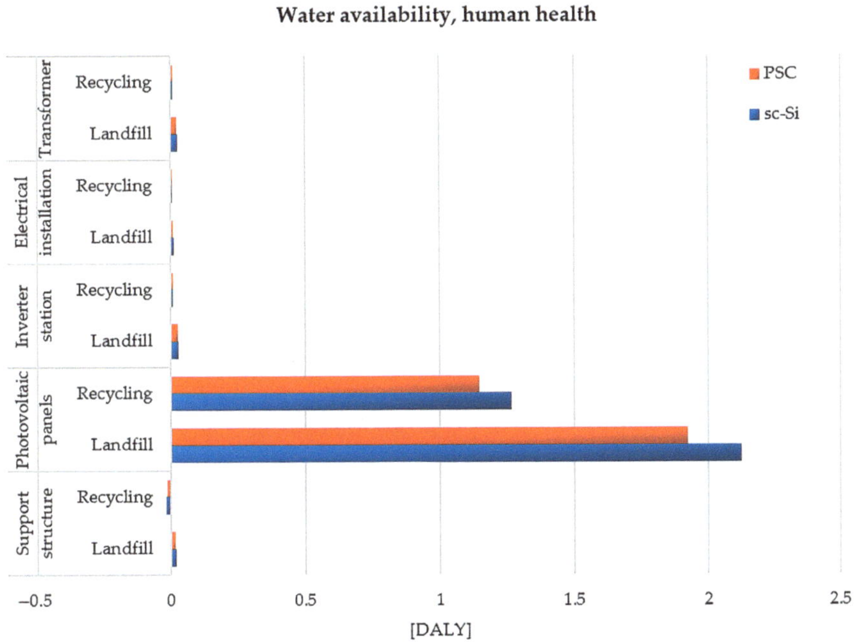

Figure 10. Characterization of environmental consequences of processes affecting water availability, affecting human health, in the life cycle of the studied photovoltaic power plants (model IMPACT World+ Endpoint ver. 1.02, unit: DALY).

Among the impact categories characterizing the impact values on the quality of ecosystems, the issue of access to water suitable for consumption is also extremely important. Within the impact category of freshwater ecotoxicity, long term, the maximum degree of negative impact was characterized by the life cycle of monocrystalline photovoltaic panels (sc-Si), which would not be recycled after the end of their use, but would be placed in a landfill (1.58×10^{-3} PDF $\times$ m^2 $\times$ yr). Post-consumer management in the form of recycling would significantly reduce the level of unsafe impact on the environment (less about 9.00×10^{-4} PDF $\times$ m^2 $\times$ yr). For each of the analyzed structural elements of both power plants, the use of recycling processes would result in a significant reduction in the value of adverse impacts on the environment. The life cycle of the power plant based on photovoltaic modules made of perovskites is characterized by a lower level of harmful impact in the analyzed area compared to the sc-Si power plant. Among the structural materials of the assessed power plants, the life cycle of which has the greatest impact on the increase in the ecotoxicity of freshwater, the following can be distinguished: copper, aluminum, iron, strontium, nickel, zinc, manganese, cadmium, and vanadium. Copper is an element that participates in several biochemical reactions. It plays an important role in photosynthesis and in the formation of proteins. However, the exploitation and processing of copper ore deposits (and other metals) results in the introduction of many harmful substances into the environment (e.g., through the discharge of mine waters, sewage from smelters and enrichment plants, dust emissions from smelters, the formation of smelter and flotation waste, etc.). The largest number of harmful emissions goes to the aquatic environment (landfill leachates, soil runoff, sewage discharge, etc.). For this reason, it is extremely important to increase the level of recycling of copper waste. Its recycling is characterized by a high level of efficiency and energy saving. It requires about 85% less energy input compared to primary production. Copper can be recovered many times, because its processing does not deteriorate the quality and functional properties. Due to the high energy consumption of the processes related to the production of copper elements and the huge interference in the environment of its mining and processing processes, despite the fact that the percentage share of copper in the mass of the studied power plants is small (about 2% for sc-Si and about 1% for PSC—Figure 6), its life cycle has a large impact on the pollution of the aquatic environment (Figure 12) [88–90].

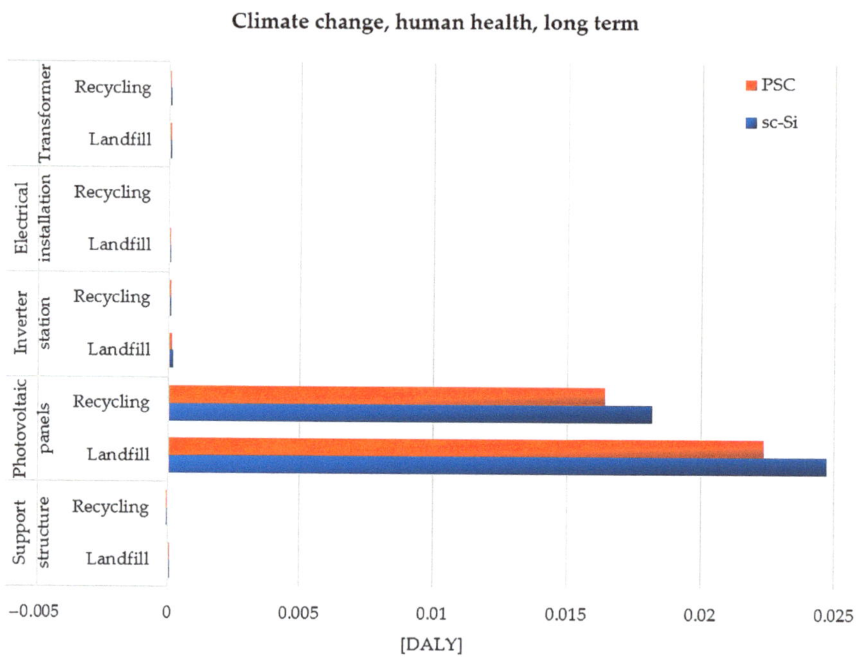

Figure 11. Characterization of the environmental consequences of emissions of climate-changing substances affecting human health (long-term perspective) over the life cycle of the studied photovoltaic power plants (model IMPACT World+ Endpoint ver. 1.02, unit: DALY).

The last of the analyzed impact categories is climate change, ecosystem quality, long term. It is therefore clear that the issue of climate change is crucial not only in the area of impact on human health, but also on the quality of the ecosystem. As in the case of the previously characterized categories, the maximum value of destructive impact was distinguished by the life cycle of photovoltaic panels of the sc-Si power plant, stored after the end of operation (6.62×10^{-5} PDF $\times$ m^2 $\times$ yr). Recycling would significantly reduce the negative impact in the assessed area (less about 2.07×10^{-5} PDF $\times$ m^2 $\times$ yr). The life cycle of a photovoltaic power plant based on monocrystalline silicon panels generates more hazardous environmental consequences compared to the life cycle of a power plant based on perovskite panels. The chemical compounds occurring in the life cycle of the considered power plants that cause the most adverse impacts in the area of climate change includes: methane, carbon dioxide, dinitrogen monoxide, sulfur hexafluoride, tetrafluoromethane (CFC-14), hexafluoroethane (HFC-116), and trifluoromethane (HFC-23). Methane is a greenhouse gas that poses a significant threat to the quality of ecosystems and human health. During the first 20 years of its presence in the atmosphere, its impact on the climate is about 85 times greater than the same mass of carbon dioxide. Its largest sources are agriculture and the energy industry (using conventional energy sources). Reducing its emissions is possible, among other things, by moving away from fossil fuels and securing abandoned oil and gas wells and closed mines against leaks. Processes related to the production and management of materials and components of photovoltaic power plants, requiring large inputs of energy and matter, are characterized by a particularly high level of methane emissions to the environment (Figure 13) [91,92].

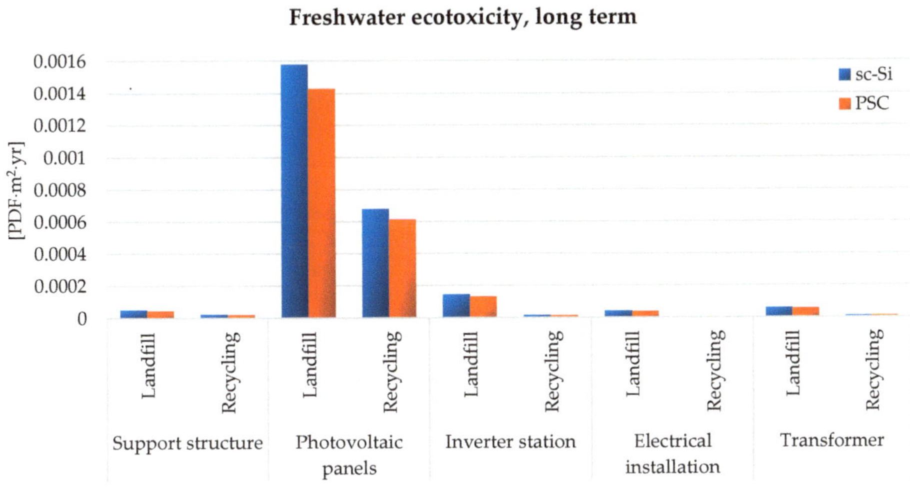

Figure 12. Characterization of the environmental consequences of the emission of substances with ecotoxic effects on freshwater ecosystems (long-term perspective) in the life cycle of the studied photovoltaic power plants (model IMPACT World+ Endpoint ver. 1.02, unit: PDF × m^2 × yr).

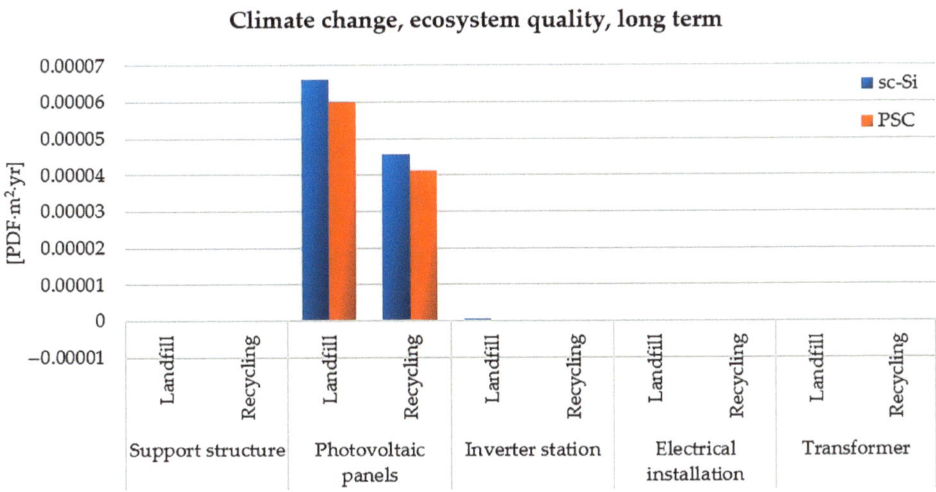

Figure 13. Characterization of the environmental consequences of emissions of substances causing climate change, affecting the quality of ecosystems (long-term perspective), in the life cycle of the studied photovoltaic power plants (model IMPACT World+ Endpoint ver. 1.02, unit: PDF × m^2 × yr).

3.1.2. Areas of Influence

As part of the analyses, the impact of the life cycles of the studied power plants on two areas of impact was also assessed—human health and the quality of the ecosystem. The highest level of the total adverse impact on human health was characteristic of the life cycle of photovoltaic panels (sc-Si: 2.16 × 10^0 DALY for landfill, and 1.29 × 10^0 DALY for recycling, PSC: 1.95 × 10^0 DALY for landfill, and 1.17 × 10^0 DALY for recycling). The remaining elements of the power plant were characterized by significantly lower values of the total harmful impact: inverter station (sc-Si: 3.21 × 10^{-2} DALY for landfill, and 1.09 × 10^{-2} DALY for recycling, PSC: 2.91 × 10^{-2} DALY for landfill, and 9.84 × 10^{-3} DALY for recycling), transformer (sc-Si: 2.26 × 10^{-2} DALY for landfill, and 1.05 × 10^{-2} DALY for recycling, PSC: 2.37 × 10^{-2} DALY for landfill, and 9.49 × 10^{-3} DALY for recycling), support structure (sc-Si: 1.96 × 10^{-2} DALY for landfill, and −1.70 × 10^{-2} DALY for recycling, PSC: 1.77 × 10^{-2} DALY for landfill, and −1.54 × 10^{-2} DALY for recycling), and electrical installation (sc-Si: 1.19 × 10^{-2} DALY for landfill, and 2.38 × 10^{-3} DALY for recycling, PSC: 1.08 × 10^{-2} DALY for landfill, and 2.16 × 10^{-3} DALY for recycling). For each element considered, for both power plants studied, the life cycle with post-consumer disposal in the form of storage had more negative consequences for human health than the cycle with the use of recycling processes. The power plant based on monocrystalline silicon (sc-Si) technology caused more destructive impacts in the assessed area compared to the perovskite (PSC) power plant (Figure 14 and Table 3).

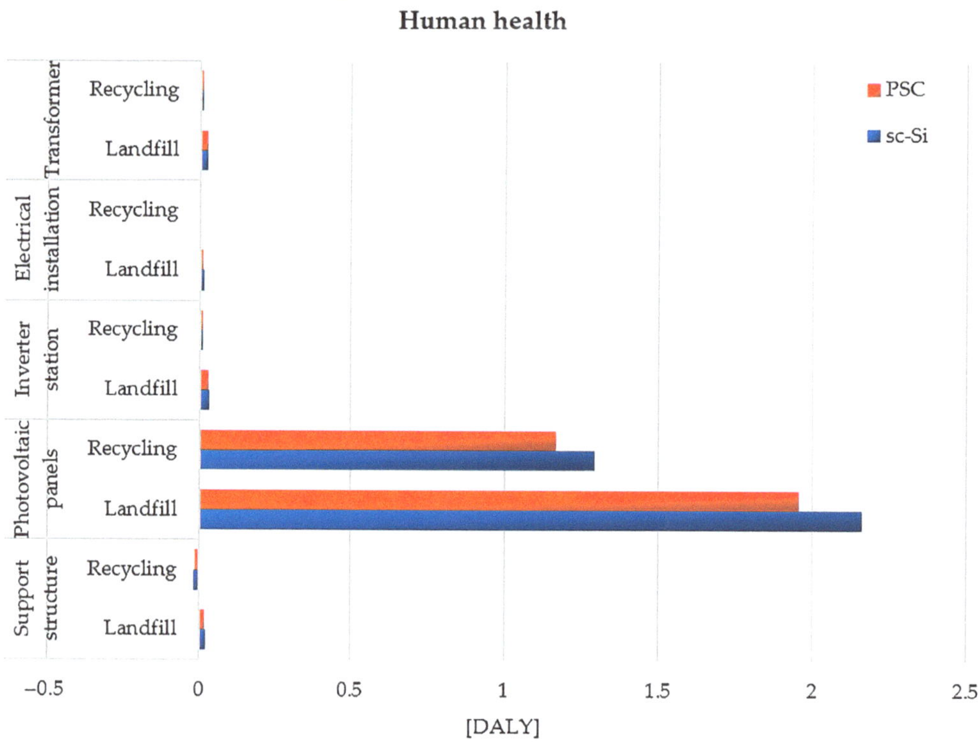

Figure 14. Characterization of the environmental consequences for human health during the life cycle of the studied photovoltaic power plants (model IMPACT World+ Endpoint ver. 1.02, unit: DALY).

A similar situation occurred in the case of the impact area covering the quality of the ecosystem—the maximum value of the total harmful impact was noted for the life cycle of photovoltaic panels (sc-Si: 1.70 × 10^{-3} PDF × m^2 × yr for landfill and

7.62×10^{-4} PDF × m² × yr for recycling; PSC: 1.54×10^{-3}, PDF × m² × yr for landfill and 6.89×10^{-4} PDF × m² × yr for recycling). Other elements were characterized by a much lower degree of combined destructive impact: inverter station (sc-Si: 1.47×10^{-4} PDF × m² × yr for landfill and 1.63×10^{-5} PDF × m² × yr for recycling; PSC: 1.33×10^{-4}, PDF × m² × yr for landfill and 1.47×10^{-5} PDF × m² × yr for recycling), transformer (sc-Si: 6.08×10^{-5} PDF × m² × yr for landfill and 9.79×10^{-6} PDF × m² × yr for recycling; PSC: 5.50×10^{-5}, PDF × m² × yr for landfill and 8.86×10^{-6} PDF × m² × yr for recycling), support structure (sc-Si: 5.18×10^{-5} PDF × m² × yr for landfill and 2.32×10^{-5} PDF × m² × yr for recycling; PSC: 4.68×10^{-5} PDF × m² × yr for landfill and 2.10×10^{-5} PDF × m² × yr for recycling), and electrical installation (sc-Si: 4.22×10^{-5} PDF × m² × yr for landfill and 1.35×10^{-6} PDF × m² × yr for recycling; PSC: 3.81×10^{-5} PDF × m² × yr for landfill and 1.22×10^{-6} PDF × m² × yr for recycling). Again, each element of both power plants had a life cycle with post-consumer management in the form of recycling causing fewer adverse effects on ecosystem quality compared to the life cycle involving landfill. The PSC power plant had fewer negative impacts in the studied range than the sc-Si power plant (Figure 15 and Table 3).

Table 3. Characterization of environmental consequences in the life cycle of the studied photovoltaic power plants for the considered impact areas and different post-consumer development scenarios (model IMPACT World+ Endpoint ver. 1.02).

Area of Influence	Power Plant Components	Support Structure		Photovoltaic Panels		Inverter Station		Electrical Installation		Transformer		Unit
	Post-Use Development Scenario	**Landfill**	**Recycling**	**Landfill**	**Recycling**	**Landfill**	**Recycling**	**Landfill**	**Recycling**	**Landfill**	**Recycling**	
Human health	**sc-Si**	$\mathbf{1.96 \times 10^{-2}}$	$\mathbf{-1.70 \times 10^{-2}}$	$\mathbf{2.16 \times 10^{0}}$	$\mathbf{1.29 \times 10^{0}}$	$\mathbf{3.21 \times 10^{-2}}$	$\mathbf{1.09 \times 10^{-2}}$	$\mathbf{1.19 \times 10^{-2}}$	$\mathbf{2.38 \times 10^{-3}}$	$\mathbf{2.62 \times 10^{-2}}$	$\mathbf{1.05 \times 10^{-2}}$	DALY
	PSC	1.77×10^{-2}	-1.54×10^{-2}	1.95×10^{0}	1.17×10^{0}	2.91×10^{-2}	9.84×10^{-3}	1.08×10^{-2}	2.16×10^{-3}	2.37×10^{-2}	9.49×10^{-3}	
Ecosystem quality	**sc-Si**	$\mathbf{5.18 \times 10^{-5}}$	$\mathbf{2.32 \times 10^{-5}}$	$\mathbf{1.70 \times 10^{-3}}$	$\mathbf{7.62 \times 10^{-4}}$	$\mathbf{1.47 \times 10^{-4}}$	$\mathbf{1.63 \times 10^{-5}}$	$\mathbf{4.22 \times 10^{-5}}$	$\mathbf{1.35 \times 10^{-6}}$	$\mathbf{6.08 \times 10^{-5}}$	$\mathbf{9.79 \times 10^{-6}}$	PDF × m² × yr
	PSC	4.68×10^{-5}	2.10×10^{-5}	1.54×10^{-3}	6.89×10^{-4}	1.33×10^{-4}	1.47×10^{-5}	3.81×10^{-5}	1.22×10^{-6}	5.50×10^{-5}	8.86×10^{-6}	

Bold data—the highest levels of negative consequences.

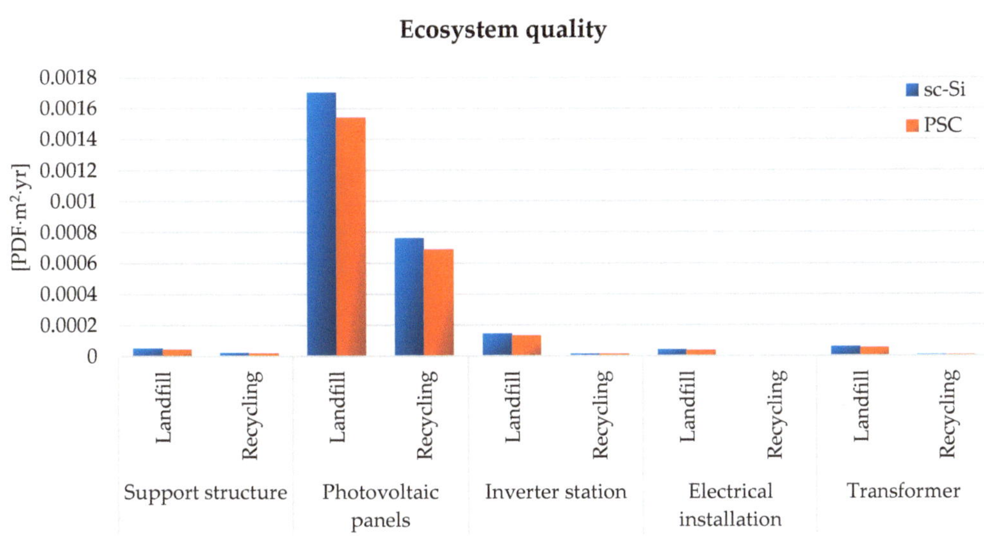

Figure 15. Characterization of environmental consequences for the quality of the ecosystem during the life cycle of the studied photovoltaic power plants (model IMPACT World+ Endpoint ver. 1.02, unit: PDF × m² × yr).

The impact areas were also analyzed in terms of potential environmental costs that the life cycles of the studied photovoltaic power plants entail (in the EUR unit—euro currency,

€). The life cycle of a power plant based on monocrystalline photovoltaic panels (sc-Si) is associated with higher environmental costs incurred as a result of deterioration of human health (total: 7.93 × 10^6 EUR for landfill and 4.25 × 10^6 EUR for recycling) and lowering the quality of the ecosystem (total: 4.36 × 10^6 EUR for landfill and 2.27 × 10^6 EUR for recycling), compared to the life cycle of a power plant based on perovskite panels (PSC). In principle, both post-consumer management processes in the form of storage and recycling cause higher environmental costs in the area of human health than in the quality of the ecosystem (Table 4).

Table 4. Grouping and weighing environmental consequences in the life cycle of the studied photovoltaic power plants for the considered impact areas and different post-consumer development scenarios (model IMPACT World+ Endpoint ver. 1.02, unit: EUR).

Element of a Technical Object		Support Structure		Photovoltaic Panels		Inverter Station		Electrical Installation		Transformer		Unit
Form of Post-Use Management		Landfill	Recycling	Landfill	Recycling	Landfill	Recycling	Landfill	Recycling	Landfill	Recycling	
Areas of Influence	Type of Power Plant											
Human health	**sc-Si**	**1.24 × 10^5**	**2.51 × 10^4**	**7.27 × 10^6**	**4.03 × 10^6**	**3.03 × 10^5**	**1.13 × 10^5**	**8.11 × 10^4**	**2.37 × 10^4**	**1.50 × 10^5**	**6.28 × 10^4**	**EUR**
	PSC	1.23 × 10^5	7.74 × 10^4	3.80 × 10^6	2.06 × 10^6	1.94 × 10^5	4.59 × 10^4	1.04 × 10^5	3.71 × 10^4	1.36 × 10^5	4.43 × 10^4	EUR
Ecosystem quality	**sc-Si**	**1.23 × 10^5**	**7.74 × 10^4**	**3.80 × 10^6**	**2.06 × 10^6**	**1.94 × 10^5**	**4.59 × 10^4**	**1.04 × 10^5**	**3.71 × 10^4**	**1.36 × 10^5**	**4.43 × 10^4**	**EUR**
	PSC	1.01 × 10^5	6.30 × 10^4	1.57 × 10^6	7.96 × 10^5	1.75 × 10^5	4.16 × 10^4	8.28 × 10^4	2.95 × 10^4	1.23 × 10^5	4.01 × 10^4	EUR

Bold data—the highest levels of negative consequences.

3.1.3. Total Impact

The element of both assessed photovoltaic power plants characterized by the highest environmental cost of the life cycle are photovoltaic panels; especially if their form of post-consumer management is landfilling (sc-Si: 1.11 × 10^7 EUR, PSC: 5.53 × 10^6 EUR). The use of recycling processes would significantly reduce the environmental cost of their life cycle (sc-Si: o 4.98 × 10^6 EUR, PSC: o 2.09 × 10^6 EUR). In the case of all other elements (support structure, inverter station, electrical installation, transformer), the environmental cost of their life cycle is lower for a photovoltaic power plant based on perovskite cell technology and for the form of post-consumer management using recycling processes. The lowest environmental cost is distinguished by the life cycle of an electrical installation, which will be recycled after the end of use (sc-Si: o 1.85 × 10^5 EUR, PSC: o 1.47 × 10^5 EUR). As mentioned earlier, this is due to the small mass of this element compared to the total mass of the power plant (Figure 16).

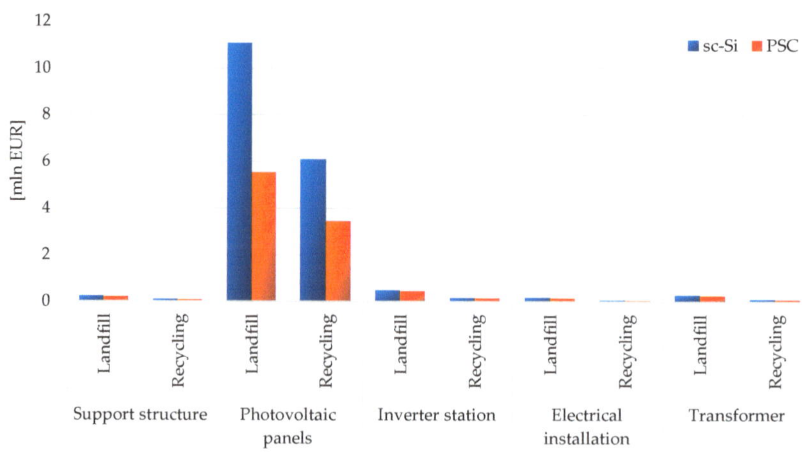

Figure 16. Grouping and weighing environmental consequences in the life cycle of the studied photovoltaic power plants, taking into account different post-consumer management scenarios (model IMPACT World+ Endpoint ver. 1.02, unit: mln EUR).

Total environmental life cycle cost of a photovoltaic power plant using monocrystalline silicon photovoltaic panels (landfill: 1.23×10^7 EUR, recycling: 6.52×10^6 EUR) was higher compared to the perovskite power plant (landfill: 6.59×10^6 EUR, recycling: 3.81×10^6 EUR). This was determined by the size of the environmental cost of photovoltaic panels, which in the case of sc-Si power plants required much higher energy and material inputs compared to those used in PSC power plants. The form of post-consumer management also had a key impact on the total value of the environmental cost of the life cycle of a photovoltaic power plant (Figure 17).

Figure 17. Grouping and weighing the environmental consequences for the ecosystem quality in the life cycle of the studied photovoltaic power plants (model IMPACT World+ Endpoint ver. 1.02, unit: mln EUR).

3.2. *IPCC 2021*

Thanks to the application of the second analytical model—IPCC 2021 GWP100 version 1.01—it was possible to determine the size of the environmental consequences for greenhouse gas emissions in the life cycle of the assessed photovoltaic power plants. The largest GHG emissions in the life cycles of both power plants occurred as a result of the use of fossil sources (fuels and mineral raw materials, impact category: GWP100—fossil). As a result, there were emissions of, among others, carbon dioxide, methane, dinitrogen monoxide, sulfur hexafluoride, tetrafluoromethane (CFC-14), trifluoromethane (HFC-23), hexafluoroethane (HFC-116), dichlorodifluoromethane (CFC-12). Their highest level was recorded in the life cycle of monocrystalline silicon photovoltaic panels (landfill: 1.01×10^6 kg CO_2 eq; recycling: 5.51×10^5 kg CO_2 eq). As mentioned earlier, the production of cells using the Czochralski method is an extremely energy-intensive process, which as a consequence generates many substances with a destructive effect on the environment (Table 5).

Table 5. Characterization of environmental consequences for greenhouse gas emissions in the life cycle of the studied photovoltaic power plants for the considered impact categories and different post-consumer management scenarios (model IPCC 2021 GWP100 ver. 1.01, unit: kg CO_2 eq).

Element of a Technical Object		Support Structure		Photovoltaic Panels		Inverter Station		Electrical Installation		Transformer		Unit
Form of Post-Use Management												
Impact Category	Type of Power Plant	Landfill	Recycling	Landfill	Recycling	Landfill	Recycling	Landfill	Recycling	Landfill	Recycling	
GWP100—fossil	**sc-Si**	**1.32×10^4**	**-3.97×10^2**	**1.01×10^6**	**5.51×10^5**	**2.32×10^4**	**6.75×10^3**	**8.76×10^3**	**1.76×10^3**	**1.83×10^4**	**6.10×10^3**	**kg CO_2 eq**
	PSC	**1.07×10^4**	**-1.55×10^3**	**5.45×10^5**	**2.69×10^5**	**2.10×10^4**	**6.11×10^3**	**6.98×10^3**	**1.40×10^3**	**1.66×10^4**	**5.52×10^3**	**kg CO_2 eq**
GWP100—biogenic	sc-Si	2.86×10^2	9.02×10^1	4.06×10^3	2.22×10^3	4.60×10^2	3.78×10^1	2.04×10^2	2.13×10^0	3.68×10^2	4.67×10^1	kg CO_2 eq
	PSC	2.33×10^2	5.58×10^1	2.28×10^3	1.10×10^3	4.16×10^2	3.42×10^1	1.62×10^2	1.70×10^0	3.33×10^2	4.23×10^1	kg CO_2 eq
GWP100—land transformation	sc-Si	4.33×10^1	1.48×10^1	1.95×10^3	9.95×10^2	6.03×10^1	9.60×10^0	3.31×10^1	1.20×10^0	4.84×10^1	1.11×10^1	kg CO_2 eq
	PSC	3.53×10^1	9.51×10^0	1.02×10^3	4.74×10^2	5.45×10^1	8.69×10^0	2.64×10^1	9.56×10^{-1}	4.38×10^1	1.01×10^1	kg CO_2 eq

Bold data—the highest levels of negative consequences.

The highest total level of greenhouse gas emissions in the life cycles of both studied power plants was distinguished by photovoltaic panels (sc-Si: landfill -1.01×10^6 kg CO_2 eq, recycling -5.54×10^5 kg CO_2 eq; PSC: landfill -5.49×10^5 kg CO_2 eq, recycling -2.71×10^5 kg CO_2 eq). For all elements of the analyzed technical facilities, the use of recycling processes is associated with lower GHG emissions compared to landfill management. The life cycles of PSC power plant elements generate less greenhouse gasses than SC power plants (Figure 18).

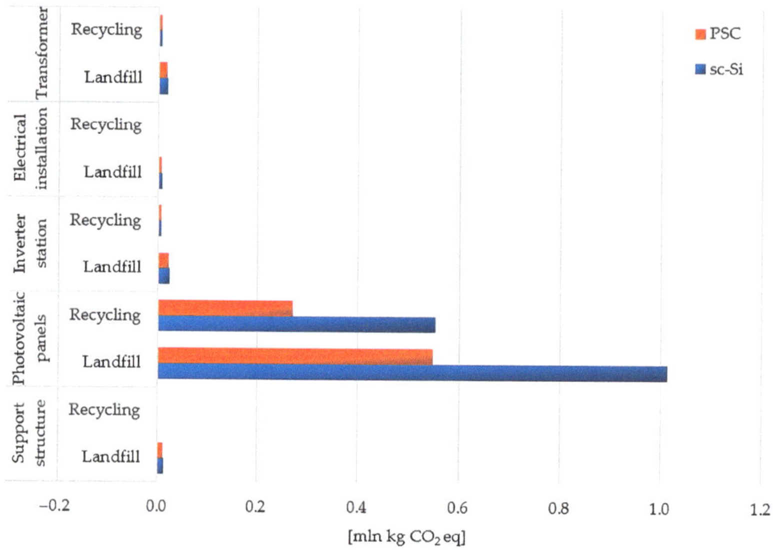

Figure 18. Characterization of environmental consequences for greenhouse gas emissions in the life cycle of the studied photovoltaic power plants, taking into account different post-consumer management scenarios (model IPCC 2021 GWP100 ver. 1.01, unit: mln kg CO_2 eq).

The total level of greenhouse gas emissions was higher for the management of materials and components of both photovoltaic power plants in the form of landfill (sc-Si: 1.08×10^6 kg CO_2 eq; PSC: 5.69×10^5 kg CO_2 eq) compared to recycling management (sc-Si: 6.05×10^5 kg CO_2 eq; PSC: 2.82×10^5 kg CO_2 eq). The life cycle of a power plant based on monocrystalline silicon panels was distinguished by a higher level of GHG emissions than the life cycle of a power plant based on perovskite panels (Figure 19).

Figure 19. Characterization of environmental consequences for greenhouse gas emissions in the life cycle of the studied photovoltaic power plants (model IPCC 2021 GWP100 ver. 1.01, unit: mln kg CO_2 eq).

3.3. Cumulative Energy Demand (CED)

The last model used in the study: CED—Cumulative Energy Demand version 1.11, allowed for determining the energy demand in the life cycle of the analyzed photovoltaic power plants. Due to the structure of the energy mix in Europe, the largest amount of energy in the life cycles of both analyzed power plants came from non-renewable sources (impact category: non-renewable fossil), mainly from crude oil, coal and gas. Among renewable sources, waterpower stood out with the highest degree of coverage of the energy demand of the research objects (impact category: renewable water). The highest level of energy demand from both non-renewable sources (total: 1.46×10^7 MJ), and renewable (total: 5.87×10^6 MJ) was noted for the life cycle of monocrystalline silicon panels, which after the end of their use were placed in a landfill. The increasingly rapid consumption of non-renewable energy sources is not only associated with the depletion of their resources; their exploitation also results in a decrease in the quality of ecosystems and the degradation of the environment. The most negative environmental consequences, and consequently, the environmental costs, are caused by opencast mining. Hence, the introduction of changes in the energy mixes of countries, toward increasing the share of renewable energy sources, is such an important problem today (Table 6).

Table 6. Characterization of energy demand in the life cycle of the studied photovoltaic power plants for the considered impact categories and different post-consumer management scenarios (model CED—Cumulative Energy Demand ver. 1.11, unit: MJ).

Element of a Technical Object		Support Structure		Photovoltaic Panels		Inverter Station		Electrical Installation		Transformer		Unit
Form of Post-Use Management												
Impact Category	Type of Power Plant	Landfill	Recycling	Landfill	Recycling	Landfill	Recycling	Landfill	Recycling	Landfill	Recycling	
Non-renewable, fossil	**sc-Si**	**1.84×10^5**	**1.69×10^4**	**1.27×10^7**	**6.96×10^6**	**2.76×10^5**	**1.23×10^5**	**1.52×10^5**	**4.98×10^4**	**2.16×10^5**	**1.08×10^5**	**MJ**
	PSC	**1.43×10^5**	**1.31×10^4**	**6.81×10^6**	**5.23×10^6**	**2.38×10^5**	**1.06×10^5**	**1.15×10^5**	**3.78×10^4**	**1.86×10^5**	**9.34×10^4**	**MJ**
Non-renewable, nuclear	sc-Si	2.55×10^4	-2.18×10^4	1.86×10^6	9.60×10^5	4.05×10^4	2.11×10^4	2.26×10^4	7.45×10^3	3.24×10^4	1.57×10^4	MJ
	PSC	1.98×10^4	-1.69×10^4	1.07×10^6	7.79×10^5	3.49×10^4	1.82×10^4	1.72×10^4	5.65×10^3	2.79×10^4	1.35×10^4	MJ
Non-renewable, biomass	sc-Si	2.11×10^1	5.08×10^0	9.71×10^2	1.90×10^2	2.95×10^1	5.27×10^0	1.58×10^1	7.61×10^{-1}	2.56×10^1	6.74×10^0	MJ
	PSC	1.64×10^1	3.94×10^0	4.85×10^2	1.21×10^2	2.54×10^1	4.54×10^0	1.20×10^1	5.77×10^{-1}	2.21×10^1	5.81×10^0	MJ
Renewable, biomass	sc-Si	3.39×10^3	1.44×10^3	5.73×10^5	3.07×10^5	4.57×10^3	2.13×10^3	2.96×10^3	1.10×10^3	3.84×10^3	2.02×10^3	MJ
	PSC	2.63×10^3	1.12×10^3	3.03×10^5	2.29×10^5	3.94×10^3	1.84×10^3	2.25×10^3	8.32×10^2	3.31×10^3	1.74×10^3	MJ
Renewable, wind, solar, geothe	sc-Si	3.69×10^3	1.95×10^3	3.72×10^5	1.53×10^5	5.95×10^3	2.71×10^3	2.80×10^3	7.28×10^2	4.20×10^3	2.29×10^3	MJ
	PSC	2.86×10^3	1.51×10^3	2.09×10^5	1.22×10^5	5.12×10^3	2.34×10^3	2.12×10^3	5.52×10^2	3.62×10^3	1.98×10^3	MJ
Renewable, water	sc-Si	3.25×10^4	-2.03×10^4	2.71×10^6	2.09×10^6	9.78×10^4	8.70×10^3	1.98×10^4	9.07×10^2	4.16×10^4	5.19×10^3	MJ
	PSC	4.19×10^4	-2.62×10^4	4.92×10^6	3.00×10^6	1.14×10^5	1.01×10^4	2.61×10^4	1.20×10^3	4.83×10^4	6.03×10^3	MJ

Bold data—the highest levels of negative consequences.

The high mass share of photovoltaic panels in the masses of the tested power plants (sc-Si—approx. 53%; PSC—approx. 48%, Figure 7) and the high demand for matter and energy in their life cycles make them the elements of the power plant with the highest cumulative demand for energy. In turn, the lowest demand in this respect, due to its small mass, was characteristic of the electrical installation with recycling as a form of post-consumer management (total: sc-Si -6.03×10^4 MJ; PSC -4.57×10^4 MJ). In the case of each of the analyzed power plant components, post-consumer management through landfilling was a source of greater energy demand in the life cycle than recycling processes, in which this energy could have been recovered (in a broader perspective) by using other materials and components in production (high energy costs of primary production are avoided) (Figure 20).

The life cycle of a photovoltaic power plant based on sc-Si technology is characterized by a higher total cumulative energy demand (landfill: 2.16×10^7 MJ; recycling: $1.172.16 \times 10^7$ MJ) compared to a power plant based on PSC technology (landfill: $1.212.16 \times 10^7$ MJ; recycling: $8.732.16 \times 10^6$ MJ). The use of recycling processes allows for a significant reduction in energy demand in the life cycles of both tested technical objects (Figure 21).

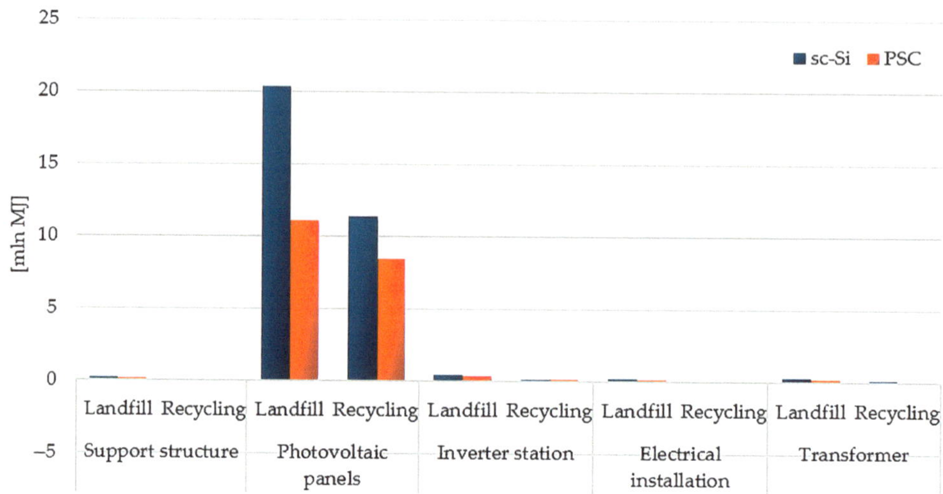

Figure 20. Characterization of energy demand in the life cycle of the studied photovoltaic power plants, taking into account different post-consumer development scenarios (model CED—Cumulative Energy Demand ver. 1.11, unit: mln MJ).

Figure 21. Characterization of energy demand in the life cycle of the studied photovoltaic power plants (model CED—Cumulative Energy Demand ver. 1.11, unit: mln MJ).

3.4. Comparison of the Total Impact on the Environment

Table 7 presents the results of the total environmental impact of the life cycle of a photovoltaic power plant based on monocrystalline silicon (sc-Si) and perovskite (PSC) modules for two post-consumer development scenarios (landfill and recycling). In terms of impact on human health, environmental quality, greenhouse gas emissions and total energy demand, the highest level of negative impact on the environment was observed in the life cycle of a monocrystalline power plant (sc-Si), which materials, components and components would be disposed of in a landfill after the end of their service life, while the lowest impact was observed in the case of a perovskite power plant (PSC), for which the post-consumer development scenario would include recycling processes. Therefore, perovskites can be considered as a construction material that better fits into the main assumptions of sustainable development, compared to the currently most commonly used monocrystalline silicon.

Table 7. Comparison of the total environmental impact of the life cycles of a photovoltaic power plant based on monocrystalline silicon (sc-Si) and perovskite (PSC) modules, for two post-consumer development scenarios (landfill and recycling).

Area of Influence	sc-Si		PSC		Unit
	Landfill	Recycling	Landfill	Recycling	
Human health	$\mathbf{2.25 \times 10^{0}}$	1.30×10^{0}	2.04×10^{0}	1.17×10^{0}	DALY
Ecosystem quality	$\mathbf{2.00 \times 10^{-3}}$	8.12×10^{-4}	1.81×10^{-3}	7.35×10^{-4}	PDF × m^2 × yr
Human health	$\mathbf{7.93 \times 10^{6}}$	4.25×10^{6}	4.36×10^{6}	2.27×10^{6}	EUR
Ecosystem quality	$\mathbf{4.36 \times 10^{6}}$	2.27×10^{6}	2.05×10^{6}	9.70×10^{5}	EUR
Total impact	$\mathbf{1.23 \times 10^{7}}$	6.52×10^{6}	6.59×10^{6}	3.81×10^{6}	EUR
GWP100	$\mathbf{1.08 \times 10^{6}}$	5.69×10^{5}	6.05×10^{5}	2.82×10^{5}	kg CO_2eq
Cumulative energy demand	$\mathbf{2.16 \times 10^{7}}$	1.17×10^{7}	1.21×10^{7}	8.73×10^{6}	MJ

Bold data—the highest levels of negative consequences.

4. Summary and Conclusions

The increase in the standard of living and the number of people in the world means that the global demand for energy is increasing year by year. Currently, most energy is obtained from fossil sources, the exploitation of which is characterized by a huge, destructive impact on the environment. The production, exploitation and post-consumer management of renewable sources also involve some use of non-renewable resources, but to a much smaller extent that that of their conventional counterparts. Photovoltaic power plants are considered "environmentally friendly", but the technologies used in their life cycles may fit into the main assumptions of sustainable development [93–95].

The main objective of the study was achieved by assessing the environmental impact of construction materials of monocrystalline and perovskite photovoltaic power plants towards their sustainable development.

The research object was the materials and construction elements of two 1 MW photovoltaic power plants. The first one was based on modules made of monocrystalline silicon, while the second one was based on perovskite (tandem type). The analyses were carried out using the Life Cycle Assessment method (IM-PACT World+ model, IPCC and CED). The assessment was carried out separately for five sets of elements: support structure, photovoltaic panels, inverter station, electrical installation and transformer. Additionally, two post-consumer management scenarios were adopted: landfill and recycling.

In the literature, there are few studies on the life cycle of photovoltaic power plants. Most often, the life cycle of photovoltaic modules or only the materials from which the cells are manufactured is assessed. The studies usually concern silicon technologies, mainly single-crystalline silicon [8–16], and multi-crystalline silicon [11,13,16–22], less often—amorphous-silicon [16,19,20], and multi-junction thin-film silicon [23]. Individual works were devoted to cells made of other materials, e.g., cadmium telluride [13,20,22], copper-indium-gallium-diselenide [23], and DSSC [24]. There are also a small number of articles addressing the life cycle of systems with large installed capacity [9,25,26]. Studies conducted so far have usually not included more detailed analyses of the impacts of photovoltaic power plants, which may pose a threat to human health, cause a decrease in the quality of ecosystems or deepen the depletion of raw material resources.

The achievement of the main objective of the study allowed for the formulation of the following conclusions:

- From the perspective of the life cycle of materials and elements of both analyzed photovoltaic power plants, the categories of impact with the highest level of harmful impact on human health include water availability, human health, and climate change, human health, long term. In terms of impact on ecosystem quality, these are the categories of freshwater ecotoxicity, long term, and climate change, ecosystem quality, long-term (Tables 1 and 2, Figures 11–14).

- Among the processes related to the reduction in water availability in the life cycles of both power plants (impact category: water availability, human health), one can distinguish, first, the consumption of water from various sources (lakes, rivers, wells, etc.) and its use for cooling turbines (during electricity generation).
- The chemical compounds affecting climate change (impact category: climate change, human health, long term, and climate change, ecosystem quality, long term) in the life cycles of both power plants, characterized by the highest harmful impact values include emissions of carbon dioxide, methane, dinitrogen monoxide, sulfur hexafluoride, tetrafluoromethane (CFC-14), hexafluoroethane (HFC-116) and trifluoromethane (HFC-23).
- Among the construction materials of the analyzed power plants, the life cycle of which has the greatest impact on the increase in freshwater ecotoxicity (impact category: freshwater ecotoxicity, long term), the following can be distinguished: copper, aluminum, iron, strontium, nickel, zinc, manganese, cadmium and vanadium.
- In the case of both assessed power plants, the highest level of total harmful impact on human health and ecosystem quality was observed during the life cycle of photovoltaic panels (Figures 14 and 15).
- The life cycle of a sc-Si power plant is associated with higher environmental costs incurred due to the deterioration of human health and the reduction in the quality of the ecosystem compared to the life cycle of a PSC power plant (Table 4).
- Both post-consumer management processes in the form of storage and recycling cause higher environmental costs in the area of human health than in the area of ecosystem quality (Table 4).
- The element of both assessed power plants with the highest environmental life cycle cost are photovoltaic panels (Figure 16).
- For all elements, the environmental cost of their life cycle is lower in the case of a photovoltaic power plant based on the use of perovskite panels and for the form of post-consumer management using recycling processes (Figure 16).
- The total environmental cost of the life cycle of a sc-Si power plant is higher than that of a PSC power plant (Figure 17).
- The highest level of greenhouse gas emissions in the life cycles of both power plants is due to the use of fossil sources (fuels and mineral raw materials, impact category: GWP100—fossil) (Table 5).
- The highest GHG emissions in the life cycles of both power plants are characteristic of photovoltaic panels (Figure 18).
- For all assessed electoral elements, the use of recycling processes is associated with lower greenhouse gas emissions compared to landfill management (Figure 18).
- The life cycle of a power plant using monocrystalline silicon panels is characterized by higher GHG emissions than a power plant using perovskite panels (Figure 19).
- The majority of energy in the life cycles of both examined photovoltaic power plants comes from non-renewable sources (impact category: non-renewable, fossil), mainly from crude oil, coal and gas (Table 6).
- Photovoltaic panels are the element of the power plant that stands out with the highest cumulative demand for energy (Figure 20).
- The life cycle of a sc-Si photovoltaic power plant is characterized by a higher total cumulative energy demand compared to a PSC power plant (Figure 21).
- The use of recycling processes enables a significant reduction in the energy demand in the life cycles of both tested technical facilities (Figure 21).
- Photovoltaic power plants based on perovskite modules fit better into the main assumptions of sustainable development than power plants using monocrystalline silicon panels.

The increasing share of renewable energy sources (including photovoltaic power plants) in the global energy balance enables more sustainable and economically efficient use of fossil fuels. It also improves the condition of the environment and increases the level of energy security. An additional advantage is more dynamic regional development, the creation of new jobs and a reduction in many environmental problems [96–98].

Continuous technological and scientific development allows for the implementation of the best available techniques (BAT). They should take into account the minimization of energy and material consumption from the perspective of the entire life cycle of photovoltaic power plants. Nowadays, it is necessary to change the way of managing environmental resources towards the rationalization of the use of natural resources. In order to achieve this goal, more sustainable technologies should be popularized, not only in the area of renewable energy sources, but in all sectors of the economy [99,100].

New methods of producing and recycling plastics, materials and components of photovoltaic power plants are being developed all over the world in order to reduce the negative impact of this type of waste on the environment and create the possibility of recovering at least some of its value (especially from photovoltaic panels that are no longer suitable for further use). Today, recycling methods allow for the recovery of only a certain part of the materials used, so there is a lot of room for progress in this area. Reducing energy and material consumption in the life cycles of photovoltaic power plants translates into a reduction in the level of their harmful impact on human health, environmental quality and the depletion of raw materials. Life cycle studies of innovative construction materials that better fit into the assumptions of sustainable development can have a significant impact on policy and future industrial practices. They provide a basis for developing recommendations in the area of sustainable, efficient development of the renewable energy sector, and, in particular,—photovoltaics. Identification of areas in the life cycle of photovoltaic power plants with the potentially greatest harmful impact on the environment will enable taking action to reduce material and energy consumption and the harmful emissions of processes related to the production, operational and post-consumer management of their construction materials. It will also facilitate work on creating innovative, more environmentally friendly materials and elements. The search for an optimal solution for a photovoltaic power plant will, therefore, primarily consist of finding the right structure that will allow obtaining the desired quality of a technical object and determining process parameters that guarantee the lowest energy and material costs at each stage of its life cycle.

Author Contributions: Conceptualization, I.P. and Z.K.; methodology, I.P. and Z.K.; software, I.P.; validation, I.P.; formal analysis, I.P. and Z.K.; investigation, I.P.; resources, I.P.; data curation, I.P.; writing—original draft preparation, I.P.; writing—review and editing, I.P. and Z.K.; visualization, I.P.; supervision, I.P. and Z.K.; project administration, I.P. All authors have read and agreed to the published version of the manuscript.

Funding: This research received no external funding.

Institutional Review Board Statement: Not applicable.

Informed Consent Statement: Not applicable.

Data Availability Statement: The data presented in this study are available on request from the author.

Conflicts of Interest: The author declares no conflicts of interest.

References

1. Li, C.; Chen, L.; Jiang, F.; Song, Z.; Wang, X.; Balvanz, A.; Ugur, E.; Liu, Y.; Liu, C.; Maxwell, A.; et al. Diamine chelates for increased stability in mixed Sn–Pb and all-perovskite tandem solar cells. *Nat. Energy* **2024**, 7, 4577–4586. [CrossRef]
2. Demirel, Y. *Energy: Production, Conversion, Storage, Conservation and Coupling*; Springer: Cham, Switzerland, 2016; pp. 441–484. [CrossRef]
3. Singh, R.; Kumar, S. *Green Technologies and Environmental Sustainability*; Springer: Cham, Switzerland, 2017; pp. 157–178. [CrossRef]
4. Monge, J. Global Solar PV Installed Capacity Will More Than Triple in the Next Ten Years. Available online: https://www.woodmac.com/news/opinion/global-solar-PV-installed-more-than-triple-in-the-next-ten-years/ (accessed on 5 September 2024).
5. Andersen, O. *Unintended Consequences of Renewable Energy*; Springer: London, UK, 2013; pp. 81–89. [CrossRef]

6. Dincer, I.; Midilli, A.; Kucuk, H. *Progress in Sustainable Energy Technologies: Generating Renewable Energy*; Springer: Cham, Switzerland, 2014; pp. 339–353. [CrossRef]
7. Leda, P.; Idzikowski, A.; Piasecka, I.; Baldowska-Witos, P.; Cierlicki, T.; Zawada, M. Management of Environmental Life Cycle Impact Assessment of a Photovoltaic Power Plant on the Atmosphere, Water, and Soil Environment. *Energies* **2023**, *16*, 4230. [CrossRef]
8. Wang, X.; Tian, X.; Chen, X.; Ren, L.; Geng, C. A Review of End-of-Life Crystalline Silicon Solar Photovoltaic Panel Recycling Technology. *Sol. Energy Mater. Sol. Cells* **2022**, *248*, 111976. [CrossRef]
9. Kato, K.; Murata, A.; Sakuta, K. An evaluation on the life cycle of photovoltaic energy system considering production energy of off-grade silicon. *Sol. Energy Mat. Sol. Cells* **2017**, *47*, 95–100. [CrossRef]
10. Golroudbary, S.R.; Lundström, M.; Wilson, B.P. Analogical Environmental Cost Assessment of Silicon Flows Used in Solar Panels by the US and China. *Sci. Rep.* **2024**, *14*, 9538. [CrossRef]
11. Dones, R.; Frischknecht, R. Life cycle assessment of photovoltaic systems: Results of Swiss studies on energy chains. *Prog. Photovolt. Res. Appl.* **2018**, *6*, 117–125. [CrossRef]
12. Heath, G.A.; Silverman, T.J.; Kempe, M.; Deceglie, M.; Ravikumar, D.; Remo, T.; Cui, H.; Sinha, P.; Libby, C.; Shaw, S.; et al. Research and Development Priorities for Silicon Photovoltaic Module Recycling to Support a Circular Economy. *Nat. Energy* **2020**, *5*, 502–510. [CrossRef]
13. Fthenakis, V.M.; Kim, H.C. Greenhouse-gas emissions from solar electric and nuclear power: A life-cycle study. *Energy Policy* **2007**, *35*, 2549–2557. [CrossRef]
14. Reich, N.H.; Alsema, E.A.; van Sark, W.G.J.H.M.; Turkenburg, W.C.; Sinke, W.C. Greenhouse Gas Emissions Associated with Photovoltaic Electricity from Crystalline Silicon Modules under Various Energy Supply Options. *Prog. Photovolt. Res. Appl.* **2011**, *19*, 603–613. [CrossRef]
15. Frankl, P.; Masini, A.; Gamberale, M.; Toccaceli, D. Simplified life-cycle analysis of PV systems in buildings: Present situation and future trends. *Prog. Photovolt. Res. Appl.* **2018**, *6*, 137–146. [CrossRef]
16. Alsema, E.A. Energy pay-back time and CO_2 emissions of PV systems. *Prog. Photovolt. Res. Appl.* **2000**, *8*, 17–25. [CrossRef]
17. Oliver, M.; Jackson, T. The evolution of economic and environmental cost for crystalline silicon photovoltaics. *Energy Policy* **2000**, *28*, 1011–1021. [CrossRef]
18. Nomura, N.; Inaba, A.; Tonooka, Y.; Akai, M. Life-cycle emission of oxidic gases from power-generation systems. *Appl. Energy* **2001**, *68*, 215–227. [CrossRef]
19. Kato, K.; Murata, A.; Sakuta, K. Energy pay-back time and lifecycle CO_2 emission of residential PV power system with silicon PV module. *Prog. Photovolt. Res. Appl.* **2018**, *6*, 105–115. [CrossRef]
20. Ito, M.; Kato, K.; Komoto, K.; Kichimi, T.; Kurokawa, K. A comparative study on cost and life-cycle analysis for 100 MW very large-scale PV (VLS-PV) systems in deserts using m-Si, a-Si, CdTe, and CIS modules. *Prog. Photovolt. Res. Appl.* **2008**, *16*, 17–30. [CrossRef]
21. Ito, M.; Kato, K.; Sugihara, H.; Kichimi, T.; Song, J.; Kurokawa, K. A preliminary study on potential for very large scale photovoltaic power generation (VLS-PV) system in the Gobi desert from economic and environmental viewpoints. *Sol. Energy Mat. Sol. Cells* **2003**, *75*, 507–517. [CrossRef]
22. Fthenakis, V.M.; Alsema, E. Photovoltaics energy payback times, greenhouse gas emissions and external costs: 2004—Early 2005 status. *Prog. Photovolt. Res. Appl.* **2006**, *14*, 275–280. [CrossRef]
23. Bravi, M.; Parisi, M.L.; Tiezzi, E.; Basosi, R. Life cycle assessment of a micro morph photovoltaic system. *Energy* **2011**, *36*, 4297–4306. [CrossRef]
24. Greijer, H.; Karlson, L.; Lindquist, S.E.; Hagfeldt, A. Environmental aspects of electricity generation from a nanocrystalline dye sensitized solar cell system. *Renew. Energy* **2001**, *23*, 27–39. [CrossRef]
25. Kato, K.; Hibino, T.; Komoto, K.; Ihara, S.; Yamamoto, S.; Fujihara, H. A life-cycle analysis on thin-film CdS/CdTe PV modules. *Sol. Energy Mat. Sol. Cells* **2001**, *67*, 279–287. [CrossRef]
26. Schaefer, H.; Hagedorn, G. Hidden energy and correlated environmental characteristics of PV power generation. *Renew. Energy* **2002**, *2*, 159–166. [CrossRef]
27. Manser, J.S.; Christians, J.A.; Kamat, V.P. Intriguing Optoelectronic Properties of Metal Halide Perovskites. *Chem. Rev.* **2016**, *116*, 12956–13008. [CrossRef] [PubMed]
28. Kojima, A.; Teshima, K.; Shirai, Y.; Miyasaka, T. Organometal Halide Perovskites as Visible-Light Sensitizers for Photovoltaic Cells. *J. Am. Chem. Soc.* **2009**, *131*, 6050–6051. [CrossRef] [PubMed]
29. Sun, K.; Wang, Y.; Xu, H.; Zhang, J.; Zhu, Y.; Hu, Z. Short-Term Stability of Perovskite Solar Cells Affected by In Situ Interface Modification. *Sol. RRL* **2019**, *3*, 1989–1996. [CrossRef]
30. *ISO 14040:2006*; Environmental Management—Life Cycle Assessment—Principles and Framework (Edition 2). International Organization for Standardization: Geneva, Switzerland, 2006.
31. *ISO 14044:2006*; Environmental management—Life Cycle Assessment—Requirements and Guidelines (Edition 1). International Organization for Standardization: Geneva, Switzerland, 2006.
32. Piotrowska, K.; Piasecka, I.; Gola, A.; Kosicka, E. Assessment of the Impact of Selected Segments of Road Transport on the Natural Environment Using LCA Analysis. In *Advances in Manufacturing IV. MANUFACTURING 2024. Lecture Notes in Mechanical Engineering*; Hamrol, A., Grabowska, M., Eds.; Springer: Cham, Switzerland, 2024; pp. 223–236. [CrossRef]

33. Muthu, S.S. *Life Cycle Sustainability Assessment (LCSA)*; Springer: Singapore, 2021; pp. 1–14. [CrossRef]
34. Schmidt, W.P. *Solutions for Sustainability Challenges. Technical Sustainability Management and Life Cycle Thinking*; Springer: Cham, Switzerland, 2024; pp. 27–96. [CrossRef]
35. Nakamura, S. *A Practical Guide to Industrial Ecology by Input-Output Analysis*; Springer: Cham, Switzerland, 2023; pp. 1–56. [CrossRef]
36. Mroziński, A.; Piasecka, I. Selected Aspects of Building, Operation and Environmental Impact of Offshore Wind Power Electric Plants. *Pol. Marit. Res.* **2015**, *22*, 86–92. [CrossRef]
37. Hesser, F.; Kral, I.; Obersteiner, G.; Hörtenhuber, S.; Kühmaier, M.; Zeller, V.; Schebek, L. *Progress in Life Cycle Assessment 2021*; Springer: Cham, Switzerland, 2023; pp. 7–40. [CrossRef]
38. Siwiec, D.; Pacana, A. Decision-Making Model Supporting Eco-Innovation in Energy Production Based on Quality, Cost and Life Cycle Assessment (LCA). *Energies* **2024**, *17*, 4318. [CrossRef]
39. Stoffels, P.; Kaspar, J.; Baehre, D.; Vielhaber, M. Holistic Material Selection Approach for More Sustainable Products. *Procedia Manuf.* **2017**, *8*, 401–408. [CrossRef]
40. Reap, J.; Roman, F.; Duncan, S.; Bras, B. A survey of unresolved problems in life cycle assessment. *Int. J. Life Cycle Assess.* **2008**, *13*, 290–300. [CrossRef]
41. Sobocińska, M.; Mazurek-Łopacińska, K.; Graczyk, A.; Kociszewski, K.; Krupowicz, J. Decision-Making Processes of Renewable Energy Consumers Compared to Other Categories of Ecological Products. *Energies* **2022**, *15*, 6272. [CrossRef]
42. Majid, S.; Zhang, X.; Khaskheli, M.B.; Hong, F.; King, P.J.H.; Shamsi, I.H. Eco-Efficiency, Environmental and Sustainable Innovation in Recycling Energy and Their Effect on Business Performance: Evidence from European SMEs. *Sustainability* **2023**, *15*, 9465. [CrossRef]
43. Kłos, Z. Ecobalancial assessment of chosen packaging processes in food industry. *Int. J. Life Cycle Assess.* **2002**, *7*, 309. [CrossRef]
44. Klemeš, J.J.; Varbanov, P.S.; Ocłoń, P.; Chin, H.H. Towards Efficient and Clean Process Integration: Utilisation of Renewable Resources and Energy-Saving Technologies. *Energies* **2019**, *12*, 4092. [CrossRef]
45. Varun; Bhat, I.K.; Prakash, R. LCA of Renewable Energy for Electricity Generation Systems—A Review. *Renew. Sustain. Energy Rev.* **2009**, *13*, 1067–1073. [CrossRef]
46. Kjaer, L.L.; Pigosso, D.C.A.; McAloone, T.C.; Birkved, M. Guidelines for evaluating the environmental performance of Product/Service-Systems through life cycle assessment. *J. Clean. Prod.* **2018**, *190*, 666–678. [CrossRef]
47. Suppipat, S.; Chotiratanapinun, T.; Teachavorasinskun, K.; Hu, A.H. *Design for Enhancing Eco-Efficiency of Energy-Related Products. The Integration of Simplified LCA Tools in Industrial Design Education*; Springer: Cham, Switzerland, 2023; pp. 15–43. [CrossRef]
48. Piotrowska, K.; Piasecka, I.; Kasner, R. Assessment of the Life Cycle of a Wind and Photovoltaic Power Plant in the Context of Sustainable Development of Energy Systems. *Materials* **2022**, *15*, 7778. [CrossRef]
49. Wang, S.; Su, D. Sustainable Product Innovation and Consumer Communication. *Sustainability* **2022**, *14*, 8395. [CrossRef]
50. Proske, M.; Finkbeiner, M. Obsolescence in LCA–Methodological Challenges and Solution Approaches. *Int. J. Life Cycle Assess.* **2020**, *25*, 495–507. [CrossRef]
51. Palousis, N.; Luong, L.; Abhary, K. *An Integrated LCA/LCC Framework for Assessing Product Sustainability Risk*; WIT Press: Southampton, UK, 2008; pp. 121–128. [CrossRef]
52. Finkbeiner, M.; Inaba, A.; Tan, R.; Christiansen, K.; Klüppel, H.-J. The New International Standards for Life Cycle Assessment: ISO 14040 and ISO 14044. *Int. J. Life Cycle Assess.* **2006**, *11*, 80–85. [CrossRef]
53. Piasecka, I.; Tomporowski, A. Analysis of Environmental and Energetical Possibilities of Sustainable Development of Wind and Photovoltaic Power Plants. *Probl. Sustain. Dev.* **2018**, *13*, 125–130.
54. Kutraleeswaran, M.; Venkatachalam, M.; Saroja, M.; Gowthaman, M.; Shankar, S. Dye sensitized solar cells—A review. *Trans. Indian Ceram. Soc.* **2017**, *4*, 26–38.
55. Tian, X.; Stranks, S.D.; You, F. Life cycle energy use and environmental implications of high-performance perovskite tandem solar cells. *Sci. Adv.* **2020**, *6*, eabb0055. [CrossRef] [PubMed]
56. Finnveden, G.; Hauschild, M.Z.; Ekvall, T.; Guinée, J.; Heijungs, R.; Hellweg, S.; Koehler, A.; Pennington, D.; Suh, S. Recent developments in Life Cycle Assessment. *J. Environ. Manag.* **2009**, *91*, 1–21. [CrossRef]
57. Guinée, J. *Handbook on Life Cycle Assessment: Operational Guide to the ISO Standards*; Springer: Dordrecht, The Netherlands, 2002; pp. 1–390. [CrossRef]
58. Fukushige, S.; Kobayashi, H.; Yamasue, E.; Hara, K. *EcoDesign for Sustainable Products, Services and Social Systems II*; Springer: Singapore, 2024; pp. 109–124. [CrossRef]
59. Bałdowska-Witos, P.; Piasecka, I.; Flizikowski, J.; Tomporowski, A.; Idzikowski, A.; Zawada, M. Life Cycle Assessment of Two Alternative Plastics for Bottle Production. *Materials* **2021**, *14*, 4552. [CrossRef]
60. Hauschild, M.Z.; Goedkoop, M.; Guinée, J.; Heijungs, R.; Huijbregts, M.; Jolliet, O.; Margni, M.; De Schryver, A.; Humbert, S.; Laurent, A.; et al. Identifying best existing practice for characterization modeling in life cycle impact assessment. *Int. J. Life Cycle Assess.* **2013**, *18*, 683–697. [CrossRef]
61. Bare, J.C.; Hofstetter, P.; Pennington, D.W.; Udo de Haes, H.A. Midpoints versus endpoints: The sacrifices and benefits. *Int. J. Life Cycle Assess.* **2000**, *5*, 319–326. [CrossRef]

62. Jolliet, O.; Müller-Wenk, R.; Bare, J.; Brent, A.; Goedkoop, M.; Heijungs, R.; Itsubo, N.; Peña, C.; Pennington, D.; Potting, J.; et al. The LCIA midpoint-damage framework of the UNEP/SETAC life cycle initiative. *Int. J. Life Cycle Assess.* **2004**, *9*, 394–404. [CrossRef]
63. Portillo, F.; Alcayde, A.; Garcia, R.M.; Fernandez-Ros, M.; Gazquez, J.A.; Novas, N. Life Cycle Assessment in Renewable Energy: Solar and Wind Perspectives. *Environments* **2024**, *11*, 147. [CrossRef]
64. Piasecka, I.; Tomporowski, A.; Piotrowska, K. Environmental analysis of post-use management of car tires. *Przem. Chem.* **2018**, *10*, 1649–1653. [CrossRef]
65. Millet, D.; Bistagnino, L.; Lanzavecchia, C.; Camous, R.; Poldma, T. Does the Potential of the Use of LCA Match the Design Team Needs? *J. Clean. Prod.* **2007**, *15*, 335–346. [CrossRef]
66. Su, D. *Sustainable Product Development. Tools, Methods and Examples*; Springer: Cham, Switzerland, 2020; pp. 15–70. [CrossRef]
67. Cooper, J.S. Specifying functional units and reference flows for comparable alternatives. *Int. J. Life Cycle Assess.* **2003**, *8*, 337–349. [CrossRef]
68. Finkbeiner, M. Product environmental footprint—Breakthrough or breakdown for policy implementation of life cycle assessment? *Int. J. Life Cycle Assess.* **2014**, *19*, 266–271. [CrossRef]
69. Sobaszek, L.; Piasecka, I.; Flizikowski, J.; Tomporowski, A.; Sokolovskij, E.; Baldowska-Witos, P. Environmentally Oriented Analysis of Benefits and Expenditures in the Life Cycle of a Wind Power Plant. *Materials* **2023**, *16*, 538. [CrossRef] [PubMed]
70. Heijungs, R. *Probability, Statistics and Life Cycle Assessment. Guidance for Dealing with Uncertainty and Sensitivity*; Springer: Cham, Switzerland, 2024; pp. 331–454. [CrossRef]
71. Bulle, C.; Margni, M.; Patouillard, L.; Boulay, A.M.; Bourgault, G.; De Bruille, V.; Cao, V.; Hauschild, M.; Henderson, A.; Humbert, S.; et al. IMPACT World+: A globally regionalized life cycle impact assessment method. *Int. J. Life Cycle Assess.* **2019**, *24*, 1653–1674. [CrossRef]
72. Jolliet, O.; Frischknecht, R.; Bare, J.; Boulay, A.M.; Bulle, C.; Fantke, P.; Gheewala, S.; Hauschild, M.; Itsubo, N.; Margni, M.; et al. Global guidance on environmental life cycle impact assessment indicators: Findings of the scoping phase. *Int. J. Life Cycle Assess.* **2014**, *19*, 962–967. [CrossRef]
73. Nalau, J.; Gilmore, E.; Howden, M. Improving adaptation assessment in the IPCC. *NJP Clim. Action* **2024**, *3*, 76. [CrossRef]
74. Masson-Delmotte, V.; Zhai, P.; Pirani, A.; Connors, S.L.; Péan, C.; Berger, S.; Caud, N.; Chen, Y.; Goldfarb, L.; Gomis, M.I.; et al. *IPCC, 2021: Climate Change 2021: The Physical Science Basis. Contribution of Working Group I to the Sixth Assessment Report of the Intergovernmental Panel on Climate Change Cambridge*; University Press: Cambridge, UK, 2021; pp. 147–1926. [CrossRef]
75. Muñoz, I.; Schmidt, J.H. Methane oxidation, biogenic carbon, and the IPCC's emission metrics. Proposal for a consistent greenhouse-gas accounting. *Int. J. Life Cycle Assess.* **2016**, *21*, 1069–1075. [CrossRef]
76. Piotrowska, K.; Piasecka, I. Specification of Environmental Consequences of the Life Cycle of Selected Post-Production Waste of Wind Power Plants Blades. *Materials* **2021**, *14*, 4975. [CrossRef]
77. Frischknecht, R.; Wyss, F.; Büsser Knöpfel, S.; Lützkendorf, T.; Balouktsi, M. Cumulative energy demand in LCA: The energy harvested approach. *Int. J. Life Cycle Assess.* **2015**, *20*, 957–969. [CrossRef]
78. Patel, M. Cumulative energy demand (CED) and cumulative CO_2 emissions for products of the organic chemical industry. *Energy* **2003**, *28*, 721–740. [CrossRef]
79. Huijbregts, M.A.J.; Hellweg, S.; Frischknecht, R.; Hendriks, H.W.M.; Hungerbühler, K.; Hendriks, A.J. Cumulative Energy Demand As Predictor for the Environmental Burden of Commodity Production. *Environ. Sci. Technol.* **2010**, *44*, 2189–2196. [CrossRef]
80. Kłos, Z.; Kalkowska, J.; Kasprzak, J. *Towards a Sustainable Future—Life Cycle Management. Challenges and Prospects*; Springer: Cham, Switzerland, 2022; pp. 87–163. [CrossRef]
81. Curran, M.A. *Interpretation, Critical Review and Reporting in Life Cycle Assessment*; Springer: Cham, Switzerland, 2023; pp. 7–50. [CrossRef]
82. Piasecka, I.; Bałdowska-Witos, P.; Piotrowska, K.; Kruszelnicka, W.; Flizikowski, J.; Tomporowski, A. Ecological life cycle assessment of the 1 MW photovoltaic power plant under Polish environmental conditions. *Przem. Chem.* **2021**, *1*, 40–46. [CrossRef]
83. Koehler, A. Water use in LCA: Managing the planet's freshwater resources. *Int. J. Life Cycle Assess.* **2008**, *13*, 451–455. [CrossRef]
84. Boulay, A.M.; Bulle, C.; Bayart, J.B.; Deschênes, L.; Margni, M. Regional Characterization of Freshwater Use in LCA: Modeling Direct Impacts on Human Health. *Environ. Sci. Technol.* **2011**, *45*, 8948–8957. [CrossRef] [PubMed]
85. Pfister, S.; Koehler, A.; Hellweg, S. Assessing the Environmental Impacts of Freshwater Consumption in LCA. *Environ. Sci. Technol.* **2009**, *43*, 4098–4104. [CrossRef]
86. Andrew, R.M. A comparison of estimates of global carbon dioxide emissions from fossil carbon sources. *Earth Syst. Sci. Data* **2020**, *12*, 1437–1465. [CrossRef]
87. Burke, P.J.; Shahiduzzaman, M.; Stern, D.I. Carbon dioxide emissions in the short run: The rate and sources of economic growth matter. *Glob. Environ. Chang.* **2015**, *33*, 109–121. [CrossRef]
88. Georgopoulos, P.G.; Roy, A.; Yonone-Lioy, M.J.; Opiekun, R.E.; Lioy, P.J. Environmental copper: Its dynamics and human exposure issues. *J. Toxicol. Environ. Health* **2011**, *4*, 341–394. [CrossRef]
89. Ibrahim, M.; El-Haes, H. Computational spectroscopic study of copper, cadmium, lead and zinc interactions in the environment. *Int. J. Environ. Pollut.* **2005**, *23*, 417–424. [CrossRef]

90. Laws, E.A. *Environmental Toxicology*; Springer: New York, NY, USA, 2013; pp. 203–238. [CrossRef]
91. Mar, K.A.; Unger, C.; Walderdorff, L.; Butler, T. Beyond CO_2 equivalence: The impacts of methane on climate, ecosystems, and health. *Environ. Sci. Policy* **2022**, *134*, 127–136. [CrossRef]
92. Miller, S.M.; Wofsy, S.C.; Michalak, A.M.; Kort, E.A.; Andrews, A.E.; Biraud, S.C.; Dlugokencky, E.J.; Eluszkiewicz, J.; Fischer, M.L.; Janssens-Maenhout, G.; et al. Anthropogenic emissions of methane. *Earth Atmos. Planet. Sci.* **2013**, *110*, 20018–20022. [CrossRef]
93. Piotrowska, K.; Bałdowska-Witos, P.; Piasecka, I.; Kasner, R.; Kruszelnicka, W.; Tomporowski, A. Identification of the most important areas of environmental impact over the life cycle of car tires. *Przem. Chem.* **2021**, *11*, 1593–1599. [CrossRef]
94. El Chaar, L.; Lamont, L.A.; El Zein, N. Review of photovoltaic technologies. *Renew. Sustain. Energy Rev.* **2011**, *15*, 2165–2175. [CrossRef]
95. Zacher, L.W. *Technology, Society and Sustainability: Selected Concepts, Issues and Cases*; Springer: Basel, Switzerland, 2017; pp. 203–221. [CrossRef]
96. Bagnall, D.M.; Boreland, M. Photovoltaic technologies. *Energy Policy* **2008**, *36*, 4390–4396. [CrossRef]
97. Khan, I. *Renewable Energy and Sustainability*; Elsevier: Amsterdam, The Netherlands, 2022; pp. 1–22. [CrossRef]
98. Leda, P.; Kruszelnicka, W.; Leda, A.; Piasecka, I.; Kłos, Z.; Tomporowski, A.; Flizikowski, J.; Opielak, M. Life Cycle Analysis of a Photovoltaic Power Plant Using the CED Method. *Energies* **2024**, *16*, 8098. [CrossRef]
99. Rigatos, G.G. *Intelligent Renewable Energy Systems: Modelling and Control*; Springer: Cham, Switzerland, 2016; pp. 339–409. [CrossRef]
100. Ross, J.R.H. *Sustainable Energy*; Elsevier: Amsterdam, The Netherlands, 2022; pp. 1–75. [CrossRef]

materials

MDPI

Article

Influence of Reclaimed Water on the Visual Quality of Automotive Coating

Piotr Woźniak and Marek Gryta *

Faculty of Chemical Technology and Engineering, West Pomeranian University of Technology in Szczecin, ul. Pułaskiego 10, 70-322 Szczecin, Poland; piotr.wozniak@zut.edu.pl

* Correspondence: marek.gryta@zut.edu.pl

Abstract: In the present study, the possibility of recovering water in a car wash station was presented. The resistance of automotive coatings to washing water recovered at 50% and 70% from wastewater generated at car wash was tested. Wastewater treatment was carried out by ultrafiltration (UF) using tubular polyvinylidene fluoride (PVDF) membranes (100 and 200 kDa) manufactured by the PCI company. The membranes retained oil contamination, suspended solids, and over 60% of surfactants. For comparison, the 0.5% Turbo Active Green solution, used at professional car washes, was also applied in paint resistance studies. The tested solutions washed the painted surfaces of samples taken from car doors for 8 days. The resistance of automotive coatings to washing solutions was assessed by measuring gloss, Log Haze, RIQ, and Rspec parameters. Scratch resistance was also assessed. The results obtained in the current study indicated that the use of water recovered from wastewater did not deteriorate the quality of the car paint coating.

Keywords: automotive coatings; car wash; water reuse; water reclamation; ultrafiltration

1. Introduction

Cars need to be washed periodically, which results in an annual worldwide water consumption of millions of cubic meters [1–3]. For this reason, regulations are increasingly being introduced to limit water consumption in washing facilities, particularly in water-scarce areas [4,5]. Obviously, the amount of washing water consumed by car washes depends on the technology used. Commercial car wash operations are realised by applying different automatic or manual systems, using approximately 150–350 L of water per car [6,7]. In Europe, some countries limit water consumption to 60–70 L per car and require to achieve a high water recovery rate [8,9].

Commercial carwash stations are typically equipped with a settling tank and an oil separator. The quality of the water treated in this way is too low to be recycled for car washing. Therefore, several additional systems are designed and tested for the treatment and reuse of the wastewater generated in carwash stations. For these purposes, multi-stage filtration through sand and activated carbon beds [5,10], membrane separation [2,11], chemical methods like coagulation and flocculation [12,13], electrocoagulation [14,15], and biological treatment [16,17] are used. Many of these methods produce purified water with parameters even better than tap water, which is usually fed to car washes.

Car washing is carried out in several stages, and at each stage, water from different purities can be used. Demineralised water (reverse osmosis separation) is used in the final rinsing stage (spot-free water) to achieve a shiny car finish without streaks or traces of salt deposits. At the initial process stage, for pre-wash and foam washing, there is no need for such pure water; hence, reclaimed water can be used for this [2,3]. Furthermore, multi-stage cleaning systems are too complex and expensive to be applied in small hand touchless car washes. Therefore, this study investigated the possibility of recycling water treated in a simple ultrafiltration (UF) system.

Small touchless car washes are popular in many countries, and there are several in every Polish city. These are self-service car washes, where the customer can choose and apply a washing programme, usually including water rinsing, foam washing, and final water rinsing. A final option, including rinsing with osmotic water and waxing for paint protection, is also available. Water demineralised by reverse osmosis is used for these final stages, but reclaimed water can be used for the initial two stages. Washing water and cleaning agents are sprayed onto the car using high-pressure nozzles (5–10 MPa), and therefore, reclaimed water must be free of suspended solids and microorganisms [2,18]. This degree of water purification can be achieved using the UF process [19,20].

The UF process makes it possible to separate the contaminants present in the effluent removed from car washing. The permeate obtained contains mainly water and part of the cleaning agents used at the car wash, which are retained in 60–80% [20]. The part of the cleaning agents remaining in the permeate can facilitate the removal of contaminants from the car surface. The reclaimed water may also contain traces of substances formed in the settling tank, e.g., aggressive products of the microorganisms' metabolisms [21], whose effects on paints are not known. Moreover, hydrolytic degradation of car coatings was found to be caused by enzymes existing in bird droppings [22]. This indicates that many components present in the environment can have an impact on the appearance of a car's body during its service life. To investigate whether such an impact is also caused by reclaimed water, the surfaces of car body components were washed for several days. Some works have presented the long-term performance of water purification systems in car wash stations [8,23]. However, the effect of reclaimed water on the durability of automotive paints has not yet been studied.

In modern cars, paint resistance is provided by the external clearcoat layer (40–50 μm) covering the coloured basecoat (15–20 μm), which is deposited on the corrosion protection layer (primer and electrocoat, 40–60 μm) [22,24]. The automotive paint forming the basecoat has little abrasion resistance, hence the scratch-free finishes provided by clearcoat. Scratch events occur from very small (micron size) to very large, and they appear white due to fractures in the clearcoat [24]. Pores and oxidation products can form on the paint surface due to negative environmental influences, resulting in reduced gloss of the paint and reduced scratch resistance [25,26].

Manufacturers design the products to have highly reflective car body panels. Gloss, distinctness of image (DOI), image fuzziness (Haze), and waviness (orange peel) measurements are used to assess surface finish quality [24,27,28]. Typical gloss meter systems are designed around standard measurement angles of 20° for high-gloss surfaces, 60° for semi-gloss surfaces, and 85° for matte surfaces [29,30]. The defined reference points of gloss unit (GU) in the standards are a 0 GU (zero reflection) and a 100 GU reference [31]. DOI quantifies the spread of light reflected at the specular angle. It gives an indication of how sharp a surface reflection is likely to be. Haze refers to a cloudy or milky appearance, also due to scattering of light. Other industry standards have been developed by Rhopoint Instruments company, two of which are peak specular reflectance (Rspec) for smoothness quantification and the reflected image quality (RIQ) parameter [29]. Rspec is very sensitive to surface texture and can be used to identify subtle differences in smooth surfaces. RIQ is used to quantify effects such as orange peel and surface waviness. This new parameter provides higher resolution results compared to DOI measurement and better mimics human perception of surface texture, especially on high-quality finishes such as automotive.

The effect of using reclaimed water on changes in the above parameters, as well as the scratch resistance of clearcoat, was investigated in the study. The tests were conducted to determine whether reclaimed water is more detrimental to the visual quality of car paint than the commercial washing solution used in the car wash.

2. Materials and Methods

2.1. Washing Water

The permeates obtained during ultrafiltration of wastewater collected from touchless car washes were used for the study. The UF process was carried out using polyvinylidene fluoride (PVDF) tubular membranes FP100 (100 kDa) and FP200 (200 kDa), manufactured by PCI (Kostrzyn, Poland). The wastewater separation process is described in work [20]. The UF process was carried out using a 50% and 75% water recovery rate, obtaining the UF permeates of the composition shown in Table 1 (Permeate 50% and Permeate 75%). In addition to the permeates, in presented studies, a 0.5% Turbo Active Green Foam solution (EuroEcol, Łódź, Poland) was also used for comparison. At the car wash stations from which the wastewater was collected, this cleaning agent was used to prepare a washing solution.

Table 1. Washing water parameters. Wastewater collected from the touchless car washes.

Parameter	Permeate 50%	Permeate 75%	T. Active Green [1]	Wastewater [2]
COD [mg/L]	535 ± 2.5	565 ± 0.5	2298 ± 13	1018 ± 10
BOD [mg/L]	237 ± 2	243 ± 2	690 ± 1	369 ± 1
total N [mg/L]	6.64 ± 0.02	8.27 ± 0.03	8.48 ± 0.02	14.4 ± 0.04
total P [mg/L]	5.26 ± 0.1	6.13 ± 0.07	0	12.5 ± 0.09
anionic [mg/L]	75.7 ± 0.7	79.6 ± 0.2	691 ± 11	135 ± 10
nonionic [mg/L]	3.46 ± 0.1	4.17 ± 0.06	30 ± 1.6	14.3 ± 1.6
pH [-]	8.5 ± 0.1	8.6 ± 0.1	8.7 ± 0.1	8.6 ± 0.1

[1] T. Active Green—0.5% solution of Turbo Active Green Foam cleaning agent; [2] wastewater—used as a feed for UF process.

The Hach cuvette tests (Hach Lange, Wrocław, Poland) were used to determine the concentration of surfactants (LCK 334-nonionic, LCK 344-anionic) and chemical oxygen demand (COD) (LCK 1014). The biological oxygen demand (BOD) was determined using test LCK 555 and the total P and total N by using LCK 348 and LCK 238 tests, respectively.

2.2. Experimental Installation

The study evaluated the effect of prolonged contact with the cleaning solution on the changes in the visual quality of the automotive paint finishes. The design of the experimental setup is shown in Figure 1. The tested car body samples mounted at an angle of approximately 40° were poured with wash water for 8 days. Washing water (2 L) was pumped by a peristaltic pump (660 mL/min) to the top of the sample and flowed in a film (5–6 cm wide, 13–15 cm long) to the tank from which it was circulated.

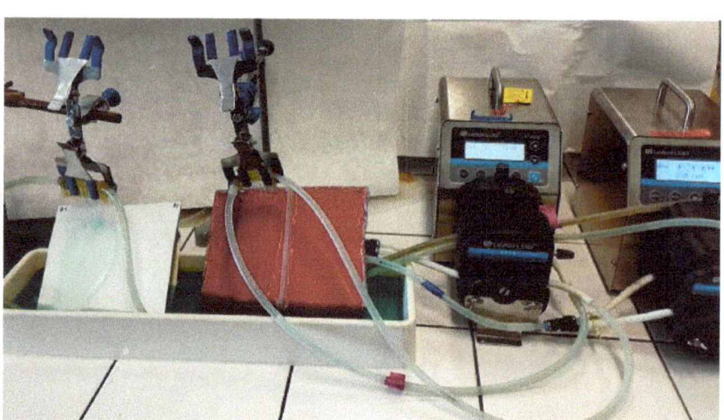

Figure 1. The image of the experimental installation.

For the study, car body samples were cut from the metal front doors of two passenger cars and painted in red (R) and cherry (CH). The CH sample had a metallic coating, which can interfere with measurements of some parameters [29]. To improve fuel economy for vehicle body construction, lighter materials such as polymers were applied in addition to steel. The polymer materials were used to make white-painted truck doors, from which samples labelled W were taken.

Several samples were used for the study, with the starting parameters (Rhopoint IQ measurements) shown in Table 2. Of the coating types tested, red paint was the least resistant to the damaging effects of the environment. For this reason, more R samples were used in the study than the others. The samples were obtained from the cars after 4–5 years of use. As cars are used over time, the amount of damage to the clearcoat layer increases, which intensifies the negative environmental impact [21]. This property also allows a better assessment of the impact of reclaimed water.

Table 2. The initial and final parameters of study samples. Average values for a sample measured 30 times. 8d—results after 8 days of washing the samples. Gloss [GU]—measured at 20°, 60° and 85°.

Sample	Gloss 20°	Gloss 60°	Gloss 85°	RIQ	Log Haze	Rspec
R#1	71.8 ± 12.2	86.3 ± 1.6	96.6 ± 2.8	57.8 ± 13.4	48.7 ± 5.8	36.5 ± 7.8
8d–T.A.G. [1]	65.9 ± 3.9	82.5 ± 0.4	96.8 ± 0.3	54.3 ± 5.4	51.5 ± 3.8	31.3 ± 3.1
R#2	66.2 ± 2.5	82.4 ± 1.6	88.8 ± 7.3	55.9 ± 5.3	47.3 ± 7.6	32.7 ± 2.3
8d–P.50% [2]	64.0 ± 8.3	82.6 ± 1.3	87.9 ± 1.9	51.8 ± 8.5	42.9 ± 2.9	30.8 ± 4.7
R#3	74.1 ± 0.7	87.1 ± 0.3	96.2 ± 0.4	58.9 ± 6.6	49.0 ± 2.9	38.2 ± 3.8
8d-P.75% [3]	69.6 ± 3.3	82.1 ± 1.5	96.7 ± 0.5	60.1 ± 5.5	60.8 ± 4.0	36.2 ± 3.4
R#4	70.1 ± 1.4	83.8 ± 0.7	96.3 ± 0.4	49.1 ± 3.5	69.7 ± 7.4	30.6 ± 1.4
8d-scratch	55.7 ± 6.4	75.9 ± 2.2	95.8 ± 0.5	49.7 ± 4.8	79.5 ± 12.6	26.1 ± 3.5
R#5	74.0 ± 2.3	87.2 ± 0.9	96.2 ± 2.5	58.9 ± 3.9	49.0 ± 3.5	38.2 ± 2.6
8d–scratch	60.4 ± 2.6	79.9 ± 1.2	97.1 ± 0.5	62.9 ± 7.0	74.2 ± 7.7	33.6 ± 3.6
R#6	64.3 ± 3.6	84.0 ± 1.4	95.8 ± 0.4	55.8 ± 3.6	77.0 ± 6.1	31.7 ± 2.1
only scratch [4]	63.1 ± 6.7	79.4 ± 2.4	96.4 ± 0.6	52.4 ± 6.4	79.4 ± 9.8	29.4 ± 3.4
W#1	78.4 ± 1.4	93.1 ± 0.9	96.2 ± 0.4	44.6 ± 4.3	48.8 ± 4.8	32.0 ± 1.9
8d–T.A.G.	81.3 ± 2.1	91.6 ± 1.0	97.5 ± 0.6	44.5 ± 6.6	46.5 ± 4.7	33.1 ± 2.8
W#2	77.8 ± 4.6	85.7 ± 1.9	93.6 ± 0.5	45.9 ± 6.2	64.9 ± 8.6	27.9 ± 1.4
8d–P.50%	81.6 ± 1.6	94.1 ± 0.4	96.8 ± 0.4	44.4 ± 4.2	49.5 ± 6.1	33.4 ± 2.6
W#3	82.3 ± 0.8	93.1 ± 0.5	96.2 ± 0.7	45.8 ± 3.8	57.4 ± 2.2	33.4 ± 1.7
8d-P.75%	82.8 ± 1.7	92.1 ± 0.4	95.8 ± 1.3	45.9 ± 5.9	56.5 ± 6.1	33.3 ± 2.7
W#4	71.7 ± 3.8	87.9 ± 1.7	96.9 ± 0.2	31.2 ± 5.8	75.2 ± 13.1	22.7 ± 3.1
CH#1	76.6 ± 16.4	86.9 ± 1.6	96.5 ± 2.5	74.4 ± 13.5	13.1 ± 4.5	51.9 ± 11.7
8d-T.A.G.	81.4 ± 1.1	91.7 ± 0.9	91.6 ± 2.5	72.3 ± 4.5	10.1 ± 1.4	52.1 ± 3.9
CH#2	83.1 ± 0.8	91.1 ± 0.7	93.2 ± 0.8	70.1 ± 7.4	11.5 ± 1.9	54.4 ± 6.6
8d–P.50%	79.6 ± 8.3	88.9 ± 1.1	88.2 ± 4.6	62.9 ± 9.8	10.1 ± 0.7	45.5 ± 4.7

[1] T.A.G.—Turbo Active Green. [2] P.50%—Permeate 50%. [3] P.75%—Permeate 75%. [4] only scratch—new sample, wiped only 10 times with paper (not exposed to wash).

Before performing the tests, the samples taken from the cars were cleaned. They were rinsed with distilled water and finely cleaned with isopropanol. The constituents of the reclaimed water could deposit on the samples; therefore, their surfaces were also rinsed with isopropanol each day before measurements were taken. Measurements with the Rhopoint IQ were taken once a day, collecting 30 datasets from a 3 × 7 cm area. The samples were divided into fields (Figure 1), and the Turbo Active Green solution flowed 3–4 cm from the surface over which the reclaimed water flowed, allowing the effect of both washing solutions to be studied for clearcoat with similar properties.

In the case of the abrasion resistance test, the alcohol-wetted samples were additionally scratched 10 times with filter paper. Alcohol was wiped off by moving the paper by hand on both sides of the sample without applying additional pressure. It was assumed that similar friction conditions were achieved and that the degree of surface scratching was only dependent on the performance of the clearcoat, which could change with prolonged contact with wash solutions.

2.3. Measuring Equipment

The quality of surface appearance was analysed using a glossmeter/goniophotometer Rhopoint IQ (Rhopoint Instruments Ltd., St Leonards on Sea, UK). The instrument uses a standard optical configuration at 60° for semi-gloss surfaces and 85° for matte surfaces. Moreover, at 20°, the Rhopoint IQ uses a diode array to measure the distribution of the reflected light at +/−7.25° from the specular angle of reflection in steps of 0.02832°. The Rhopoint IQ uses a 512-element linear diode array, which profiles reflected light in a large arc from 14° to 27°. The measures at 20° allow the instrument to calculate Haze, DOI, RIQ and Rspec parameters.

Haze describes the milky halo or bloom seen on high gloss surfaces. When measuring haze values, higher numbers indicate a lower-quality surface. The IQ instrument can display the natural haze value or Log Haze value. Log Haze is commonly quoted for paints and coatings as this scale has better corroboration with human perception of surface quality [29]. The Log Haze unit is calculated from the relationship:

$$\text{Log Haze} = 1285\ (\log 10((\text{Haze}/20) + 1)) \quad (1)$$

The Rhopoint IQ measures the DOI (or improved RIQ parameter) of a surface by quantifying the way a reflected measurement beam is spread and distorted around the specular angle. The DOI value of a surface is a number between 01 and 100; a surface that exhibits a perfect undistorted image returns a value of 100, and as the value decreases, the image becomes less discernible. However, two highly reflective surfaces that have very small changes in orange peel or texture will show very little or no change in DOI due to the way that it is calculated but will appear quite different visually. When using the RIQ option, a greater differentiation is achieved. The RIQ value of a surface is also a number between 0 and 100; a surface that exhibits a perfect undistorted image returns a value of 100; as the values decrease, higher surface texture is present, and the image sharpness is reduced [29].

Rspec is the peak reflectance measured over a very narrow angle in the specular direction (+/−) 0.0991°. Waviness or rippling on a surface acts as a concave or convex reflector deflecting light around the specular angle. When Rspec is equal to the gloss, the surface is smooth. The RSpec parameter drops as the texture becomes apparent.

The pH of each solution was measured using a 6P Ultrameter (Myron L Company, Carlsbad, CA, USA).

A Stemi 508 optical microscope (Carl Zeiss, Oberkochen, Germany) and an OLS 5100 (Olympus, Hamburg, Germany) laser scanning microscope (LSM) were used to observe the surfaces of the tested samples. The LSM microscope software was used to measure the surface roughness parameters: root mean square height (Sq) and arithmetic mean height (Sa). The surface of filter paper was observed using an SU8020 (Hitachi High Technologies Co., Tokyo, Japan) scanning electron microscope (SEM).

3. Results

3.1. Long-Term Washing

In the visual assessment, the samples taken from the car doors were glossy and showed little difference in light images reflected from a surface. The samples covered with metallic paint (CH) were the most glossy and, in reflection, showed a clear image. The light image on the surface of the sample taken from the polymer door (W) was more fuzzy, and an orange peel was visible. LSM microscope observations showed that several scratches were present on the surface of the samples after 4–5 years of use of the cars, especially on the sample from the red door (Figure 2a). A similar scratch structure in the clearcoat layer was presented in work [26].

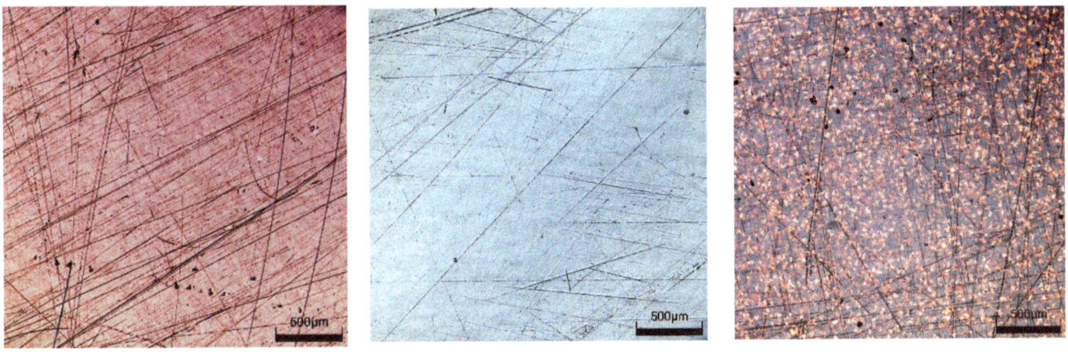

Figure 2. LSM images of tested non-washed samples: (**a**) R#1, (**b**) W#1, (**c**) CH#1.

The presence of scratches affects the roughness values and paint parameters tested in the study [29]. It has been indicated that with an increase in roughness, the results of gloss values decrease rapidly [31]. The roughness values obtained for the samples tested are shown in Table 3. Higher values were obtained before the washing process, especially for the CH sample coated with metallic paint (Sq = 8.51 μm and Sa = 5.27 μm). The roughness values decreased after 8 days (190 h) of washing with the test liquids. During car operation, due to weathering degradation, i.e., hydrolytic and photodegradation, small molecules and shorter polymeric chains are produced, which increases roughness [21,32]. The washing liquids were probably able to wash away these products, and, as a result, lower roughness values for washed samples were recorded.

Table 3. Tortuosity parameters: Sq–root mean square height, Sa–arithmetic mean height.

Parameter	Non washed	Permeate 50%	Permeate 75%	T. Active Green
R-Sq	3.07 ± 0.33	2.81 ± 0.22	1.63 ± 0.47	2.93 ± 0.41
R-Sa	2.33 ± 0.27	2.12 ± 0.14	1.41 ± 0.32	1.99 ± 0.25
W-Sq	4.94 ± 0.15	3.51 ± 0.50	3.42 ± 0.17	3.24 ± 0.11
W-Sa	3.81 ± 0.19	2.49 ± 0.48	2.41 ± 0.12	2.08 ± 0.09
CH-Sq	8.51 ± 2.74	2.81 ± 0.22	-	3.16 ± 0.03
CH-Sa	5.27 ± 1.34	2.12 ± 0.14	-	1.74 ± 0.02

Despite significant changes in roughness, microscopic observations did not show significant changes in surface structure. The image of the CH samples before and after washing was similar (Figure 3). The presence of scratches on the clearcoat surface could facilitate the degradation of polymer coatings [28]. However, LSM images did not show significant changes in the scratch structure (Figure 4). This result indicates that both the Turbo Active Green solution and the water recycled from the car wash wastewater did not damage the clearcoat or increase the scratches. In contrast, after 300 h of exposure to biological substances, several local defects, as well as a decreased appearance on the clearcoat surface, were observed [22]. This result confirmed that it is important to wash the car to remove bird droppings and insects.

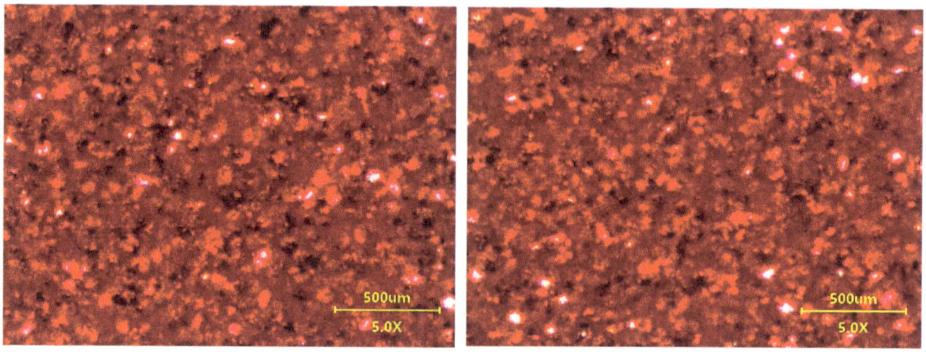

(a) (b)

Figure 3. Microscopic images of surface CH#2 sample covered with metallic coating: (**a**) non-washed, (**b**) after 8 days washing with Permeate 50%.

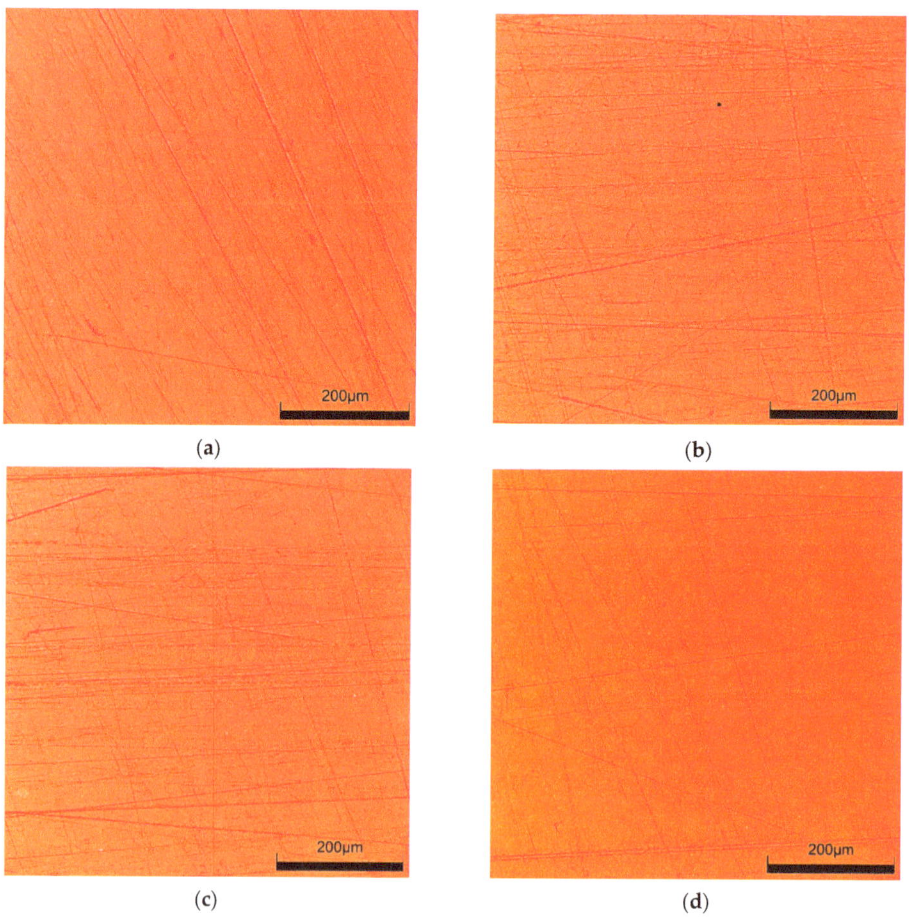

(a) (b) (c) (d)

Figure 4. LSM images of R samples surface: (**a**) R#1-non washed, (**b**) R#1-washed by Turbo Active Green, (**c**) R#2-washed by Permeate 50%, (**d**) R#3-washed by Permeate 75%.

The gloss values obtained for measurements at 60° were above 70 GU (Table 2), indicating that the samples tested are of the high gloss type. In this case, it is recommended that the analysis be performed at 20°, which improves accuracy and resolution on high gloss and metallic samples [29,33]. The results of the measurements for the samples taken from the different door locations differed by more than 10 per cent; hence, the relative changes in the values of the measured parameters (X/X_0) are shown in Figures 5–7.

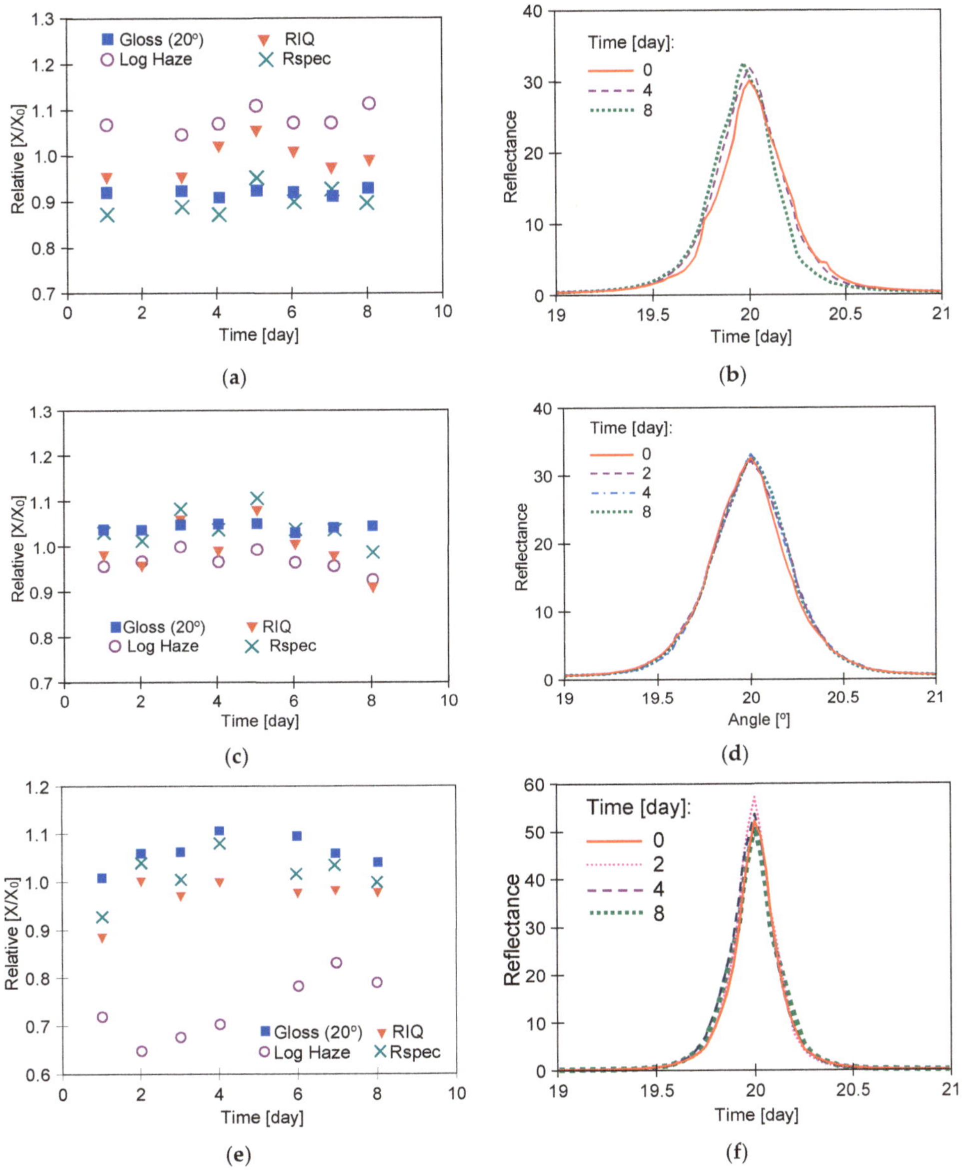

Figure 5. The result of car body samples washing with 0.5% Turbo Active Green solution. Samples: (**a**,**b**)—R#1, (**c**,**d**)—W#1, and (**e**,**f**)—CH#1. Goniophotometric profiles: (**b**,**d**,**f**).

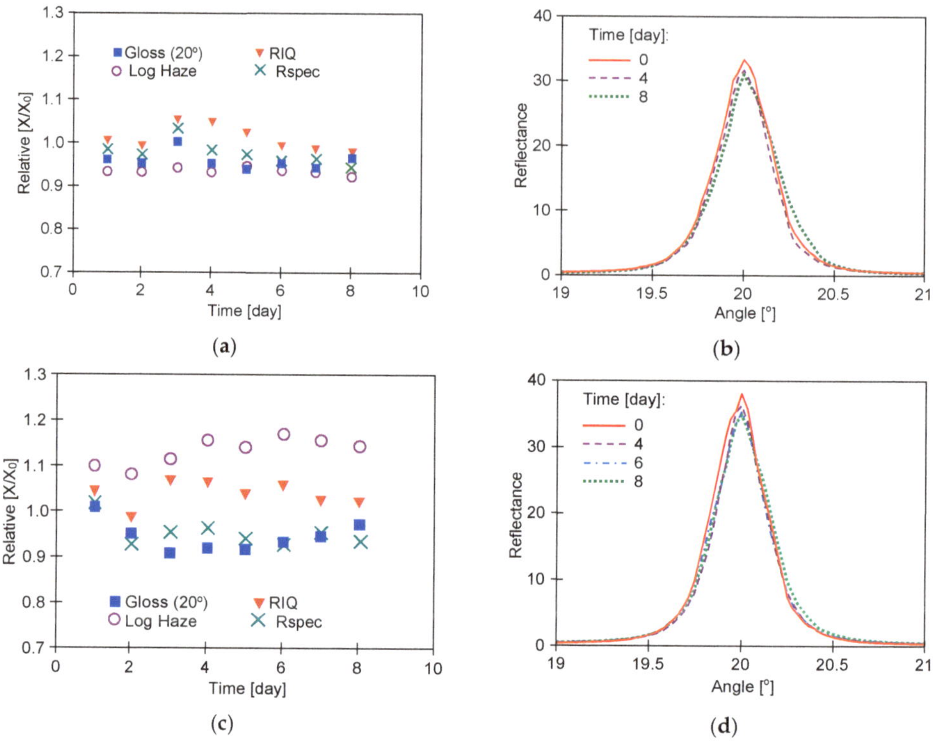

Figure 6. The results R samples washing with the UF permeates: (**a**,**b**) R#2-Permeate 50%, (**c**,**d**) R#3-Permeate 75%. Goniophotometric profiles: (**b**,**d**).

Coatings of car bodies are frequently modified with nanoparticles, and the acrylic base coating or the metalised interlayer coating is popularly used. Unfortunately, exposing cars to environmental impacts resulted in the ageing of the coating material, which caused the corrosion of polymer coatings [28]. The acrylic coatings aged climatically for 2 years showed significant destruction of their surface in the form of microcrackings, craters, and etchings [21]. These processes influenced the surface topography, resulting in a gloss decrease from 92 to 72 GU. Similar gloss values, in the range of 70–80 GU, were obtained for samples taken from cars after more than 4 years of use (Table 2).

Obviously, car washing can cause both mechanical and chemical damage to polymer coatings [24,25,33]. This is related to the fact that cleaning agents contain many chemicals in addition to detergents [34]. The active foams commonly used at car washes are alkaline (Table 1, pH values), which helps to wash away oils and grease. However, alkalis can accelerate the degradation of automotive coatings [21]; hence, active foams deposited on the car surface should be rinsed off after a few minutes.

In the present work, the samples were in contact with the 0.5% Turbo Active Green solution for 8 days, which resulted in minor changes in the measured parameters, such as an improvement in haze (Figure 5). The measured gloss values for samples R#1 and W#1 on consecutive days were almost constant, with progressively small changes in the other parameters. In most cases of the samples tested, the larger changes occurred after the first day of testing. This was probably related to the leaching of the preservatives from the clearcoat (e.g., car wax), which was applied several times during car service life. Car waxes are used not only to protect the body of the car but also to improve the gloss and visual appearance of the car coating. Hence, their removal results in a decrease in gloss values ($X/X_0 < 1$) and an increase in the Log Haze parameter ($X/X_0 > 1$). In the case of Turbo Active Green solution, the image blurring increased slightly when washing sample R#1 (Figure 5a); however, gloss improved as a result of reduced roughness (Table 3). Alkaline liquids have a strong effect on polymers [34]; hence, they can remove the resulting degradation products from their surface. This was probably the reason for the slight improvement in the surface quality parameters of samples W#1 (Figure 5c) and CH#1 (Figure 5e). Nevertheless, the progressive changes in surface structure were small and similar shapes of goniophotometric profiles were obtained in each case (Figure 5).

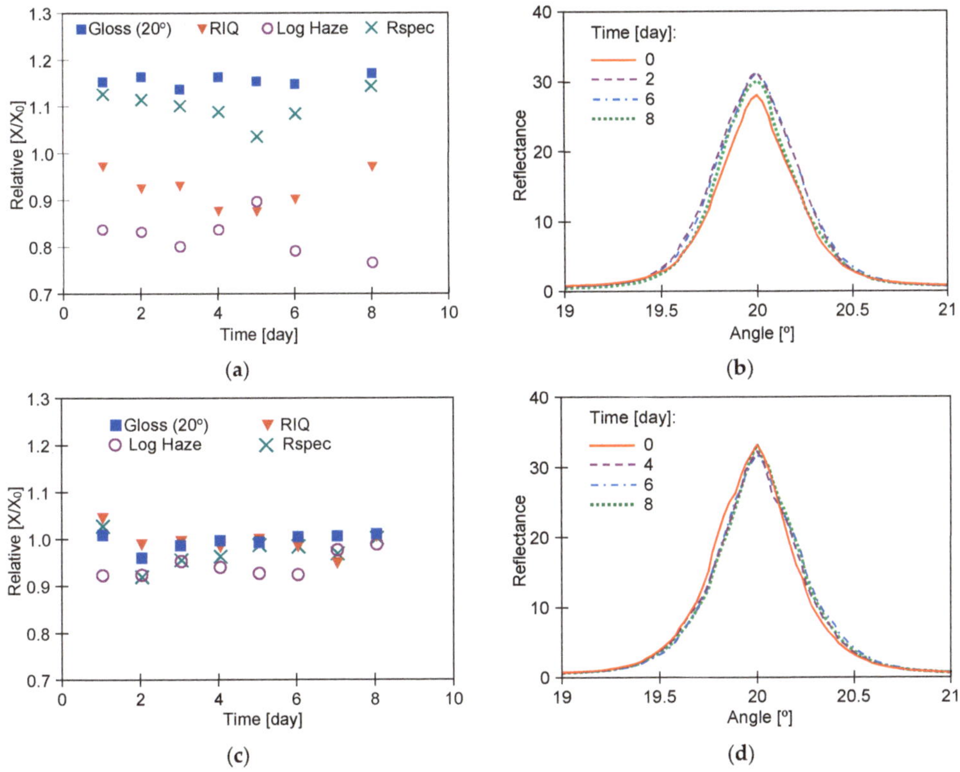

Figure 7. The results of W samples washing with the UF permeates: (**a**,**b**) W#2—Permeate 50%, (**c**,**d**) W#3—Permeate 75%. Goniophotometric profiles: (**b**,**d**).

The testing of a sample covered with a metallic coating (CH) showed the lowest values of Log Haze (Table 2), indicating the good performance of this coating. Reflection haze is caused by micro-texture on a surface, which causes a small amount of light to be reflected adjacent to the gloss angle. For metallic paint, a certain amount of diffuse light is reflected from metallic particles within the material. This diffuse light exaggerates the haze signal for these surfaces, causing higher-than-expected readings. In Rhopoint IQ, this error was eliminated by a compensation procedure [29]. The better surface quality of the CH sample is also evidenced by the higher-shaped reflectance profiles (Figure 5f). The more narrow and higher the goniophotometric profile is, the closer the gloss is to 100%. A widening of this profile means that the influence of haze increases [29].

In the next stage of the current study, the UF permeate obtained from the wastewater (Table 1) was used to wash the samples. Wastewater was taken from car washes where alkaline Turbo Active Green Foam was used. The UF membranes do not retain low molecular weight compounds like NaOH; hence, the pH values of the tested solutions were similar (pH = 8.5–8.7). The membranes used for UF retained more than 60% of the surfactants, and their residues were present in the permeates. The P content in the collected wastewater was 12–14.5 mg/L [34]. The Turbo Active Green solution contained no phosphate; hence, the P detected in the permeates came from the removed car pollutants. To determine whether such compositional differences could have caused changes in the structure of the clearcoat layer, the 8-day washing tests were carried out. The effects of washing the samples with water recycled from wastewater (50% and 75%) were presented in Figures 6–8.

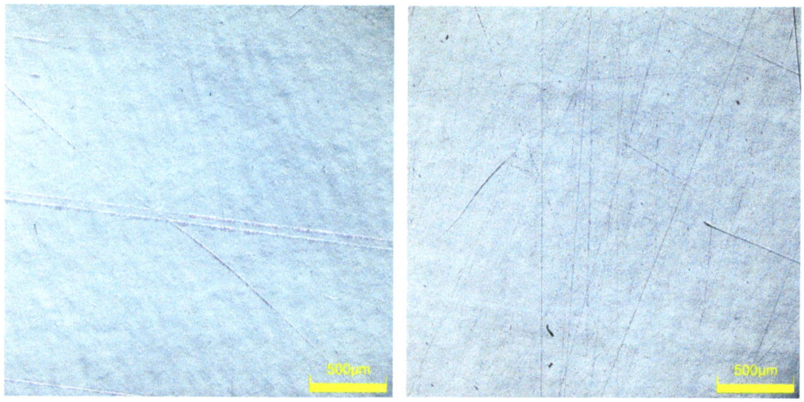

(**a**) (**b**)

Figure 8. LSM images surface of W#2 sample: (**a**) non-washed, (**b**) after 8 days washing with Permeate 50%.

For samples called 'R', greater changes in the parameters studied were observed for the Permeate 75% (Figure 6c), while the opposite results were obtained for W samples (Figure 7a). Moreover, Permeate 75% had a similar composition to the Permeate 50% (Table 1). These results indicate that the changes in clearcoat properties were not significantly affected by the degree of water recovery from the wastewater. In addition, the LSM images obtained for both permeate types showed no differences, and the observed surface structure of the washed 'R' samples was similar to that of the unwashed samples (Figure 4). It can, therefore, be assumed that the differences in the measurements obtained are due to the differences in the initial surface properties of the samples used (Table 2).

The samples called 'W' taken from the polymer door showed the greatest distortion of the reflected image. Under the influence of long-term washing, the surface properties improved, and the Log Haze parameter decreased, especially for sample W#2, which also increased the goniophotometric shape (Figure 7b). LSM microscope observations confirmed that, as a result of sample W washing, the orange peel decreased significantly (Figure 8), which resulted in a reduction of the Log Haze value, as also shown in another paper [29].

Almost constant values of the parameters were obtained for the sample with metallic paint (CH#2), confirming the good resistance of this type of paint (Figure 9). As a result, the surface image before and after 8 days of washing was similar (Figure 3).

Due to the alkaline properties (pH > 8.5) of active foams, it is recommended that they should only be in contact with the car body for a few minutes during washing. Therefore, extending the contact time in the tests to 8 days could have caused damage to the car paint. Despite this, no significant damage to the clearcoat was found in the performed tests, and even some of the parameters tested, such as Log Haze, improved slightly. During the study, each measurement was repeated several times, moving the Rhopoint IQ apparatus by 3–5 mm across the measured area. Despite such small steps, the values obtained often differed by more than 10%, resulting in significant standard deviation values calculated for 30 repetitions (Table 2). These discrepancies were observed for all types of samples tested. Therefore, despite slight differences in individual measurements, it can be concluded that the solutions tested had a similar effect on the samples used. Therefore, it can also be concluded that the results obtained for samples washed with reclaimed water were similar to those obtained for samples washed with Turbo Active Green solution.

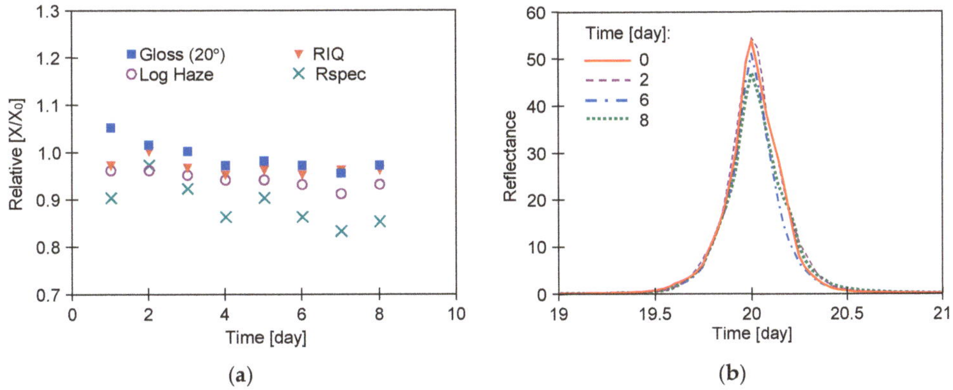

Figure 9. The results of CH#2 sample washing with Permeate 50%. (**a**) changes during 8 days of washing, (**b**) goniophotometric profiles.

3.2. Scratch Resistance Test

It has been documented that the protective coating (clearcoat) applied to improve appearance also provides scratch resistance of automotive coatings [24,32]. Degradation of the clearcoat layer reduces abrasion and scratch resistance [22]. It was assumed that if the reclaimed water tested accelerates paint degradation, it also reduces scratch resistance. For comparison, samples taken from car doors were also washed with a 0.5% Turbo Active Green solution. The tested samples were wetted with isopropanol, and then their surface was rubbed several times with the use of filter paper. Its surface is uneven (Figure 10) and contains various minerals in addition to cellulose [21], which can cause scratches on the clearcoat surface. Rubbing the non-washed sample (Figure 11a) through the filter paper several times resulted in significant scratching of the clearcoat surface (Figure 11b). This confirms the well-known fact that automotive coatings can only be wiped with very soft materials. The damage effects of R samples treated with the use of filter paper for 8 days of washing are shown in Figure 11c (Turbo Active Green) and Figure 11d (Permeate 50%).

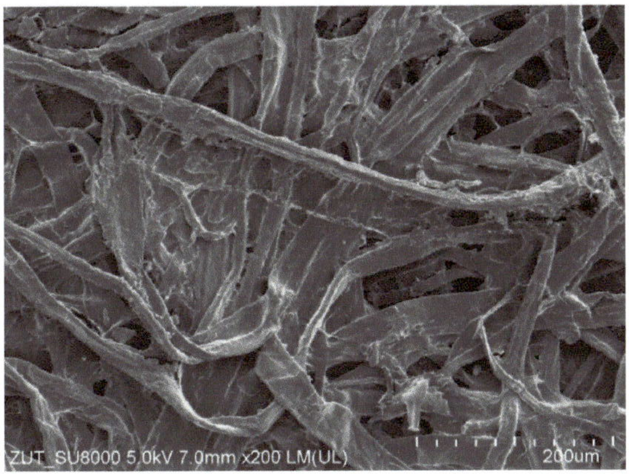

Figure 10. SEM image surface of filter paper applied for scratch resistance tests.

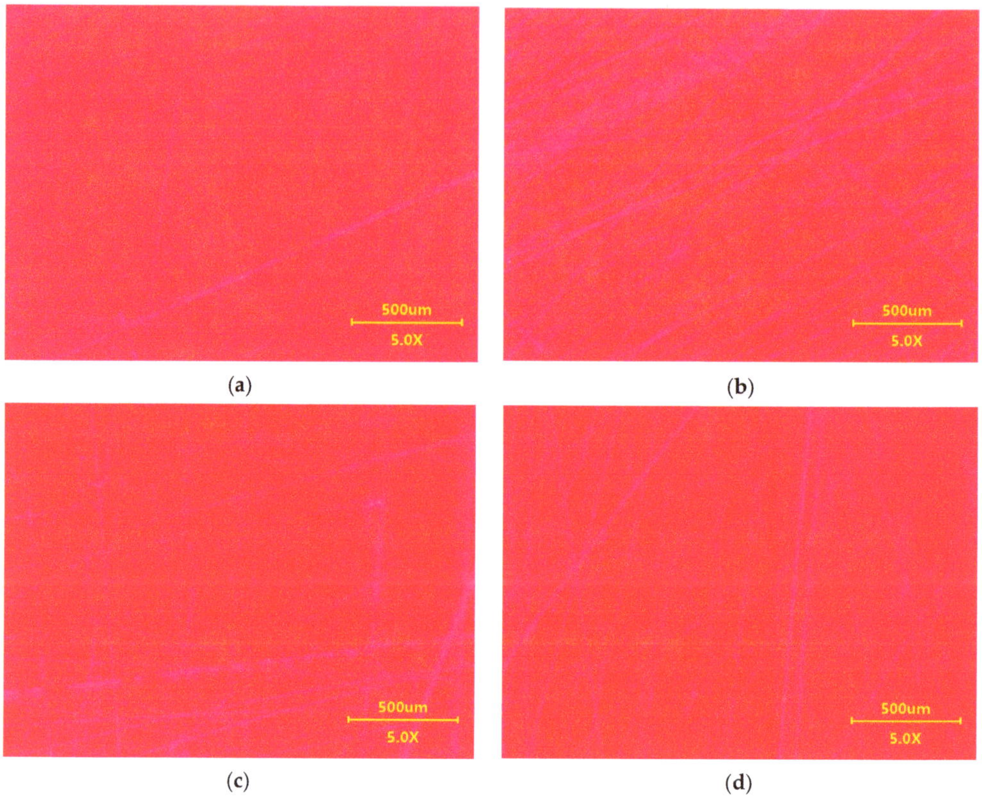

Figure 11. Microscopic images of the surface of R samples used for the scratch resistance test: (**a**) R#6 new, (**b**) R#6-rubbed 10 times through filter paper, (**c**) R#4–8 days washed with Turbo Active Green solution and rubbed daily through filter paper, (**d**) R#5–8 days washed with Permeate 50% solution and rubbed daily through filter paper.

The results of microscopy observations did not show that the use of reclaimed water increased the amount of scratching (Figure 11). Furthermore, the degree of surface damage on the washed samples was similar to the scratching level noted for the non-washed sample (Figure 11b). However, measurements with the Rhopoint IQ apparatus showed that there was a decrease in the quality of the clearcoat surface on the following test day (Figure 12). As a result of the resulting scratches, the gloss value for the sample washed with Turbo Active Green solution decreased by 30 per cent and for the reclaimed water by 15 per cent. The height of the goniophotometric curves decreased with measurement time, which also indicates an increase in clearcoat turbidity (Figure 12b,d). Although both samples were taken from the same car door, the reflectance profiles for samples R#4 washed by Turbo Active Green were lower and wider (Figure 12b) than observed in Figure 12d (Permeate 50%). This was due to the fact that the surface of sample R#4 was of poorer quality, as indicated by the slightly poorer initial parameters (Table 2).

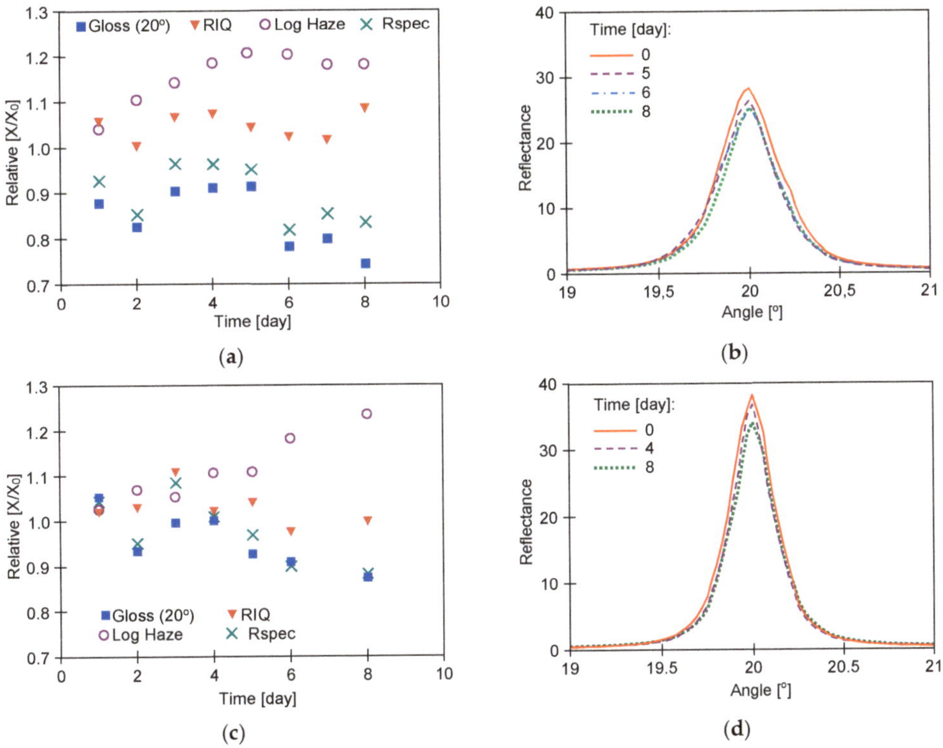

Figure 12. Changes of the clearcoat parameters during the scratch test with 8 days washing by 0.5% solution of Turbo Active Green-sample R#4 (**a**,**b**) and Permeate 50%-sample R#5 (**c**,**d**).

4. Conclusions

The alkaline solution of Active Turbo Green (pH = 8.7), used at car washes, and UF permeates (pH = 8.5–8.6) obtained from wastewater collected at the car wash, were used to test the resistance of automotive coatings to washing water. In the study conducted, car body samples were in contact with the washing water for 8 days. The results showed that the effects of the reclaimed water (UF permeates) were similar to those obtained for the washing solution used at professional car washes.

The gloss of new car coatings is generally in excess of 90 GU. After 4–5 years of car use, due to environmental degradation, it has decreased to a level of 70–80 GU. The actions of the reclaimed washing water tested only slightly altered this value and, in some cases, improved the gloss and haze parameters. Assuming that the dosing time of the solutions at the car wash does not exceed 20 min, a period of 8 days corresponds to washing the car 600 times. In reality, during service life, cars are washed much less frequently; thus, it can be assumed that washing cars at a touchless car wash does not impair the quality of the automotive coatings, even when reclaimed water is used.

Author Contributions: Conceptualization, P.W. and M.G.; methodology, P.W.; validation, M.G.; investigation, P.W.; resources, M.G.; data curation, P.W.; writing—original draft preparation, M.G. and P.W.; writing—review and editing, M.G.; visualization, P.W. and M.G.; supervision, M.G.; project administration, M.G.; funding acquisition, M.G. All authors have read and agreed to the published version of the manuscript.

Funding: The publication was financed from the Polish budget within a framework of the program of the Minister of Education and Science in Poland entitled "Science for the Society", project No. NdS/538617/2021/2022; amount of funding 352 135 PLN; total amount of funding 352 135 PLN.

Institutional Review Board Statement: Not applicable.

Informed Consent Statement: Not applicable.

Data Availability Statement: The original contributions presented in the study are included in the article, further inquiries can be directed to the corresponding author.

Conflicts of Interest: The authors declare no conflicts of interest. The funders had no role in the design of the study; in the collection, analyses, or interpretation of data; in the writing of the manuscript; and in the decision to publish the results.

References

1. Kuan, W.-H.; Hu, C.-Y.; Ke, L.-W.; Wu, J.-M. A Review of On-Site Carwash Wastewater Treatment. *Sustainability* **2022**, *14*, 5764. [CrossRef]
2. Moazzem, S.; Wills, J.; Fan, L.; Roddick, F.; Jegatheesan, V. Performance of ceramic ultrafiltration and reverse osmosis membranes in treating car wash wastewater for reuse. *Environ. Sci. Pollut. Res.* **2018**, *25*, 8654–8668. [CrossRef] [PubMed]
3. Monney, I.; Donkor, E.A.; Buamah, R. Clean vehicles, polluted waters: Empirical estimates of water consumption and pollution loads of the carwash industry. *Heliyon* **2020**, *6*, e03952. [CrossRef] [PubMed]
4. Angelakis, A.N.; Gikas, P. Water reuse: Overview of current practices and trends in the world with emphasis on EU states. *Water Util. J.* **2014**, *8*, 67–78.
5. Elgaali, E.; Akram, M. Recycling and Reuse of Wastewater Generated in Car-Washing Facilities. *Adv. Sci. Technol. Eng. Syst. J.* **2021**, *6*, 521–525. [CrossRef]
6. Mujumdar, M.M.; Rajagolkar, S.P.; Jadhav, P. Treatment of vehicle washing waste water for maximum reuse of treated water and reduce fresh water consumption. *Int. J. Recent Res. Asp.* **2020**, *7*, 1–5. Available online: https://www.academia.edu/43768187 (accessed on 17 September 2024).
7. Hu, C.-Y.; Kuan, W.-H.; Ke, L.-W.; Wu, J.-M. A study of car wash wastewater treatment by cyclo-flow filtration. *Water* **2022**, *14*, 1476. [CrossRef]
8. Zaneti, R.; Etchepare, R.; Rubio, J. Car wash wastewater reclamation. Full-scale application and upcoming features. *Resour. Conserv. Recycl.* **2011**, *55*, 953–959. [CrossRef]
9. Uçar, D. Membrane processes for the reuse of car washing wastewater. *J. Water Reuse Desalin.* **2018**, *8*, 169–175. [CrossRef]
10. Li, T.; Xue-Jun, T.; Fu-Yi, C.; Qi, Z.; Jun, Y. Reuse of carwash wastewater with hollow fiber membrane aided by enhanced coagulation and activated carbon treatments. *Water Sci. Technol.* **2007**, *56*, 111–118. [CrossRef]
11. Pinto, A.C.S.; de Barros Grossi, L.; de Melo, R.A.C.; de Assis, T.M.; Ribeiro, V.M.; Amaral, M.C.S.; de Souza Figueiredo, K.C. Carwash Wastewater Treatment by Micro and Ultrafiltration Membranes: Effects of Geometry, Pore Size, Pressure Difference and Feed Flow Rate in Transport Properties. *J. Water Process Eng.* **2017**, *17*, 143–148. [CrossRef]
12. Veit, M.T.; Novais, Í.G.V.; Juchen, P.T.; Palácio, S.M.; Da Cunha Gonçalves, G.; Zanette, J.C. Automotive Wash Effluent Treatment Using Combined Process of Coagulation/Flocculation/Sedimentation–Adsorption. *Water Air Soil Pollut.* **2020**, *231*, 494. [CrossRef]
13. Fayed, M.; Shewitah, M.A.; Dupont, R.R.; Fayed, M.; Badr, M.M. Treatability Study of Car Wash Wastewater Using Upgraded Physical Technique with Sustainable Flocculant. *Sustainability* **2023**, *15*, 8581. [CrossRef]
14. El-Ashtoukhy, E.-S.Z.; Amin, N.K.; Fouad, Y.O. Treatment of real wastewater produced from Mobil car wash station using electrocoagulation technique. *Environ. Monit. Assess.* **2015**, *187*, 628. [CrossRef] [PubMed]
15. Gönder, Z.B.; Balcıoğlu, G.; Vergili, I.; Kaya, Y. Electrochemical treatment of carwash wastewater using Fe and Al electrode: Techno-economic analysis and sludge characterization. *J. Environ. Manag.* **2017**, *200*, 380–390. [CrossRef]
16. Do, K.-U.; Kim, J.-H.; Chu, X.-Q. Sludge Characteristics and Performance of a Membrane Bioreactor for Treating Oily Wastewater from a Car Wash Service Station. *Desalin. Water Treat.* **2018**, *120*, 166–172. [CrossRef]
17. Boluarte, I.A.R.; Andersen, M.; Pramanik, B.K.; Chang, C.-Y.; Bagshaw, S.; Farago, L.; Jegatheesan, V.; Shu, L. Reuse of car wash wastewater by chemical coagulation and membrane bioreactor treatment processes. *Int. Biodeterior. Biodegrad.* **2016**, *113*, 44–48. [CrossRef]
18. Woźniak, P.; Dubicki, M.; Gryta, M. Microbiological Hazard Analysis of Car Wash Wastewater. *Pol. J. Environ. Stud.* **2023**, *32*, 3871–3882. [CrossRef]
19. Tomczak, W.; Woźniak, P.; Gryta, M.; Grzechulska-Damszel, J.; Daniluk, M. Cleaning of Ultrafiltration Membranes: Long-Term Treatment of Car Wash Wastewater as a Case Study. *Membranes* **2024**, *14*, 159. [CrossRef]
20. Woźniak, P.; Gryta, M. Application of Polymeric Tubular Ultrafiltration Membranes for Separation of Car Wash Wastewater. *Membranes* **2024**, *14*, 210. [CrossRef]
21. Kotnarowska, D. The Influence of Battery Acid on the Destruction of Acrylic Coatings of Car Bodies. *Coatings* **2021**, *11*, 967. [CrossRef]

22. Dahatonde, B.; Kadam, M.; Gupta, S. Review: Degradation of automotive clear coat caused by bird droppings. *Paintindia* **2020**, *70*, 78–90.
23. Zaneti, R.N.; Etchepare, R.; Rubio, J. Car Wash Wastewater Treatment and Water Reuse—A Case Study. *Water Sci. Technol.* **2013**, *67*, 82–88. [CrossRef] [PubMed]
24. Seubert, C.; Nietering, K.; Nichols, M.; Wykoff, R.; Bollin, S. An Overview of the Scratch Resistance of Automotive Coatings: Exterior Clearcoats and Polycarbonate Hardcoats. *Coatings* **2012**, *2*, 221–234. [CrossRef]
25. Sawyer-Beaulieu, S.; Tam, E.; Hussein, A. Measuring Corrosion on Vehicles, in Real-Time, Using Digital Imaging and Analysis Techniques. *Materials* **2022**, *15*, 3053. [CrossRef]
26. Kotnarowska, D.; Wojtyniak, M. Influence of Ageing on Mechanical Properties of Epoxy Coatings. *Sol. St. Phen.* **2009**, *147–149*, 825–830. [CrossRef]
27. Gruber, D.P.; Buder-Stroisznigg, M.; Wallner, G.; Strauss, B.; Jandel, L.; Lang, R.W. A novel methodology for the evaluation of distinctness of image of glossy surfaces. *Prog. Org. Coat.* **2008**, *63*, 377–381. [CrossRef]
28. Perrin, F.X.; Irigoyen, M.; Aragon, E.; Vernet, J.L. Evaluation of accelerated weathering tests for three paint systems: A comparative study of their aging behaviour. *Polym. Degrad. Stab.* **2001**, *72*, 115–124. [CrossRef]
29. The Rhopoint IQ. Available online: https://www.rhopointamericas.com/download/rhopoint-iq-goniophotometer-datasheet-english/ (accessed on 17 September 2024).
30. Redford, J.; Mullany, B. Classification of Visual Smoothness Standards Using Multi-Scale Areal Texture Parameters and Low-Magnification Coherence Scanning Interferometry. *Materials* **2024**, *17*, 1653. [CrossRef]
31. Frankhuizen, N. Measuring Gloss? *Paint Coat. Ind.* **2015**, *31*, 70–73.
32. Yari, H.; Moradian, S.; Tahmasebi, N. The weathering performance of acrylic melamine automotive clearcoats containing hydrophobic nanosilica. *J. Coat. Technol. Res.* **2014**, *11*, 351–360. [CrossRef]
33. Edited by Streitberger, H.-J.; Dössel, K.-F. *Automotive Paints and Coatings*, 2nd ed.; WILEY-VCH Verlag GmbH & Co. KgaA: Weinheim, Germany, 2008.
34. Tomczak, W.; Gryta, M. The Application of Polyethersulfone Ultrafiltration Membranes for Separation of Car Wash Wastewaters: Experiments and Modelling. *Membranes* **2023**, *13*, 321. [CrossRef] [PubMed]

materials

MDPI

Article

Mechanical and Durability Properties of Rubberized Sulfur Concrete Using Waste Tire Crumb Rubber

Okpin Na [1] and Giyeol Lee [2,*]

1 Hyundai E&C, Technology Research Center, Seoul 03058, Republic of Korea; nao@hdec.co.kr
2 Department of Landscape Architecture, College of Agriculture and Life Science, Chonnam National University, Gwangju 61186, Republic of Korea
* Correspondence: gylee@jnu.ac.kr

Abstract: The use of rubber crumbs provides a viable solution for alleviating the disposal problem of waste tires. In this study, rubberized sulfur concrete (RSC) was researched to investigate the optimal mixture proportion and to improve the mixing process in terms of compressive strength and durability performance. For the mixture of the RSC, sand, rubber particles, and micro-filler were adopted as aggregates and sulfur was used for the binding material. Moreover, two mixing processes were applied: the dry mixing process and the wet mixing process. Based on the test results, the increment of rubber particles in the mixture led to a decrease in the compressive strength for both the dry and wet mixing processes. To minimize the voids between the sand and rubber particles, the micro-filler was used at 5% of the total volume. The amount of sulfur varied slightly depending on the mixing process: 30% sulfur for the dry mixing process and 34% sulfur for the wet mixing process, respectively. Consequently, compared to the dry mixing process, the wet mixing process increased the bonding force between sulfur and rubber powder due to the simultaneous heating and combining. In toughness, the wet mixing process demonstrates a 40% higher energy absorption capability compared to the dry mixing process. For the durability performance of the RSC, the mixture with 20% rubber particles produced using the wet mixing process exhibited better corrosion and freeze–thaw resistance.

Keywords: rubberized sulfur concrete; dry and wet mixing process; toughness index; durability

1. Introduction

The use of rubber crumbs provides a viable solution for alleviating the disposal problem of waste tires. Recently, the number of registered domestic automobiles in Korea has approached one per two people, leading to a high interest in recycling waste tires. Waste tires are reused through three reproducing methods: original form, pyrolysis, and the crumb powder form. In the construction field, the crumb powder manufacturing method, particularly using cryogenic grinding, is prevalent, producing a 200 μm fine powder. The composition of these waste tires, excluding a minuscule amount of moisture, consists of volatile matter and fixed carbon. Moreover, the primary raw material in tire manufacturing is styrene–butadiene rubber (SBR). The sulfur material, a source of harmful gas, comprises about 0.97% [1].

The utilization of waste tires has been extensively researched in Korea and abroad since the mid-1990s. For enhancing the performance of concrete pavement in Korea, approximately 7% waste tire powder in the concrete mix was used [2]. Also, to evaluate the deformation resistance of the asphalt concrete pavement, the research of waste tire particles was conducted. The results concluded that the addition of CRM (Crumb Rubber Modifier) and the use of polymer-modified asphalt mixtures have a significant effect on improving the deformation resistance of the materials. Furthermore, although the tensile strength of conventional asphalt concrete was superior, the other properties, such as the ductility, the maximum vertical deformation, and the resilience modulus, were much better [3]. However, despite the overall remarkable material properties, the bonding issue between the crumbed waste tires and the cementitious matrices of the asphalt concrete have come out. To enhance the adhesion between the cement mortar and the crumb rubber powder, styrene–butadiene rubber latex (SBR) and resin powder were used, and their mechanical and durability properties were investigated [4–6]. Recently, to reduce the spalling of high-strength concrete under a fire condition, waste tire particles with metakaolin were applied. In the research results, metakaolin improved the fire resistance of the concrete itself and the waste tires had significant effects on the explosion reduction due to the absorption of the internal expansion [7,8]. Furthermore, rubberized concrete has been studied in the utilization of surface-treated rubber particles with silane or organosulfur to improve the mechanical properties [9–18].

On the other hand, some researchers have studied sulfur concrete. For sulfur to be used as a binder, it needs to be melted down at temperatures ranging from 130 to 146 °C. The manufacturing process of sulfur concrete is similar to that of asphalt concrete. The basic mixture materials of sulfur concrete consist of sulfur, aggregates, and mineral admixture. Sulfur concrete has excellent properties such as high strength, friction resistance, and chemical resistance, along with thermoplastic properties. In addition, it can be pulverized for recycling as well as molded without the degradation of its mechanical properties [19–24]. Currently, sulfur concrete is used in chemical plants and port facilities that require durability and chemical resistance. Another application of sulfur concrete could be in extreme construction environments such as the polar region and the lunar surface, due to its non-exothermic reaction [25,26].

Recently, to improve the fire resistance and ductility of sulfur concrete, glass fiber was adopted [27]. However, even though many researchers have studied sulfur concrete, the effect of incorporating waste rubber particles has not been sufficiently investigated.

The purpose of this study is to investigate the optimal mixture proportion and to improve the mixing process in terms of compressive strength and durability performance. To optimize the mixture proportion of the RSC, sand, rubber particles, and micro-filler as aggregates were adopted, and sulfur was used for the binding material. For the optimal mix design, initially, the effects of the sulfur amount, the rubber content, and the micro-filler were investigated in accordance with the dry mixing processes. Next, the optimal mix design was adjusted with the wet mix process, and then, the toughness and durability performances were evaluated.

2. Experimental Detail

2.1. Materials

Concrete can consist of three physical phases including aggregate, paste, and porosity. Sulfur, sand, rubber particles and micro-fillers such as fly ash were used as constituents for the sulfur concrete. Therefore, we refer to the concrete as rubberized sulfur concrete. It is important to note that sulfur served as the binder in the mixing process, and no hydration reaction occurred because no cement or water was used in the mixture. The details of the dry process and the wet process will be introduced later. The specific gravity of river sand as a fine aggregate was 2.6 g/cm^3 with a fineness modulus of 2.63 and water absorption of 2.5%, respectively. The rubber particle was produced by JaiTire Industries Inc. (Denver, CO, USA). The particle size of the rubber is illustrated in Figure 1. The bulk density and specific density of the rubber particles were 0.5 g/cm^3 and 1.0 g/cm^3, respectively. The sulfur was from Test Mark Industries, Pennsylvania. Figure 1 shows the size distributions of the sand and rubber particles, and Table 1 indicates the physical and mechanical properties of the sulfur, which was used for the binder in this study.

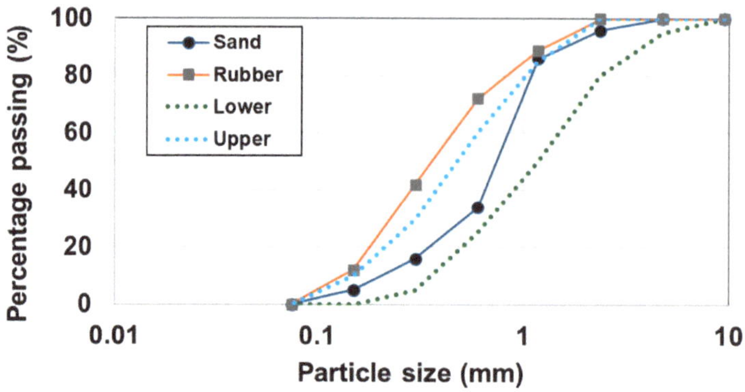

Figure 1. Sieve analysis of sand and rubber particles.

Table 1. Physical and mechanical properties of sulfur binder [28–34].

Properties	Curing	Temp. (°C)	Value	Test Method
Compressive Strength (MPa)	2 h	25	34.5	ASTM C617
	48 h	25	48.3	ASTM C579
Tensile Strength (MPa)	48 h	25	4.8	ASTM C307
Flexural Strength (MPa)	48 h	25	712.4	ASTM C580
Density (g/cm^3)	-	-	1.9	ASTM C905
Coefficient of Thermal Expansion (cm/cm/°C)	-	-	3.8×10^{-5}	ASTM C531
Water Absorption	-	-	0.2%	ASTM C413
Color	Dark Gray			

2.2. Mix Design

To determine the optimized mixing proportion for rubberized sulfur concrete (RSC), the mixture proportion was indicated according to the volume ratio of the materials as shown in Table 2. Firstly, the initial volume proportion of the sulfur, 16%, was determined based on the apparent workability of N1 and N2. Then, to obtain further workability, the sulfur content was increased to 26%, as shown in N3 to N8, and the sand was replaced by rubber particles up to 50% of the volume ratio of the sand. The micro-filler in the asphalt mix plays an important role in the stiffening and toughening of a binder [35,36].

Table 2. Preliminary mixture proportion of RSC (ratio by volume, %).

No.	Sulfur (%)	Total Aggregate (%)			
			Micro-Filler	Rubber	Sand
N1	16	84	0	0	84
N2	16	84	0	8	76
N3	26	74	0	0	74
N4	26	74	0	7	66
N5	26	74	0	15	59
N6	26	74	0	22	52
N7	26	74	0	29	44
N8	26	74	0	37	37

Table 2. *Cont.*

No.	Sulfur (%)	Total Aggregate (%)			
			Micro-Filler	Rubber	Sand
N9	30	70	7	0	63
N10	30	70	7	7	56
N11	30	70	7	14	49
N12	30	70	7	21	42
N13	30	70	7	28	35
N14	30	70	7	35	28
N15	33	67	7	20	40
N16	37	63	6	19	38

The micro-filler was also used to improve the mechanical properties in the mix designs N9 to N16. The initial volume of the micro-filler was set at 10% of all the compounds, except the sulfur. In N9 to N14, the amount of rubber powder in the place of sand increased up to 50% by volume. In addition, in N15 and N16, an amount of sulfur was added to investigate the effect of the binder and the micro-filler. In Figure 2, the process for selecting the optimal mix design conditions in this study are illustrated in three stages, with detailed variables considered at each step.

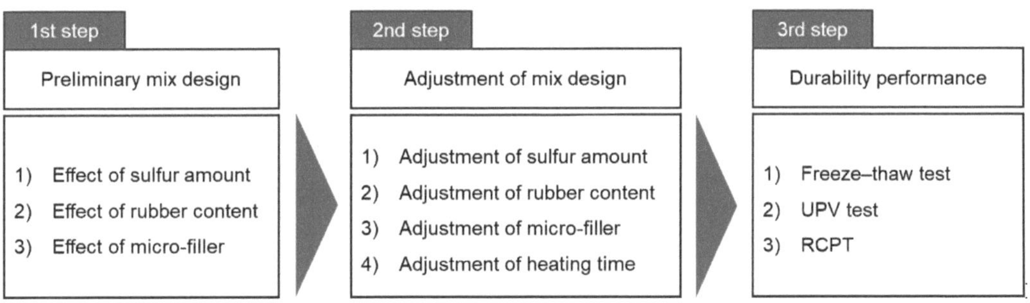

Figure 2. Decision-making process for optimal mix design.

2.3. Mixing Process

The mixing process of the RSC can be divided into a dry mixing process and a wet mixing process as shown in Figure 3. For the dry process, the sulfur was placed into a pot and heated first. The suppliers recommended a melt temperature range for the sulfur from 135 °C to 146 °C. The heating temperature of the mixing process was suitable at 146 °C and the heating time for the dry process was about 30 min until the sulfur had completely melted to a slurry. The other materials (the rubber particles, the sand, and the Portland cement) were mixed uniformly with a hand paddle. They were then preheated to about 95 °C in an oven for 30 min and then were added to the hot pot. For the wet process, the sulfur and rubber particles were put together into the pot and heated. The heating time varied to find the specific time that produced the highest strength in this study. The other materials (other than the sulfur and the rubber particles) were preheated to about 102 °C in an oven for 30 min before being added to the pot.

One can see that the main difference between the wet and the dry process is that only the sulfur is pre-heated in the dry process, while the sulfur and the rubber particles are pre-heated together in the wet process. Two similar processes are commonly used in the asphalt industry, where the asphalt can be heated first (dry process) or heated together with aggregates (wet process). It was observed that the two processes have significantly different effects on various properties of the asphalt concrete. The two processes were used in the present study to examine if they have any significant effect on the properties of the RSC.

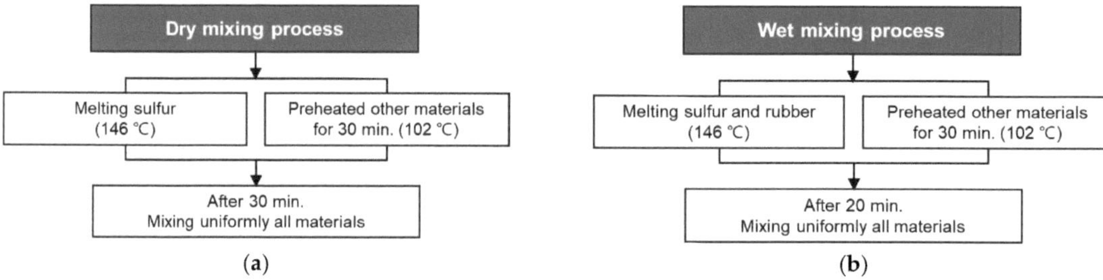

Figure 3. Two types of mixing processes. (**a**) Dry mixing process. (**b**) Wet mixing process.

After adding all the materials to the pot, the actual temperature of the mixture was lower than the original melting temperature. The real temperature of the heated mixture in the pot was hard to measure because the mixture was in a semi-fluid and semi-solid state. The mixture was continually heated in the pot until the mixture became visually uniform and semi-fluid.

2.4. Specimen Preparation and Test Methods

Making sulfur concrete is not a common practice in the concrete industry. Therefore, the specimen preparation procedures used in this study are described in detail. Referring to ASTM C 1312-97 (Making and Conditioning Chemical-Resistant Sulfur Polymer Cement Concrete Test Specimens in the Laboratory), all the materials were weighed accurately [37]. Then, two different procedures, a dry process and a wet process, were used to prepare the rubberized sulfur concrete. The molds used for making the RSC samples were 75 mm diameter and 150 mm height cylinders. If the size of a specimen is bigger than 75 mm by 150 mm, the preparation of the specimen is not easy because the sulfur is hardened very fast. One batch contained six specimens for each design mix. The inside surfaces of the molds were coated slightly with mineral oil. For measuring the axial deformation, the gauge length of the axial extensometer was 100 mm. Before testing for the compressive strength, the RSC specimens were capped like regular Portland cement concrete cylinders. Compressive strength tests were performed using a universal testing machines after seven days.

3. Results and Discussion

3.1. Effect of Sulfur Amount with Rubber Particles and Micro-Filler

There have not been sufficient research results for the RSC used for sulfur as a binder and rubber particles obtained from used tires as an aggregate. Therefore, the proper range of sulfur for the RSC has not been revealed in the literature. Thus, in this study, the initial sulfur amount of 16% was adopted for the mixture proportion of the RSC and the sulfur amount was increased to 30%, according to the visual inspection of the workability and mixing appearance.

Table 3 and Figure 4 show the effect of the sulfur amount on the compressive strength of the RSC, in terms of rubber particles and micro-filler. N1, N3, and N9 contained no rubber and had three different sulfur contents, 16%, 26%, and 30%, respectively (see Table 3). N2, N4, and N10 contained the same amount of rubber particles, a 10% volumetric ratio of rubber to the aggregate (aggregate includes rubber, micro-filler, and sand), as well as had three different sulfur contents.

Table 3. Mixture proportion of the RSC regarding the effect of sulfur amount.

No.	Sulfur (%)	Total Aggregate (%)			Ratio of Total Aggregate (%)		
		Micro-Filler	Rubber	Sand	Micro-Filler	Rubber	Sand
N1	16	0	0	84	0	0	100
N3	26	0	0	74	0	0	100
N9	30	7	0	63	10	0	90
N2	16	0	8	76	0	10	90
N4	26	0	7	67	0	10	90
N10	30	7	7	56	10	10	80

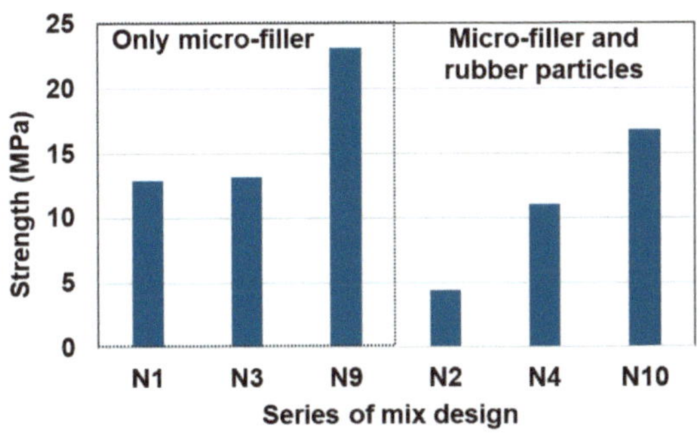

Figure 4. Effect of sulfur amount on rubber particles and micro-filler on compressive strength.

From N1 and N2, one can see that with the same amount of sulfur binder, when a part of the natural aggregates was replaced by rubber particles, the strength reduced by about a third. In fact, when the specimens of N1 and N2 were mixed, it was observed that the rubber aggregates were only partially covered by the sulfur cement and many voids existed in the matrix, which caused difficulty in mixing and placement. Therefore, sulfur content is a very important mix design parameter for the RSC. Without a sufficient sulfur amount, the strength of the RSC will be reduced by the soft rubber particles (as shown in Figure 4) as well as by the low-quality sulfur paste matrix. N9 illustrates the better mixture with a sufficient sulfur amount and micro-filler.

In order to study the effect of sulfur cement content on the compressive strength of the RSC, the volumetric sulfur amount of N3 and N4 was increased to 63%, compared to the sulfur amount of N1 and N2. The volumetric ratio of rubber to sand for N1 and N3 was the same (0:100) and so was the ratio of N2 and N4 (10:90), as shown in Table 3. From Figure 4, one can see that with the same amount of rubber replacement (about 7%), the strength reduction from N3 to N4 with a 26% sulfur amount is not as high as that from N1 to N2 with 16% sulfur. This means that there must be a sufficient amount of sulfur as the binder, and 16% is not sufficient. Based on the present test data, the sulfur content of N3 and N4 may be considered sufficient, which is 26%. With sufficient binder, the addition of rubber particles still reduces the strength of the RSC, which is the strength reduction observed in N3 and N4. As shown in Table 3 and Figure 4, in case the rubber particles and micro-filler are mixed additionally as aggregates, the sulfur binder shall be increased for workability.

Based on the results of Figure 4, the effect of the rubber particles is investigated in Table 4. The test series of N3 to N8 is demonstrated for the increment of rubber particles and that of N9 to N14 is added with a micro-filler of 7%. In Figure 5, it is very clear that, with the increase in rubber content, the compressive strength of the RSC decreases considerably. This result is expected because the strength of the rubber particles is much lower than that of the sulfur concrete (without rubber). One important result shown in Figure 5 is that with the addition of micro-filler, the compressive strength of the RSC is increased dramatically.

In particular, for the high rubber contents from 30 to 50%, the compressive strengths of the RSC with micro-filler are doubled compared to those without the micro-filler. This means that the micro-filler reduces the porosity of the RSC and improves the strength of the material significantly.

Table 4. Mixture proportion of RSC regarding the effect of rubber particles.

No.	Sulfur (%)	Total Aggregate (%)			Ratio of Total Aggregate (%)		
		Micro-Filler	Rubber	Sand	Micro-Filler	Rubber	Sand
N3	26	0	0	74	0	0	100
N4	26	0	7	66	0	10	90
N5	26	0	15	59	0	20	80
N6	26	0	22	52	0	30	70
N7	26	0	29	44	0	40	60
N8	26	0	37	37	0	50	50
N9	30	7	0	63	10	0	90
N10	30	7	7	56	10	10	80
N11	30	7	14	49	10	20	70
N12	30	7	21	42	10	30	60
N13	30	7	28	35	10	40	50
N14	30	7	35	28	10	50	40

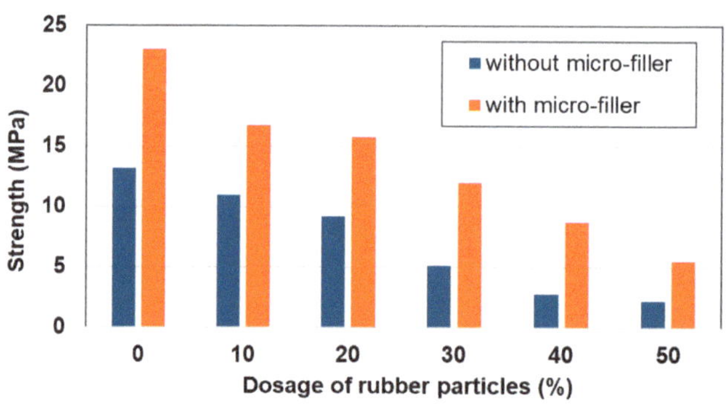

Figure 5. Effect of rubber particles with and without micro-filler on compressive strength.

The two test series in Figure 5 are almost parallel with almost the same negative slope, which means that the micro-filler can improve only the properties of the sulfur matrix and cannot compensate the strength loss due to the addition of the rubber particles. It also means that the improvement of the matrix by the micro-filler is not affected by the rubber particles. In other words, the strengthening of the matrix by micro-filler and the reduction in strength by the added rubber particles are two independent mechanisms.

Based on the results of N9 in Figures 4 and 5, additional tests were prepared to investigate the effect of sulfur content with micro-filler and rubber particles. Figure 6 indicates the strength comparison among specimens N12, N15 and N16 in Table 5. For these test series, the volumetric ratio of rubber to aggregate (aggregate includes cement, rubber, and sand) was kept as a constant 30%, while the sulfur contents were different, from 30% to 37%. The sulfur contents of N12, N15, and Nl6 were 1:1.1:1.23, respectively. The strength ratios of the three specimens were 1:0.84:0.78. It is important to see that the advantages of increasing the sulfur content are the compressive strength and workability of the RSC only in a certain range. If the sulfur content is beyond a critical value, the strength of the RSC will decrease. This indicates that the sulfur amount should be controlled within a proper range because excessive sulfur in the mixture has an adverse effect on its strength. Based on the results obtained in this study, the starting value for the sulfur proportion is about 30% of the total volume. A sulfur content that is less than the critical value may result in low workability, difficulty in mixing and placement, and low strength; a sulfur content that is higher than the critical value may lead to a reduction in the strength of the RSC.

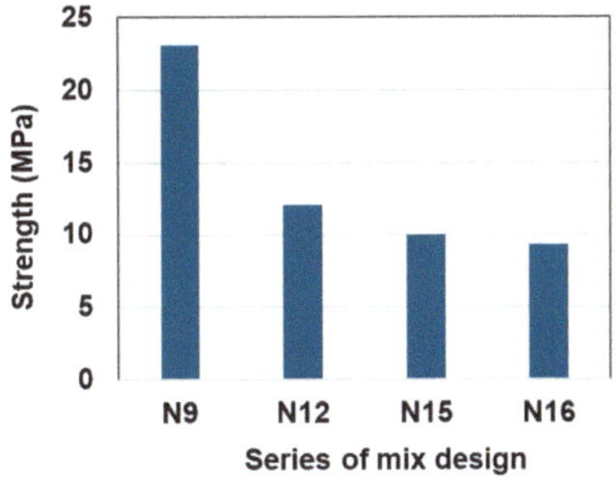

Figure 6. Effect of sulfur amount with and without rubber particles on compressive strength.

Table 5. Mixture proportion of the RSC regarding the effect of the increased sulfur amount and the rubber particles.

No.	Sulfur (%)	Total Aggregate (%)			Ratio of Total Aggregate (%)		
		Micro-Filler	Rubber	Sand	Micro-Filler	Rubber	Sand
N9	30	7	0	63	10	0	90
N12	30	7	21	42	10	30	60
N15	33	7	20	40	10	30	60
N16	37	6	19	38	10	30	60

3.2. Effect of Micro-Filler

Table 6 was designated for the effect of micro-filler on the compressive strength of RSC. The content of the micro-filler was fixed at 7% of the total volume. In order to find the optimal micro-filler content, several micro-filler contents of 3% and 5% were applied to additional RSC mixture series and tested for their compressive strength. The mixture proportions of N12, N12_MF1, and N12_MF2 with different micro-filler contents are listed in Table 6 and the test results are shown in Figure 7. It can be seen that N12_MF2 is 13% higher than N12 in compression strength. The micro-filler content of N12_MF2 is about 5% of the total volume. Therefore, 5% micro-filler by volume can be considered the best content in this study.

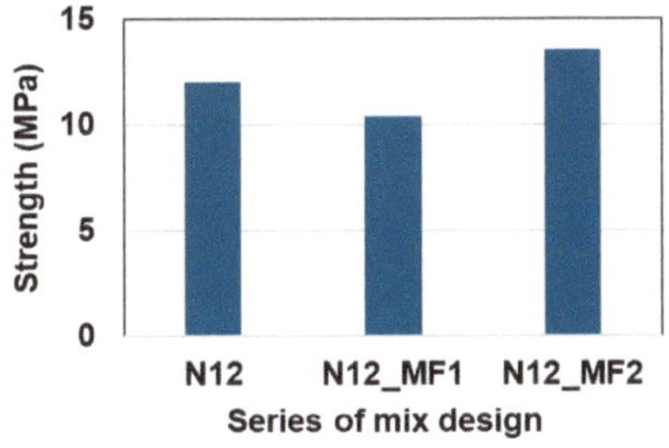

Figure 7. Effect of micro-filler on compressive strength.

Table 6. Mixture proportion of RSC regarding the effect of micro-filler.

No.	Sulfur (%)	Total Aggregate (%)			Ratio of Total Aggregate (%)		
		Micro-Filler	Rubber	Sand	Micro-Filler	Rubber	Sand
N12	30	7	20	40	10	30	60
N12_MF1	30	3	20	46	4	30	66
N12_MF2	30	5	20	44	7	30	63
N12_MF3	30	5	0	65	7	0	93
N12_MF4	30	5	7	58	7	10	83
N12_MF5	30	5	15	51	7	20	73
N12_MF6	30	5	28	37	7	40	53
N12_MF7	30	5	35	30	7	50	43

While fixing the micro-filler content to 5%, the effects of the rubber content on RSC strength with increments of rubber content were investigated. The mixture series from N12_MF2 to N12_MF7 had the same sulfur and micro-filler contents and different rubber contents ranging from 0% to 50% of the rubber to aggregate ratio. Figure 8 shows the effect of the rubber dosage on the compressive strength of the RSC. It can be seen that the strength decreases when the rubber content increases, similar to the trend shown in Figure 5. The difference between Figures 5 and 8 is the micro-filler content. In Figure 8, among the mixture proportions of N12_MF2 to N12_MF7, the lowest reduced strength ratio is N12_MF5, with 20% rubber particle and 5% micro-filler.

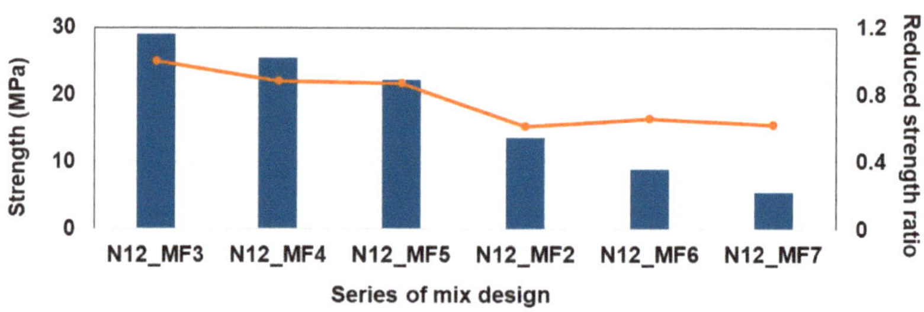

Figure 8. Effect of sulfur amount on rubber particles and micro-filler on compressive strength and reduced strength ratio.

3.3. Effect of Wet Mixing Process

As previously mentioned regarding RSC, the main factor influencing compressive strength is the micro-filler, which fills the voids and gaps between the aggregates such as the sand and rubber particles. Another significant factor affecting mechanical properties is the mixing process, which can be either dry or wet. The differences between these two mixing processes were described in the previous chapter. To explore the differences between the dry and wet mixing processes, the baseline mixtures of N12 and N12_MF2 were used for comparison. For the wet mixing process, a series of mixtures from W1 to W5 was proposed, with the total aggregate content fixed at 70% and the sulfur content increased, as shown in Table 7.

First of all, Figure 9 illustrates the test results comparing the dry mixing process of N12 and N12_MF2 with the wet mixing process of W2 and W3_MF1. Based on these results, the wet mixing process demonstrates better performance than the dry mixing process when using the same mixture proportions. That is, the compressive strength of N12 is 64% higher than that of W2 because the melted sulfur and rubber particles were mixed together more effectively, resulting in better bonding with fewer voids and gaps between the sulfur matrix and the rubber particles.

Table 7. Mixture proportion of RSC regarding the effect of the wet mixing process.

No.	Sulfur (%)	Total Aggregate (%)			Ratio of Total Aggregate (%)		
		Micro-Filler	Rubber	Sand	Micro-Filler	Rubber	Sand
N12	30	7	20	40	10	30	60
N12_MF2	30	5	20	42	7	30	63
W1	32	7	21	41	10	30	60
W2	33	7	20	41	10	30	60
W3	34	7	20	40	10	30	60
W3_MF1	34	5	20	42	7	30	63
W4	36	7	19	39	10	30	60
W5	37	7	19	38	10	30	60

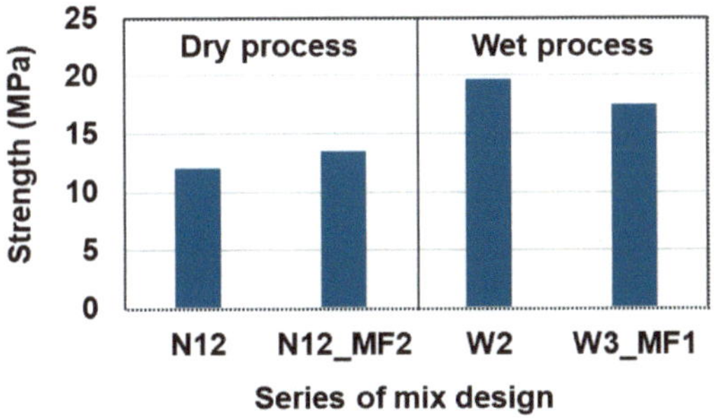

Figure 9. Difference in compressive strength depending on the dry and wet mixing process.

In Figure 10, it can be seen that as the sulfur content increased from 32% to 37%, the compressive strength initially increased and then decreased. The maximum compressive strength, just above 20 MPa, can be observed in W3 in Figure 10. This mixture has a sulfur content of approximately 34% by total volume, which can be considered the optimal sulfur content and will be used in the following sections.

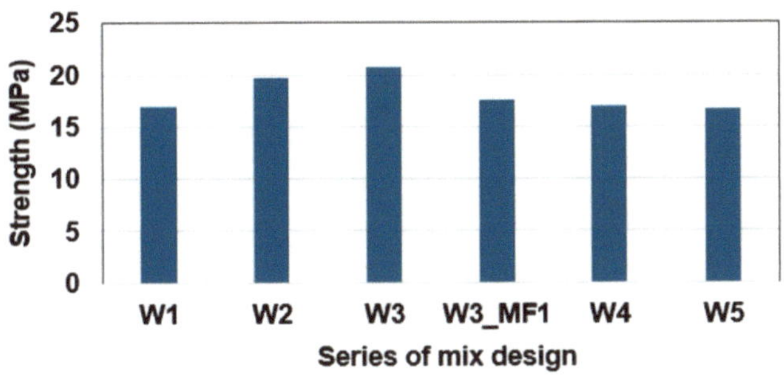

Figure 10. Effect of compressive strength regarding the dosage of rubber particles.

3.4. Modified Mixing Process

In the wet mixing process, heating time is a crucial factor, since sulfur is melted together with rubber. The optimal temperature for melting sulfur powder ranges from 135 °C to 146 °C; in this study, 146 °C was used. The heating durations were set to 30, 45, and 60 min. The compressive strength was measured for each heating duration at 146 °C with the mix conditions of W3. Figure 11 illustrates how strength varies with heating duration, showing that the highest strength was achieved with a heating duration of 45 min. Prolonged heating beyond this duration led to sulfur burning, which resulted in a decrease in strength. Consequently, a modified wet mixing process with the optimal heating duration was recommended, and the detailed procedure is illustrated in Figure 12.

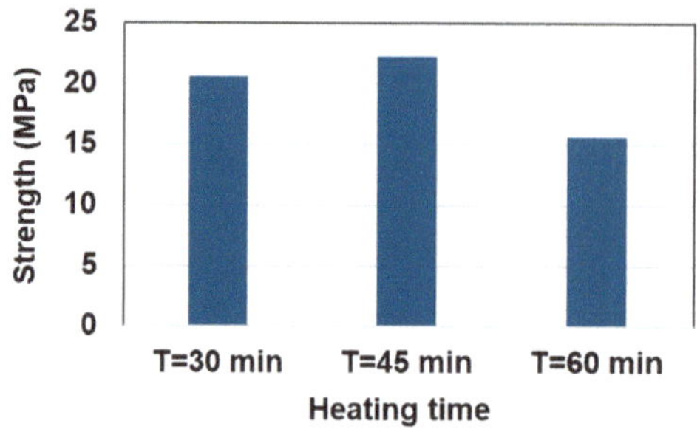

Figure 11. Effect of compressive strength depending on the heating time.

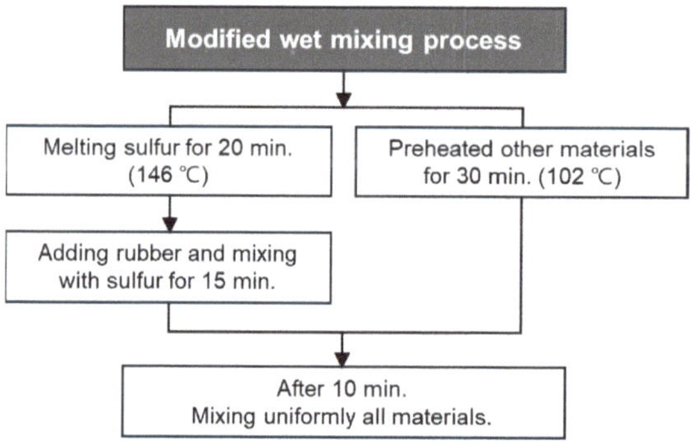

Figure 12. Proposed modified wet mixing process.

Table 8 presents the results of compressive strength measurements with varying amounts of rubber particles from 0% to 50% under the optimal wet mixing conditions. The data indicate that the strength decreased almost linearly with the increasing rubber powder content. The relationship between the strength and the rubber powder content can be predicted using the formula shown in Figure 13.

Table 8. Mixture proportion of RSC regarding the effect of rubber particles via the wet mixing process.

No.	Sulfur (%)	Total Aggregate (%)			Ratio of Total Aggregate (%)		
		Micro-Filler	Rubber	Sand	Micro-Filler	Rubber	Sand
W3_R0	34	7	0	60	10	0	90
W3_R1	34	7	7	53	10	10	80
W3_R2	34	7	13	47	10	20	70
W3_R3	34	7	20	40	10	30	60
W3_R4	34	7	27	33	10	40	50
W3_R5	34	7	33	27	10	50	40

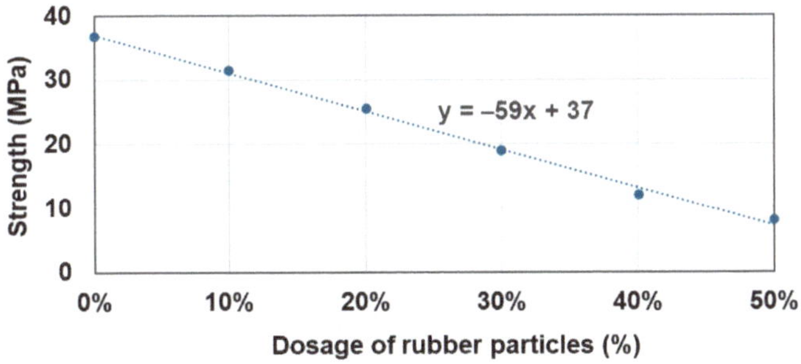

Figure 13. Regression analysis of mix design for RSC regarding the dosage of rubber particles.

3.5. Toughness of RSC According to Two Mixing Process

The toughness of a material can be determined from the area under the stress–strain curve. This allows for an understanding of the material's brittleness and ductility, as well as its resistance to fatigue fracture. Figure 14 illustrates the setting up for a compressive strength test, using MTS axial extensometers to obtain the stress–strain curve. Each extensometer is 114 mm long and it is attached to both sides of the specimen. Figures 15 and 16 show the stress–strain curves for RSC specimens with varying rubber powder content (10% to 50%), tested under the optimal mix condition, using both dry and modified wet mixing processes. Under the same mixture proportion, the RSC specimens produced by the wet mixing process generally exhibited higher stress and longer strain.

Figure 14. Set-up of RSC specimen with strain gauges.

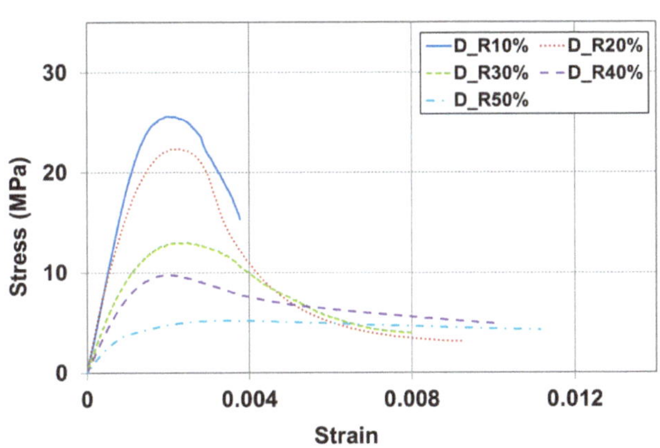

Figure 15. Stress–strain curve on dry mixing process.

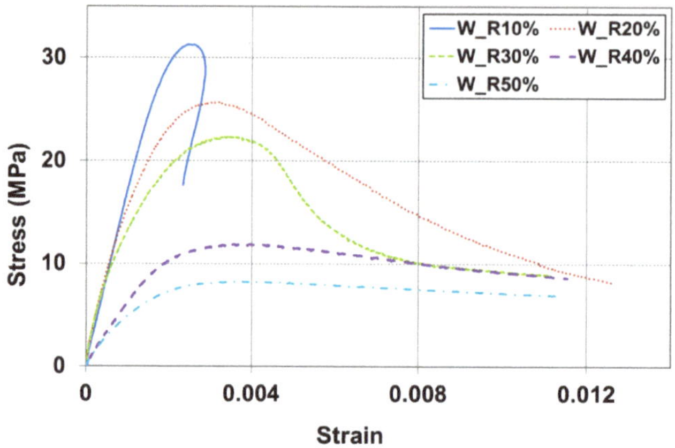

Figure 16. Stress–strain curve on wet mixing process.

Figure 17 illustrates the method for calculating the toughness index. The toughness index can be expressed as the ratio of the area OAB up to the initial crack to the area OACD up to the comparative deformation [38]. Figure 18 shows only the comparative result where 20% rubber particle was mixed, for the purpose of determining the toughness index of the RSC according to their manufacturing process, as seen in Figures 15 and 16. In Figure 18, the toughness index is represented as the area ratios for strains of 0.8% and 1.2%, which are four times the yield strain. For the calculation of the area underneath the strain-stress curve, Image J (ver. 1.46), a public domain Java-based image processing and analysis program, was used [39]. As shown in Table 9, it can be observed that the method using the wet mixing process exhibited a higher energy absorption capability of 40% and better resistance to fatigue impact compared to the dry mixing process.

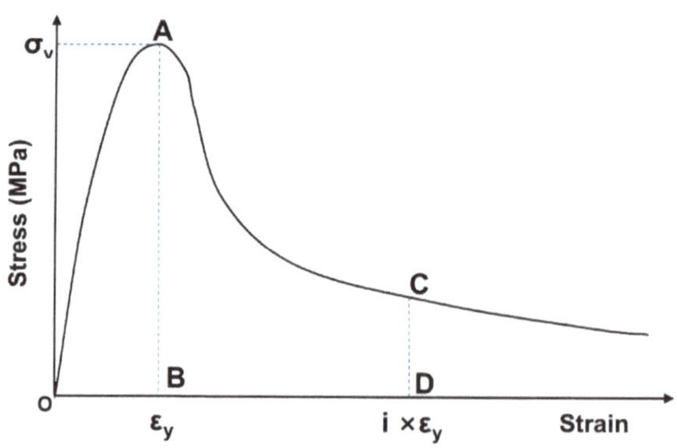

Figure 17. Toughness index from stress (σ)–strain (ε) curve.

Table 9. Toughness indices of different mixing processes.

No.	B		D		OAB	OACD	Toughness Index
	εy	σy	4εy	σ			
D_R20%	0.0022463	22.12	0.008978	3.16	77,021	205,830	2.67
W_R20%	0.0036945	25.52	0.012141	8.63	125,880	470,602	3.74

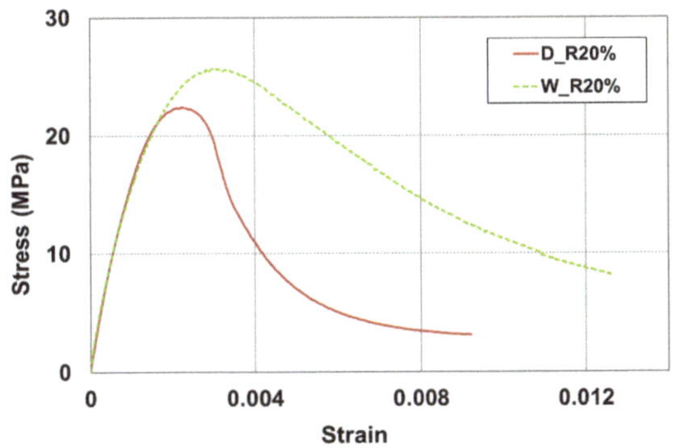

Figure 18. Comparison of toughness between D_R20% and W_R20%.

3.6. Durability Performance of RSC

3.6.1. Freezing–Thawing Resistance Test

The purpose of this test is to evaluate the resistance of freezing and thawing in water, as outlined by ASTM C666 [40]. The freezing-and-thawing chamber is equipped with refrigeration and heating systems, along with controls, to continuously and automatically generate reproducible cycles that meet the specified temperature requirements. In the chamber, cylindrical RSC specimens with 75 mm diameter by 100 mm height were used and they were cast with a modified wet mixing process. For monitoring the temperatures inside the chamber and in the center of the specimen, temperature sensors were installed as shown in Figure 19a,b. The temperature range in the center of the specimen was from 0 °C to −13 °C and the ambient temperature range was about 20 °C to −12 °C as shown in Figure 19d.

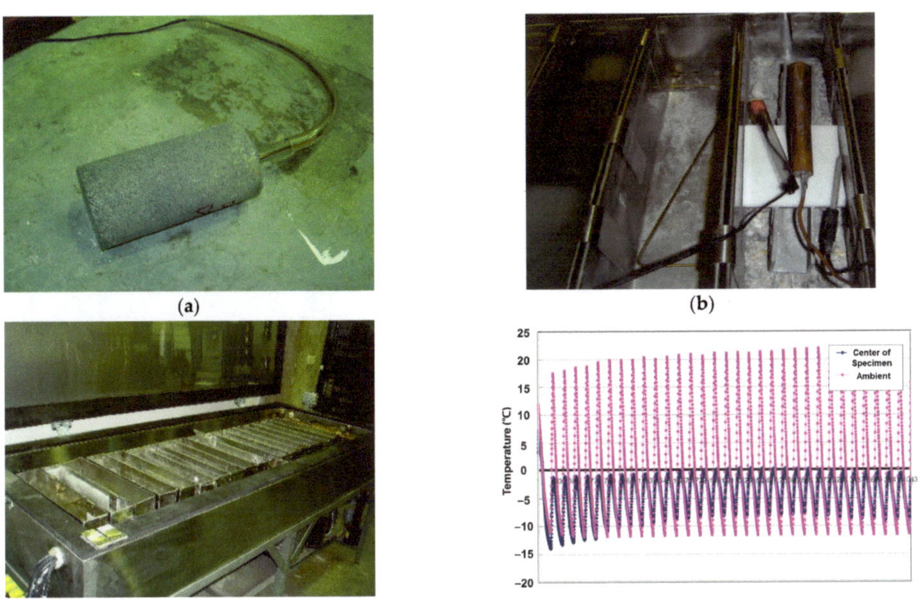

(c) (d)

Figure 19. Temperature sensors and test equipment. (**a**) Set-up the sensor in the center of the specimen (**b**) Sensor for measuring the ambient temperature inside the chamber (**c**) Freezing–thawing chamber (**d**) Freezing-thawing temperature cycles.

This freezing–thawing test (F-T test) was conducted for 300 cycles and one complete cycle took approximately 3 h. After 30 cycles, the transmission time was measured with an ultrasonic pulse velocity (UPV) tester as shown in Figure 20a. According to the UPV results shown in Figure 20b, the transmission time through the RSC specimen increased slightly. Specifically, for the case with 10% rubber particle replacement of the total aggregate, the variation in transmission time was significantly higher compared to the other cases. The accumulation of the freezing–thawing cycles resulted in damage to the specimen, particularly with lower rubber dosages. In other words, rubber particles are a primary factor contributing to increased micro-damage in RSC.

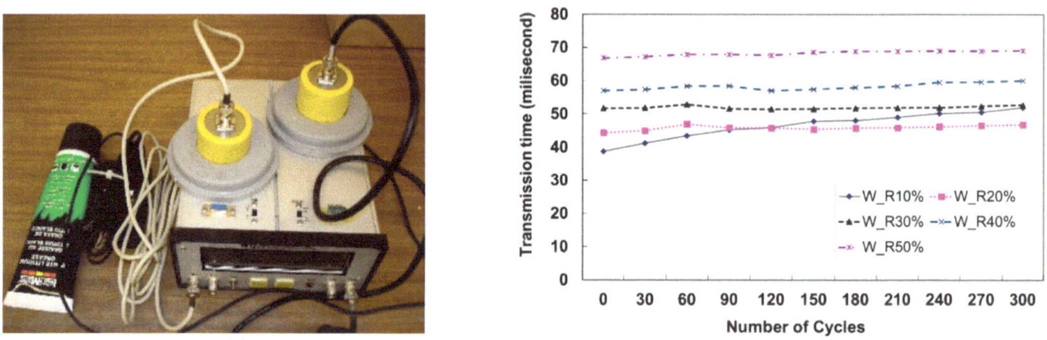

(**a**) (**b**)

Figure 20. UPV test after F-T cycles. (**a**) Ultra-sonic pulse velocity tester (**b**) traveling time with UPV.

To investigate the effect of the mechanical properties under freezing–thawing conditions, a compression test was conducted before and after 300 freezing–thawing cycles. As shown in Figure 21a, the maximum reduction in compressive strength was 40% for the W_R10% mixture. Conversely, the strength reduction for the W_R20% mixture was lower compared to the other mixtures, indicating that rubber particles were effective in mitigating the contraction and expansion of the RSC under freezing and thawing conditions. Furthermore, Figure 21b shows that the elastic modulus of the RSC decreased significantly, by up to about 80%. This suggests that rubber particles may enhance the durability performance of RSC.

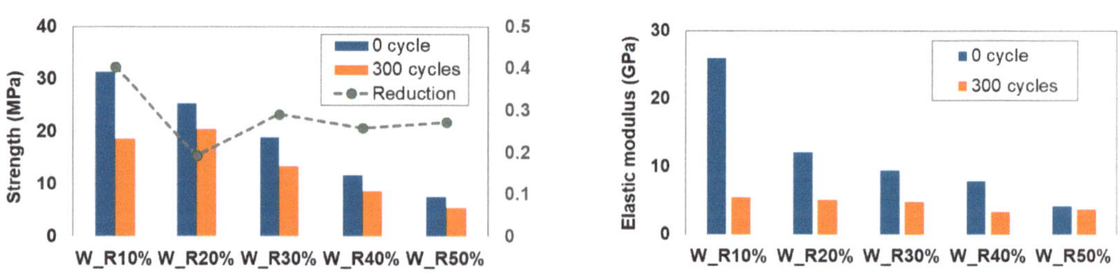

(a) (b)

Figure 21. Reduction in mechanical properties. (**a**) Compression strength after F-T cycles. (**b**) Elasticity after F-T cycles.

3.6.2. Rapid Chloride Penetration Test

A rapid chloride penetration test was conducted to evaluate the resistance of rubberized sulfur concrete to chloride ion penetration in moist and chemically aggressive environments, in accordance with ASTM C1202 [41]. The RCP test was conducted using cylindrical specimens with a diameter of 100 mm and a height of 50 mm, which were cured for 7 days after casting. The experimental apparatus known as the PROOVE-It tester was used, as illustrated in Figure 22e.

(a) (b) (c) (d) (e)

Figure 22. Procedure of RCPT. (**a**) Preparation of specimens. (**b**) Set-up with pump. (**c**) Vacuuming. (**d**) Assembling test cell. (**e**) Proove-It tester.

Before conducting the RCP test, the cylindrical specimens were prepared by sealing the cylinder walls with silicone and allowing it to dry for 1 h. This coating was intended to prevent fluid leakage from the sidewalls and to ensure a linear flow of electrical current through the specimen. The specimens were then subjected to a vacuum for three hours and saturated for 18 h according to ASTM standards. Following this, each specimen was immediately loaded into the Proove-It device. Once loaded, solutions were added to the two test cell containers: one containing a 3% NaCl solution and the other containing 0.3 N NaOH. The testing device then applied a 60-volt potential across the faces of the specimen for six hours. All the procedures of the RCP test are demonstrated in Figure 22a–e. During this period, the apparatus measured the impedance in coulombs throughout the entire duration of the test.

This test method measures the electrical conductivity of the specimen, which indicates the permeability of sulfur-binding materials. Permeability is determined by the total porosity of the RSC and the connectivity of its pores. For optimal resistance to moisture and aggressive chemicals, the permeability should be minimized. In Figure 23a, the permeability of D_R20% and W_R20% was compared. As noted in the previous chapter, W_R20% was produced using a wet mixing process with 20% replacement of the total aggregate. Based on the test results, the ion penetrabilities of D_R20% and W_R20% were 2200 and 1650 coulombs, respectively. According to Figure 23b, the level of D_R20% was moderate and that of W_R20% was low. This is because the internal organization of W_R20% is likely to be more sophisticated due to the different mixing process.

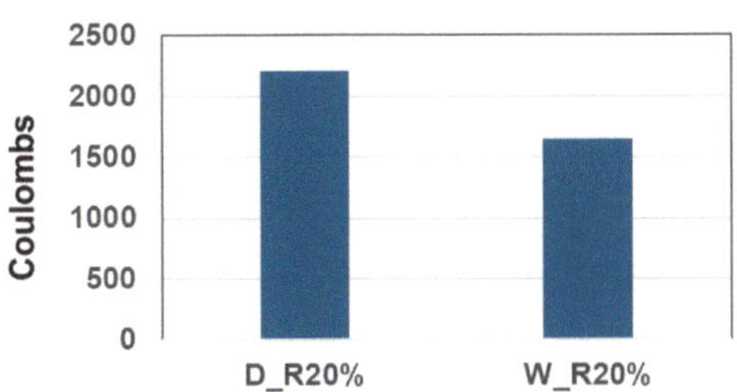

(a)

Charge passed	Chloride ion penetrability
> 4000	High
> 2000~4000	Moderate
> 1000~2000	Low
100~1000	Very Low
<100	Negligible

(b)

Figure 23. Result of RCPT. (**a**) Comparison of permeability between D_R20% and W_R20%. (**b**) Ion penetrability index as per charge passed.

4. Conclusions

This study aimed to determine the optimal mixture proportion for rubberized sulfur concrete (RSC) in terms of compressive strength. For the optimal mix design, initially, the effects of the sulfur amount, the rubber content, and the micro-filler were investigated in accordance with the dry mixing processes. Next, the optimal mix design was adjusted with the wet mix process. Then, the toughness and durability performances were evaluated. For optimizing the mixture proportion of the RSC, sand, rubber particles, and micro-filler were adopted as aggregates and sulfur was used for the binding material. Moreover, two mixing processes were applied: the dry mixing process and the wet mixing process.

Based on the test results, the increment of rubber particles in the mixture led to a decrease in the compressive strength for both the dry and wet mixing processes. To minimize the voids between the sand and rubber particles, the micro-filler was used at 5% of the total volume. The amount of sulfur varied slightly depending on the mixing process: 30% sulfur for the dry mixing process and 34% sulfur for the wet mixing process.

In the wet mixing process, the sulfur was not completely mixed with the rubber particles during the heating time, as the rubber particles tended to be burned before the sulfur melted. Thus, a modified wet mixing process was proposed. Consequently, compared to the dry mixing process, the wet mixing process increased the bonding force between the sulfur and the rubber powder due to their simultaneous heating and combining.

For the toughness of the RSC, the stress–strain profiles of the dry and wet mixing processes were compared using a mixture with 20% rubber particles. The toughness index was represented by the area ratios for strains of 0.8% and 1.2%, which were four times the yield strain. The results show that the wet mixing process demonstrates a 40% higher energy absorption capability compared to the dry mixing process. Therefore, it is determined that RSC can be applied to construction sites where low-strength high-toughness materials are required, such as in road paving or lightweight walls.

For the durability performance of the RSC, a freezing–thawing (F-T) test and a rapid chloride ion penetration (RCP) test were conducted. According to the results of the transmission time, the increment of the F-T cycles led to damage in the RSC specimens, but the rubber particles were able to enhance the durability of the RSC by alleviating the micro-damage. The RCP test was assessed to determine the permeability of the sulfur-binding materials. According to the results of the RCPT, the mixture with 20% rubber particles produced using the wet mixing process exhibited better permeability compared to the dry mixing process.

These findings can be applied to precast products such as bricks, sound-absorbing barriers, and lightweight wall panels. Additionally, RSC is suitable for 3D-printing materials due to its quick setting and early strength development.

Author Contributions: Conceptualization, G.L. and O.N.; data curation, O.N.; formal analysis, O.N.; funding acquisition, G.L.; investigation, G.L. and O.N.; methodology, O.N.; supervision, G.L.; validation, O.N.; visualization, O.N.; writing—original draft, O.N.; writing—review and editing, O.N. All authors have read and agreed to the published version of the manuscript.

Funding: This research received no external funding.

Institutional Review Board Statement: Not applicable.

Informed Consent Statement: Not applicable.

Data Availability Statement: The raw data supporting the conclusions of this article will be made available by the authors on request.

Conflicts of Interest: The author Okpin Na was employed by the company Hyundai E&C. The remaining authors declare that the research was conducted in the absence of any commercial or financial relationships that could be construed as a potential conflict of interest.

References

1. Korea Tire Manufacturers Association. Available online: www.kotma.or.kr (accessed on 13 October 2024).
2. Bae, J.S.; Moon, J.S.; Ko, Y.Z.; Kim, J.W. Performance Evaluation of Wasted Tire—Chip Added Concrete Road Pavement. *KSCE* **1996**, *16*, 235–247.
3. Kim, N. Experimental Performance Characteristics of Crumb Rubber-Modified (CRM) Asphalt Concrete. *KOSHAM—J. Korean Soc. Hazard Mitig.* **2003**, *3*, 89–97.

4. Song, H.; Jo, Y.K.; Soh, Y.S. A Study on the Fundamental Properties of Cement Mortar Using Polymer Coated Crumb Rubber. *Mag. Korea Concr. Inst.* **1996**, *8*, 163–172.
5. Park, S.K. Development of Green Cement Type Grouting Materials with High Toughness and Non-Shrinkage Including Powder of Waste Tire and Resin. *J. Korea Concr. Inst.* **2007**, *19*, 623–630.
6. Epps, J.A. *Uses of Recycled Rubber Tires in Highways: Synthesis of Highway Practice 198*; Transportation Research Board, National Research Council: Washington, DC, USA, 1994.
7. Lee, M.H.; Lee, S.H. Fundamental Study of Fire-Proof Characteristics of High Strength Concrete Using Meta-Kaolin and Waste Tire Chip. *J. Korea Concr. Inst.* **2008**, *20*, 89–97. [CrossRef]
8. Güneyisi, E.; Gesoğlu, M.; Özturan, T. Properties of rubberized concretes containing silica fume. *Cem. Concr. Res.* **2004**, *34*, 2309–2317. [CrossRef]
9. Biel, T.D.; Lee, H. Use of recycled tire rubbers in concrete. In Proceedings of the ASCE Third Materials Engineering Conference, Infrastructure: New Materials and Methods of Repair, San Diego, CA, USA, 13–16 November 1994; pp. 351–358.
10. Eldin, N.N.; Senouci, A.B. Rubber-tire particles as concrete aggregate. *J. Mat. Civ. Eng.* **1993**, *5*, 478–496. [CrossRef]
11. Topcu, I.B. The properties of rubberized concretes. *Cem. Concr. Res.* **1995**, *25*, 304–310. [CrossRef]
12. Topcu, I.B. Assessment of the brittleness index of rubberized concretes. *Cem. Concr. Res.* **1997**, *27*, 177–183. [CrossRef]
13. Chung, K.H.; Hong, Y.K. Introductory Behavior of Rubber Concrete. *J. Appl. Polym. Sci.* **1999**, *72*, 35–40. [CrossRef]
14. Li, G.; Stubblefield, M.A.; Garrick, G.; Eggers, J.; Abadie, C.; Huang, B. Development of waste tire modified concrete. *Cem. Concr. Res.* **2004**, *34*, 2283–2289. [CrossRef]
15. Chou, L.H.; Chang, J.R.; Lee, M.T. Use of Waste Rubber as Concrete Additive. *Waste Manag. Res.* **2007**, *25*, 68–76. [CrossRef] [PubMed]
16. Chou, L.H.; Lin, C.N.; Lu, C.K.; Lee, C.H.; Lee, M.T. Improving Rubber Concrete by Waste Organic Sulfur Compounds. *Waste Manag. Res.* **2010**, *28*, 29–35. [CrossRef] [PubMed]
17. Chou, L.H. Mortars with Partially Oxidized Waste Rubber Crumbs. *Adv. Mater. Res.* **2011**, *156–175*, 1421–1424. [CrossRef]
18. Chou, L.H. On Improvements of Rubberized Concrete with Organosulfur Compounds. *Adv. Mater. Res.* **2011**, *156–175*, 1459–1462. [CrossRef]
19. Gillott, J.E.; Jordann, I.J.; Loov, R.E.; Shrive, N.G. Effect of Different Aggregates on Durability of Sulphur Concretes. *Durab. Build. Mater.* **1983**, *1*, 255–273.
20. Mohamed, A.M.O.; Gamal, M.E. Hydro-Mechanical Behavior of a Newly Developed Sulfur Polymer Concrete. *Cem. Concr. Compos.* **2009**, *31*, 186–194. [CrossRef]
21. Vlahovic, M.M.; Martinovic, S.P.; Boljanac, T.D.; Javanic, P.B. Durability of Sulfur Concrete in Various Aggressive Environments. *Constr. Build. Mater.* **2011**, *25*, 3926–3934. [CrossRef]
22. Mohammed, S.; Poornima, V. Strength and durability study of sulphur concrete with replaced fine aggregate. *Mater. Today Proc.* **2018**, *5*, 23888–23897. [CrossRef]
23. Ghasemi, S.; Nikudel, M.R.; Zalooli, A.; Khamehchiyan, M.; Alizadeh, A.; Yousefvand, F.; Ghasemi, A.M.R. Durability Assessment of Sulfur Concrete and Portland Concrete in Laboratory Conditions and Marine Environments. *J. Mater. Civ. Eng.* **2022**, *34*, 1–14. [CrossRef]
24. Romadhon, E.S.; Hanif, A. Analysis of Compressive Strength of Sulfur Concrete. *J. Mech. Civ. Ind. Eng.* **2022**, *3*, 7–16. [CrossRef]
25. The Sulphur Institute. Available online: https://www.sulphurinstitute.org/about-sulphur/sulphur-construction-materials/ (accessed on 13 October 2024).
26. Grugel, R.N. *Sulfur 'Concrete' for Lunar Applications-Environmental Considerations*; Marchall Space Flight Center: Huntsville, AL, USA, 2008.
27. Siddharthan, P.; Sukumar, S.; Sivaraja, M. Comparative Study on Sulphur Concrete with Modified Glass Fiber. *J. Cardiovasc. Dis. Res.* **2021**, *12*, 2234–2239.
28. *ASTM C 617*; Standard Practice for Capping Cylindrical Concrete Specimens. ASTM International: West Conshohocken, PA, USA, 2015.
29. *ASTM C 579*; Standard Test Methods for Compressive Strength of Chemical-Resistant Mortars, Grouts, Monolithic Surfacings, and Polymer Concretes. ASTM International: West Conshohocken, PA, USA, 2018.
30. *ASTM C 307*; Standard Test Method for Tensile Strength of Chemical-Resistant Mortar, Grouts, and Monolithic Surfacings. ASTM International: West Conshohocken, PA, USA, 2018.
31. *ASTM C 580*; Standard Test Method for Flexural Strength and Modulus of Elasticity of Chemical-Resistant Mortars, Grouts, Monolithic Surfacings, and Polymer Concretes. ASTM International: West Conshohocken, PA, USA, 2018.
32. *ASTM C 905*; Standard Test Methods for Apparent Density of Chemical-Resistant Mortars, Grouts, Monolithic Surfacings, and Polymer Concretes. ASTM International: West Conshohocken, PA, USA, 2018.
33. *ASTM C 531*; Standard Test Method for Linear Shrinkage and Coefficient of Thermal Expansion of Chemical-Resistant Mortars, Grouts, Monolithic Surfacings, and Polymer Concretes. ASTM International: West Conshohocken, PA, USA, 2018.
34. *ASTM C 413*; Standard Test Method for Absorption of Chemical-Resistant Mortars, Grouts, Monolithic Surfacings, and Polymer Concretes. ASTM International: West Conshohocken, PA, USA, 1997.
35. Remisova, E. Study of mineral filler effect on asphalt mixtures properties. In Proceedings of the 6th International Conference on Bituminous Mixtures and Pavements, Thessaloniki, Greece, 10–12 June 2015.

36. Chen, Y.; Xu, S.; Tebaldi, G.; Romeo, E. Role of mineral filler in asphalt mixture. *Road Mater. Pavement Des.* **2022**, *23*, 1–40. [CrossRef]
37. *ASTM C 1312*; Standard Practice for Making and Conditioning Chemical-Resistant Sulfur Polymer Cement Concrete Test Specimens in the Laboratory. ASTM International: West Conshohocken, PA, USA, 2018.
38. Bae, J.S.; Lim, J.H.; Kim, K.S. A Study on Existing Evaluation Method and TES Method about Toughness of Fiber Reinforced Concrete. *Mag. Korea Concr. Inst.* **1998**, *10*, 247–286.
39. Ferreira, T.; Rasband, W. ImageJ User Guide. Available online: https://imagej.net/ij/docs/guide/user-guide.pdf (accessed on 13 October 2024).
40. *ASTM C666*; Standard Test Method for Resistance of Concrete to Rapid Freezing and Thawing. ASTM International: West Conshohocken, PA, USA, 2015.
41. *ASTM C1202*; Standard Test Method for Electrical Indication of Concrete's Ability to Resist Chloride Ion Penetration. ASTM International: West Conshohocken, PA, USA, 2019.

materials

MDPI

Review

Valorization of Residue from Aluminum Industries: A Review

Andrie Harmaji *, Reza Jafari * and Guy Simard

Department of Applied Sciences, University of Québec in Chicoutimi (UQAC), 555, Boulevard de l'Université, Saguenay, QC G7H 2B1, Canada
* Correspondence: aharmaji@etu.uqac.ca (A.H.); reza_jafari@uqac.ca (R.J.)

Abstract: Recycling and reusing industrial waste and by-products are topics of great importance across all industries, but they hold particular significance in the metal industry. Aluminum, the most widely used non-ferrous metal globally, generates considerable waste during production, including dross, salt slag, spent carbon cathode and bauxite residue. Extensive research has been conducted to recycle and re-extract the remaining aluminum from these wastes. Given their varied environmental impacts, recycling these materials to maximize residue utilization is crucial. The components of dross, salt slag, and bauxite residue include aluminum and various oxides. Through recycling, alumina can be extracted using processes such as pyrometallurgy and hydrometallurgy, which involve leaching, iron oxide separation, and the production of alumina salt. Initially, the paper will provide a brief introduction to the generation of aluminum residues—namely, dross, salt slag, and bauxite residue—including their environmental impacts, followed by an exploration of their potential applications in sectors such as environmental management, energy, and construction materials.

Keywords: aluminum; valorization; waste heat; dross; salt slag; spent carbon cathode; bauxite residue

1. Introduction

Aluminum (Al) and aluminum alloys are highly valued metals with diverse applications because of their superior thermal characteristics, electrical conductivity, low weight, and corrosion resistance [1]. As one of the most recycled materials, aluminum's significant value and extensive use in the construction, automotive, packaging, and aerospace industries highlight its importance. Notably, aluminum used in building and automotive parts is recycled at rates up to 90% at the end of its useful life [2], and 75% of all aluminum ever produced remains in use today [3]. Ranking second only to steel in the amount of non-ferrous metal produced, aluminum's production volume surpasses that of all other non-ferrous metals combined [4]. In particular, aluminum alloys are crucial in the aerospace and automotive industries, where they often replace steel due to their ability to naturally form a passive oxide layer when exposed to air, thereby preventing corrosion under various conditions [5].

The aluminum industry generates both non-solid residues (e.g., waste heat) and solid waste, including aluminum dross, salt slags, spent carbon cathode, and bauxite residues. These residues from both primary and secondary aluminum production pose a global challenge. The recycling and reuse of industrial waste and by-products are paramount in construction technology. Traditional industrial by-products like fly ash, granulated blast furnace slag, and silica fume have been extensively researched for their reactivity and effectiveness as active additions in cement and concrete manufacturing [6]. Less reactive wastes serve as additional cementitious materials for cement production or as inert aggregates for concrete and mortar. The residues from aluminum industries, rich in metal oxides and containing valuable rare earth elements, underscore the importance of valorizing aluminum production residues for their incorporation and reuse in everyday applications. This review begins with a brief overview of aluminum production and the residues it generates—namely, waste heat, aluminum dross, salt slag, spent carbon cathode, and bauxite residues. We then explore the valorization potential of each residue and present opportunities for large-scale residue reuse.

2. Overview of Aluminum

In the following sections, we describe the Hall–Héroult process and explain the technology applied during the production process to fully comprehend the operations of the global aluminum industry. Opportunities for the use of residues can then be determined, and we can identify constraints on integrating novel procedures for minimizing and reusing waste products.

2.1. The Hall–Héroult Process

The current method of producing primary aluminum was established in 1886 when Paul Héroult and Charles Hall independently discovered that the reduction of molten aluminum oxide in cryolite (Na_3AlF_6) served as a more affordable means of aluminum production. As both shared the patent, this production technique is known as the Hall–Héroult process (Figure 1). In 1887, Karl Joseph Bayer discovered that alumina (Al_2O_3) could be extracted from bauxite, an ore named after the French province of Les Baux, and used as a cost-effective feedstock for the Hall–Héroult process.

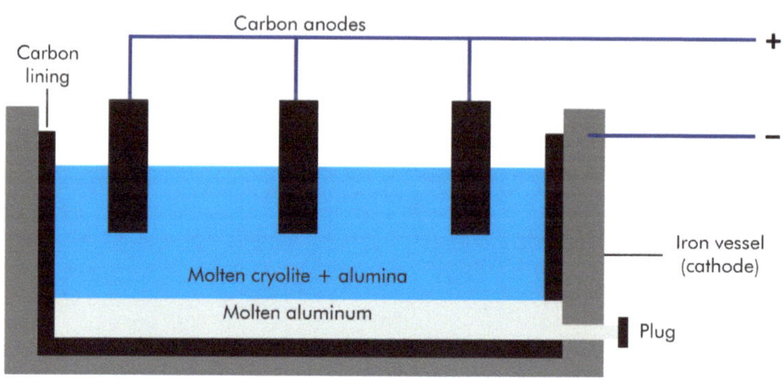

Figure 1. Illustration of the Hall–Héroult process.

The Bayer process separates alumina from bauxite. Bauxite is dissolved in a hot caustic solution to break it down into a green liquor. This green liquor is then separated from the undissolved material and allowed to cool, where it precipitates pure alumina hydrate. The hydrate is subsequently calcined to produce alumina. The high-purity petroleum coke used to make the carbon anodes for aluminum reduction is ground, calcined, and combined with ground used carbon anodes and pitch, which serves as a binder, to create a paste. Although the Hall–Héroult process consumes a significant amount of electrical energy, manufacturers have gradually increased the efficiency of this technique. Nonetheless, the extraction process produces significant CO_2 emissions from the anodes and relatively high heat loss from the electrolytic cells.

2.2. Current Production of Aluminum

Aluminum is the most abundant metallic element, constituting 7.96% of the Earth's crust [7]. It is seldom found in its elemental form because of its high reactivity, especially when in contact with oxygen, necessitating its extraction from minerals in ores. Bauxite is the principal ore used for aluminum extraction, accounting for more than 99% of primary aluminum production [8]. Open-cast mines on the Earth's surface are the main sites for bauxite extraction [9]. Bauxite contains three main types of aluminum hydroxide compounds: gibbsite ($Al(OH)_3$), diaspore (α-AlO(OH)), and boehmite (γ-AlO(OH)) [10]. Typically, alumina constitutes 40–60% of bauxite's weight, along with trace amounts of iron, silicon, and titanium compounds [3], necessitating refinement because of these trace compounds.

Advancements in the metallurgy of aluminum alloys, along with population growth and increased economic activity [11], have led to a rise in bauxite extraction. In 2023, the world's aluminum production increased marginally to 70 million metric tons (MT) from 68.4 million MT in 2022. Of this total, 41 million metric tons, or more than half, came from China, followed by India and Russia [12]. According to official data, China's primary aluminum output increased by 7.10% year over year to 14.24 Mt between January and April of this year. An unsustainable increase in alumina imports (+75% y/y to 1.06 Mt), the intermediate product between bauxite and metal in the primary aluminum production chain, is the main cause of this growth [13]. In this country, alumina is produced with an estimated 82 million metric tons produced, followed by Australia with about 19 million metric tons that year [14]. The largest producer of aluminum in India, Vedanta Aluminum, ranks among the top three global companies, with a production capacity of 3 million tonnes per annum (MTPA) [15]. The company announced the new 1.5 MTPA expansion at its alumina refinery in Lanjigarh, Odisha in early 2024. As a result, the alumina refinery's production capacity has grown from its previous 2 MTPA to 3.5 MTPA. Global demand for aluminum is expected to double over the next decade [16], though more conservative estimates suggest a potential doubling or tripling of demand by 2050 [17]. Australia, China, Guinea, Brazil, India, and Indonesia are the six largest bauxite producers currently [18]. While bauxite ore is plentiful, the growth in secondary production and recycling of aluminum has also surged; hence, the estimated 55–75 Gt of global bauxite reserves are likely to meet demand for many centuries [19].

Primary aluminum is produced exclusively from mined ore. The refined ore undergoes electrolytic reduction (Figure 2) before being shaped through casting, rolling, or extrusion to produce bulk material. For products with specific requirements, manufacturers process this bulk material further. Secondary aluminum production involves scrap pretreatment as well as smelting and refining. Pretreatment operations include the processing, cleaning, and sorting of scrap. The smelting and refining processes encompass cleaning, melting, refining, alloying, and pouring of aluminum recovered from scrap.

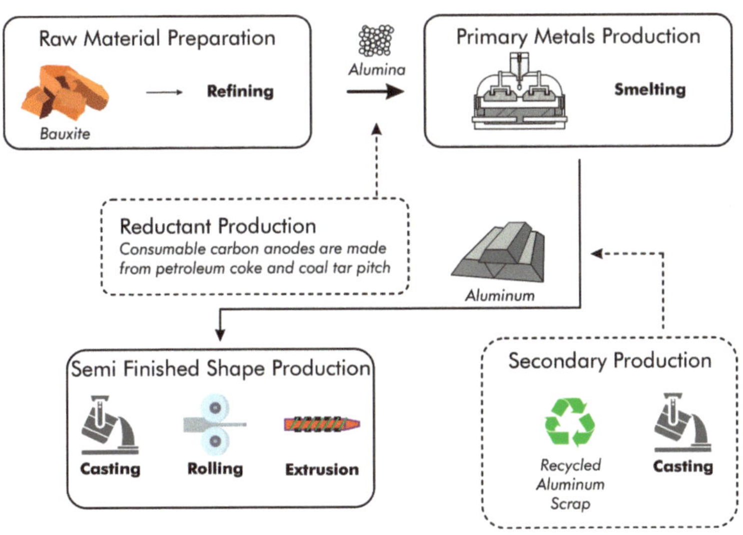

Figure 2. Flow diagram of primary and secondary aluminum processing.

2.3. The Bayer Processes

The Bayer process is the most widely recognized refining method, though the nepheline-based process and the combined or parallel Bayer–Sinter process are used in some regions. The nepheline-based process uses nepheline syenite as an alternative to bauxite [20], prized for its high aluminum content. However, it presents challenges because of (a) the slow dissolution of aluminum-bearing phases and (b) the formation of silica gel at high pulp densities and low treatment temperatures [21,22].

In the Bayer process, bauxite ore undergoes processing to produce purified alumina, necessitated by the impurities within the ore (Figure 3). High-porosity and reactive activated alumina are produced in situ from the bauxite rocks with minimal processing, without direct purification from the ore [23]. This process involves adding ground and blended bauxite and sodium hydroxide (NaOH) to a pressure vessel at temperatures ranging from 150 to 265 °C, depending on the concentration of aluminum compounds present [24]. The compounds dissolve, creating an equilibrium as described by Equations (1) and (2) [25], with conditions favoring the reaction's shift to the right to produce hydrated $NaAl(OH)_4$ in a step known as digestion.

$$\text{Gibbsite}: Al(OH)_3 + Na^+ + OH^- \rightleftharpoons Al(OH)_4{}^- + Na^+. \quad (1)$$

$$\text{Diaspore and Boehmite}: Al(OH)_3 + Na^+ + OH^- \rightleftharpoons Al(OH)_4{}^- + Na^+. \quad (2)$$

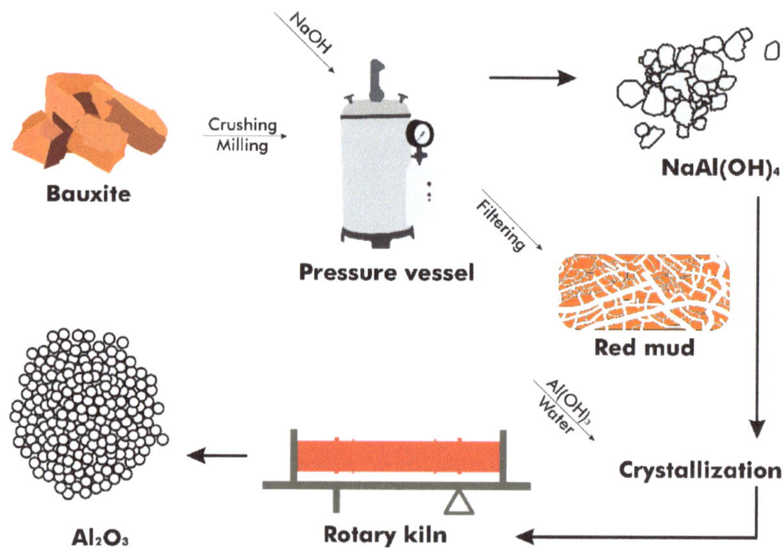

Figure 3. Flow diagram of the Bayer process.

A supersaturated sodium aluminate liquor known as a slurry is produced by chemically treating the solution. The equilibrium is then shifted to the left by reversing Equation (3) and by cooling the mixture in the presence of seeded crystals of $Al(OH)_3$. Aluminum hydroxide is heated at approximately 1000 °C in a rotary kiln or calciner to remove water and produce anhydrous aluminum oxide, while the sodium hydroxide is recycled [26].

$$2Al(OH)_3 \rightarrow Al_2O_3 + 3H_2O \quad (3)$$

The primary smelters or other industries then acquire the anhydrous aluminum oxide for processing. Figure 4 presents the stages of a typical refining procedure.

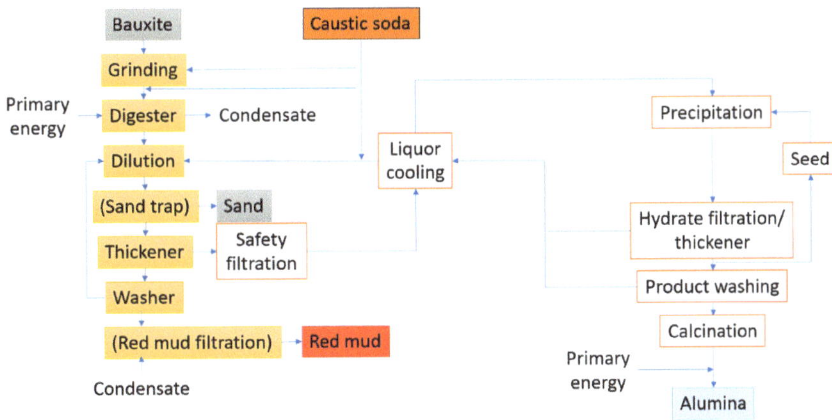

Figure 4. Overview of the stages of refining bauxite to alumina.

3. Impact of Aluminum Production on the Environment

The manufacturing of aluminum involves numerous additional supporting activities and resources, all of which must be considered in any life cycle analysis of aluminum production. The impacts of aluminum production include the construction of roads and infrastructure, as well as the extraction and disposal of a large amount of rock. Once the mine site is depleted, the post-restoration soils exhibit low water-holding capacities, rendering them unsuitable for crop cultivation. Furthermore, aluminum production is associated with the damage and even loss of ecosystems through photochemical ozone formation, acidification, eutrophication, and ecotoxicity [27]. The aluminum industry emits between 0.45 and 5 Gt of carbon dioxide (CO_2) equivalent annually [28]. Industrial facilities converting alumina into aluminum often rely on power plants powered by fossil fuels, thereby increasing greenhouse gas emissions. Power generation may also include hydroelectricity, which reduces carbon emissions but still poses environmental impacts related to dam construction and the flooding of large areas for reservoirs. Particulate and gaseous emissions during aluminum reduction processes include carbon monoxide (CO), CO_2, sulfur oxides (SOx), aluminum fluoride (AlF_3), calcium fluoride (CaF_2), and volatile organic compounds. Polycyclic aromatic hydrocarbons (PAHs), pollutants created during the electrolytic process, are particularly concerning given their carcinogenic potential. Hence, any future sustainable plant design must prioritize the efficient use of natural resources, minimize the input of residues into the environment, and explore alternative, recyclable energy sources such as waste heat.

3.1. Waste Heat

Primary aluminum production consumes significantly more energy and water than secondary production. The efficiency of existing plants can be improved using modern technologies. Recovering, reclaiming, and using generated waste heat are methods for reducing energy consumption. Heat recovery techniques, including the use of electrolytic cells, can help minimize heat loss and reduce the need for fuel and electricity. Most heat is lost through the cell sidewalls, with a smaller amount lost through off-gassing. In secondary aluminum production, scrap melting in high-temperature furnaces, a crucial step, only employs waste heat recovery in about one-third of existing high-capacity furnaces. Nevertheless, several methods for recovering waste heat are available (Figure 5).

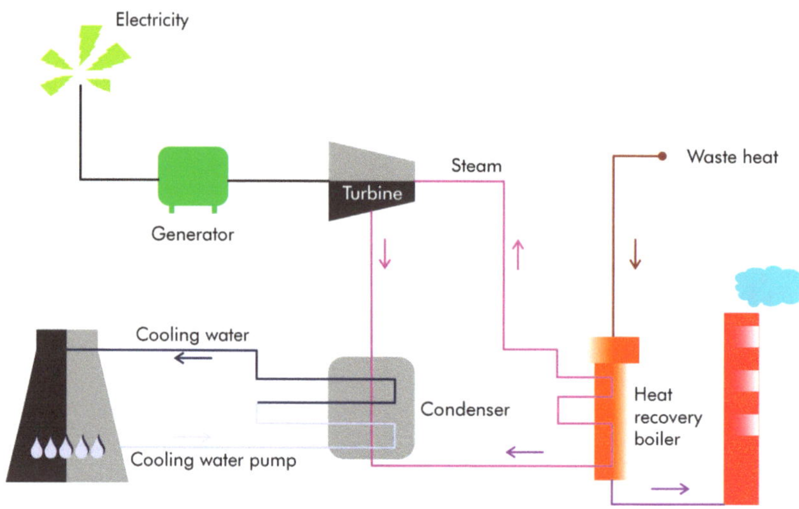

Figure 5. A heat recovery model. The steam is fed into the highly efficient turbine/generator unit, which generates electrical energy.

3.2. Aluminum Dross

A significant challenge in producing pure aluminum is the generation of other solid wastes. Primary aluminum production generates considerably more atmospheric emissions and solid waste than secondary production. Materials with less than 45% aluminum content are classified as "dross", whereas residues with more than 45% aluminum content are termed "skimmings" [29]. Dross can be further divided into "white dross", from primary smelters without salt covers, and "black dross", originating from secondary smelters. Metal beads, crystallized salt, and solid nonmetallic particles are found in the nonmetallic waste products from dross smelting processes. White dross comprises a fine powder obtained by skimming the molten aluminum, containing between 20% and 45% recoverable metallic aluminum. In the second stage of aluminum production, the flux-containing dross resulting from the melting operation of aluminum scrap in reverberatory furnaces is referred to as "black dross", distinguished by its darker color [30]. Black dross typically contains aluminum oxide (20–50%), a salt–flux mixture (40–55%), and aluminum metal, with the recoverable aluminum content usually ranging between 10% and 20%. It has a much higher salt content (typically greater than 40%) than white dross [31]. In the separation process of aluminum production, white dross requires more energy and water than black dross and also produces more waste [32]. Globally, more than 95% of the four million tonnes of white dross and over a million tonnes of black dross produced annually are landfilled.

3.3. Salt Slag

The nonmetallic residue resulting from dross smelting operations is known as salt slag and typically contains 3–7% residual metallic aluminum, 15–30% aluminum oxide, 30–55% sodium chloride, and 15–30% potassium chloride, along with other materials that vary in abundance depending on the type of scrap [33]. Salt slag, also referred to as salt cake, is classified as toxic and hazardous waste with variable harmful properties [34]. The scrap/dross and salt flux are re-added to an oil or gas-fired furnace during the rotary salt furnace process. The salt facilitates the agglomeration and separation of the metal, thereby increasing metal recovery by shielding it from the atmosphere's reactivity [35]. Additionally, it enhances heat transfer to the metal, prevents metal oxidation, and absorbs contaminants including oxides, nitrides, carbides, and other substances found in scrap or created by reactions during melting [36]. Aluminum metal and salt slag are removed from the furnace after melting. After tapping and cooling, the salt cake is formed, containing all the nonmetallic components from the raw mix [37]. The oxides in the dross (from the raw mixture) form a net-like structure, trapping aluminum. This structure is disrupted by the molten flux, which also facilitates the coalescence and sinking of aluminum droplets into the aluminum bath [38]. Carbon typically remains in the salt slag following the decomposition of organic contaminants. If insufficient amounts of salt are present, a high concentration of oxides and other contaminants may cause the molten salt to become highly viscous. In practice, a more viscous slag results in significant metal loss because of the entrapment of metal droplets.

Carbides, nitrides, and phosphates bound with aluminum, along with the oxides of alloying elements (e.g., Cu, Fe, Si, Zn), make up most of the nonmetallic compounds. When liquid aluminum comes into contact with finely dispersed carbon, Al_4C_3 is formed, which originates from organic contaminants in scrap such as paints, plastic coatings, and hybrid-sandwich components. The AlN-containing dross is fed into rotary furnaces, where it is captured by the salt slag. The high concentration of NaCl and KCl in salt slag causes the slag to release chlorides into water when in contact with a water source or groundwater. The presence of toxic, harmful, explosive, poisonous, and foul-smelling gases, e.g., NH_3, CH_4, PH_3, H_2, and H_2S, means that the gaseous emissions resulting from the contact of the salt slag with water may also have a significant negative impact on the environment [39]. Depending on the type of scrap, it may contain carbides, nitrides, phosphides, and sulfides [40].

3.4. Spent Carbon Cathode

Another solid waste produced from aluminum is the spent carbon cathode. This carbon-rich material produced when electrolysis cells in the aluminum manufacturing process are overhauled [41]. Most aluminum electrolytic cells only last 5 to 8 years because high-temperature electrolytes continuously erode the cathode carbon during production [42]. The main way that SCC is currently disposed of is in landfills [43], which may release highly toxic cyanide and soluble fluoride into the environment, endangering ecological and environmental safety [44]. Moreover, it also has a significant effect on the wellbeing and ecological equilibrium of plants and animals [45]. While carbon recovery is relatively easy, recovering fluorine resources is typically difficult, primarily for innocuous detoxification [46]. Among the techniques used to treat spent SCC are high-temperature heating [47], chemical leaching [48], and physical separation [49]. The drawbacks of chemical leaching and physical separation are their lengthy processing times, complicated procedures, and high wastewater production. The combustion and boiling points of dangerous compounds, like fluorides, serve as the foundation for the high-temperature heating treatment. Moreover, these techniques have a number of drawbacks, including low leaching rates, lengthy leaching cycles, high energy consumption, low recovery rates, insufficiently pure carbon powder, and the production of new pollutants throughout the treatment process [50]. Therefore, finding a treatment method that is both effective and pollution-free is essential.

3.5. Bauxite Residue

The most significant waste created in the primary aluminum process is bauxite residue or "red mud", generated during the Bayer process. This red-colored sludge is produced in the second stage of alumina production. The high concentration of Fe_2O_3 in bauxite residue gives it its red-colored appearance. To produce one ton of aluminum, four tons of bauxite are required, and two tons of bauxite residue are generated. Red mud consists primarily of iron (Fe), silicon (Si), and titanium (Ti) oxides, sodium hydroxide, unextracted aluminum oxide, zinc (Zn), phosphorus (P), nickel (Ni), and vanadium (V), as well as other oxides. The composition of red mud varies among aluminum production facilities (Table 1).

Table 1. Oxide compositions of red mud produced globally.

Country	Plant	Location	Major Constituents (wt%)						Ref.
			Fe_2O_3	Al_2O_3	SiO_2	TiO_2	Na_2O	CaO	
Australia	Boyne Smelters	Gladstone	23–36	20–30	13–17	4–8	5–11	1–3	[51]
	Rio Tinto	Gove	30	20.8	17.1	8.3	8.1	2	[51]
	ALCOA	Kwinana	27–31	12–20	27–54	1–3	1–3.3	0–3	[51]
Brazil	Alunorte	Barcarena	45.6	15.1	15.6	4.29	7.5	1.16	[52]
Bosnia	Zvornik	Birac	48.5	14.14	11.53	5.42	7.5	–	[53]
Canada	Rio Tinto	Vaudreuil	31.60	20.61	8.89	6.23	10.26	1.66	[54]
China	Weiqiao	Shandong	59.37	16.16	9.11	–	2.78	2.17	[55]
France	Pechiney	Gardanne	42	14	6	11	2	–	[56]
Germany	FRG Baudart	Stade	38.75	20	13	5.5	8.16	–	[57]
Greece	Aluminum of Greece	Agios Nikolaos	44.60	23.6	10.2	5.7	2.5	11.2	[58]
Hungary	MAL	Ajka	38.75	15.2	10.15	4.6	8.12	–	[57]
India	Nalco	Damanjodi	47.85	22.64	12.51	3.58	10.25	1.86	[59]
Indonesia	ICA	Tayan	35.10	29.5	2.96	–	4.89	0.18	[60]
	BAI	Bintan	44.66	28.87	20.21	3.03	–	0.28	[61]
Iran	Jajarm	North Khosaran	22.17	13.98	13	7.17	2.01	24.25	[62]
Ireland	RUSAL	Askeaton	47	17	7	12	5	–	[63]
Italy	Eurallumina SpA	Sardinia	18	26	20	6	12	6.7	[64]
Jamaica	ALPART	Nain, St. Elizabeth	50.9	14.2	3.4	6.87	3.18	–	[57]
Russia	Bogoslovsky	Urals	42.1	12.7	9.4	4.3	4.8	7.8	[65]
	RUSAL	Kamensk-Uralsky	12.2	25.5	2.5	-	28.3	2.5	[66]
	Ural	Kamensk-Uralsky	36.9	11.8	8.71	3.54	0.27	23.8	[67]
	Tatarka Deposit	Krasnoyarsk Krai	6.86–34.05	41.44–59.45	1.48–7.97	1.17–4.08	0.05–0.39	0.01–0.4	[68]
Suriname	ALCOA	Paranam	24.81	19	11.9	12.15	9.2	–	[57]
Spain	Alcoa	Guinea	37.5	21.2	4.4	11.45	3.6	5.51	[52]
Taiwan	Sigma Group	Kaohsiung	41.3	20.21	17.93	2.9	3.8	–	[57]
Turkey	Eti Alüminyum	Seydizehir	37.84	20.24	15.27	6.15	9.43	2.23	[57]
USA	Alcoa	Arkansas	55.6	12.15	4.5	4.5	2–5	–	[57]
UK	Alcan	Lynemouth	46.0	20.0	5.0	6.0	8.0	1.0	[69]
Ukraine	Nikolaev	Halytsynove	39–43	17–19	9.5–11.1	4.4–5.6	6.2–6.9	7.6–9.5	[70]

The bauxite residue becomes extremely alkaline, with pH values from 9 to 13, when bauxite is treated with concentrated NaOH at high temperature and pressure [71]. Bauxite residue waste management typically involves controlled landfill disposal. Because sodium hydroxide is added during the Bayer process, the residue is extremely alkaline [72]. In Canada, red mud is considered a Class 8 hazardous good (corrosive material) under the dangerous goods transport regulation, Sections 2.40–2.42, in accordance with UN code No. 3244. Thus, specific packaging and the use of UN-compliant containers are required for all shipments. Transportation via airfreight is quite difficult, except in limited quantities. Meanwhile, sea freight is easier, and road transportation packaging standards are normally used. Moreover, given the high alkalinity of the water and soil (pH 10–13), the presence of heavy metals, and even traces of radioactive elements, there is a significant contamination risk [73]. Extensive treatment involving maintenance and monitoring is necessary, as is a sizable storage space for its disposal.

4. Valorization of Residue from Aluminum Production

Given its high pH and trace amounts of heavy metals, aluminum residue poses a serious environmental risk. The pursuit of producing cleaner bauxite residue has become a key focus for the aluminum industry, necessitating that governments and producers adopt a circular economy model for aluminum production, use, and reuse [74]. The circular economy model facilitates waste reduction by adding value to products [75]. Aluminum is one metal that benefits significantly from the circular economy; it can be recycled numerous times without losing its mechanical, physical, or chemical properties. The following subsections discuss strategies to promote the valorization of waste heat, white and black dross, salt slag, and bauxite residue from aluminum manufacturing.

4.1. Waste Heat Recovery

Several studies have focused on the use of waste heat from the aluminum industry, which could benefit from potential sources of recoverable heat, such as those from refining, smelting, recycling, and secondary melting processes. Heat recovery is also feasible from calciners and hot alumina. For calciners, various technologies are under consideration to recover latent heat from water vapor [25]. There is potential to use recoverable heat from hot alumina to produce heated water for generating electricity or for other processes within aluminum processing plants. However, the availability of hot water for plant operations is often limited.

Electrolysis pots have shown good potential for heat recovery. The steel-shell-protected wall can be fitted with a thermoelectric device. In Europe, heat exchangers on the exhaust gas stream are used to recover heat from aluminum smelting cells or reactors [26]. Hot exhaust gases from fuel combustion, produced during anode baking processes, contain significant amounts of tar vapor and volatiles, which are problematic. The primary cast furnace produces exhaust gas flows and temperatures. The gas may contain hydrogen fluoride [27] at concentrations of less than 1 to 10 ppm, which limits the use of conventional heat exchangers for waste heat recovery. The melting furnace emits a considerable quantity of exhaust gases at temperatures around 870 °C, which may include flux material vapors, organic vapors, and particulates. An option is to treat the gases with absorbents, eliminating the need to cool the gases and producing cleaner gases suitable for heat recovery in existing systems.

The gases from crucible heaters, gas generators, and reverberatory (reverb) furnaces—with temperatures ranging from 790 to 1090 °C—contain natural gas combustion products and are free of major contaminants. These gases can account for up to 60% of the total heat input for heating systems and are viable candidates for heat recovery. Rotary furnaces produce cold gases laden with a large number of contaminants, and the industry has yet to develop an approach for their recovery. Consequently, research is underway to develop high-temperature polymers or composites for gas cleaning applications in waste heat recovery. Delacquering systems remove volatile materials, such as paint coatings, from used beverage cans and preheat them to about 480 °C before releasing them from a rotary kiln. The unit emits relatively clean exhaust gases at a temperature of about 340 °C, yet these gases do not undergo any heat recovery. Nonrecoverable heat is also found in crucible heaters. Nowicki et al. presented a mapping of heat demands at Alcoa Deschambault Quebec (ADQ) [76]. Their conservative assumptions indicated that the ADQ facility had the potential to produce approximately 10 MW of electricity using extractable waste heat. An aluminum foundry in Romania evaluated a waste heat recovery system [77]. Uses for the waste heat in the foundry included direct heating during winter, conversion to steam during summer, and electricity generation during spring and autumn when both heating and cooling demands are low. Wang et al. analyzed the performance of the organic Rankine cycle (ORC) in recovering low-temperature waste heat from aluminum reduction cells [78], observing that the variation in heat source temperature significantly affects net power output. When wind and solar power are used as a major energy supply for industrial aluminum electrolysis, the heat-exchanging system is also suitable for aluminum cells. Waste heat can be used for the aluminum die-casting industry [79]. Egilegor et al. analyzed the potential of heat recovery in aluminum low-pressure die casting to ensure effective operation and prolonged lifetime under high temperatures [80]. They concluded that given the working temperature/pressure and fluids, selecting appropriate materials for the heat pipe shells was crucial.

4.2. Aluminum Dross

Given the significant variability in the composition of aluminum dross among batches, greater effort is required to identify potential applications for this material. Various techniques for recycling and reusing aluminum dross show promise for managing aluminum dross and its reuse in diverse applications, such as the development of construction and building materials, separation media and agents, alumina extraction, and hydrogen production (Figure 6). We explore several of these applications in the following subsections.

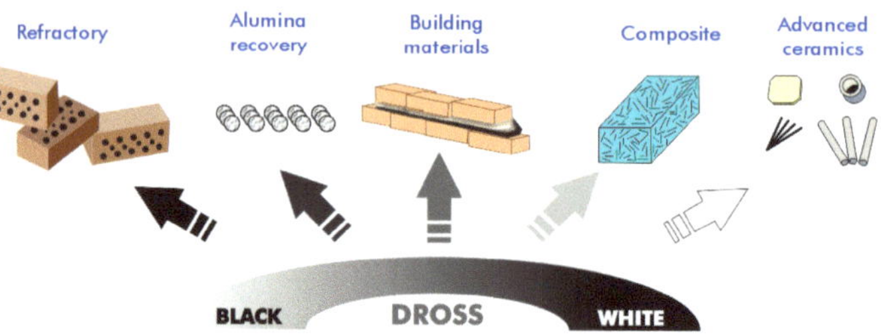

Figure 6. Potential application of aluminum dross residues.

4.2.1. Aluminum Metal Recovery

Aluminum metal can be recovered using cost-effective recovery processes, and aluminum oxides find applications in the metallurgical and construction industries. The recycling and reuse of materials are crucial for reducing waste generation and decreasing the economy's dependence on the extraction of primary (virgin) raw materials. This circular economy conserves resources and energy while slowing the depletion of virgin natural resources. During the Hall–Héroult process, aluminum metal can be reclaimed from white dross, which can be recycled through a rapid slurry-forming (RSF) process (Figure 7).

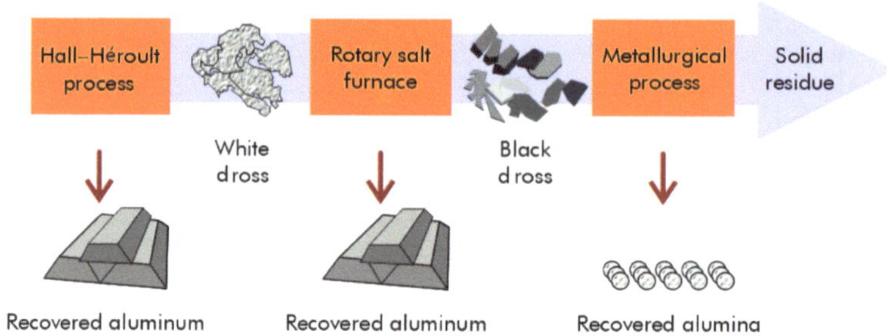

Figure 7. Recovery of aluminum and alumina from white and black dross using RSF and metallurgical processes.

Several methods have been developed to recover dross while retaining the maximum amount of aluminum metal, such as spreading hot dross to accelerate cooling, combining dry pressing and alkaline roasting processes, and using an inert gas like argon to prevent oxidation [81]. The standard procedures for processing aluminum dross involve initially cooling the dross to room temperature using methods such as rotary drum coolers, stationary inert gas coolers, and floor spreading. The remaining metal can be recovered by reheating the raw dross (or a milled concentrate) with salt flux in a rotary furnace or, less commonly, in a side-well reverberatory furnace [82].

Manual separation and density-based separation methods like hydraulic and pneumatic classifiers are less effective. Zuo et al. demonstrated a method for recovering salts and ammonium hydroxide from black aluminum dross using catalytic hydrolysis, followed by filtration and drying before calcination [83]. Aluminum and other valuable elements can be recovered in the form of alloys, such as aluminum nitride (AlN), by adding alumina in electrochemical reduction. Salts are also reclaimed during the filtration of leachate followed by crystallization.

4.2.2. Applications in Construction Materials

The use of white or black dross as a filler in the construction industry is common. Both white and black dross can serve as fillers in asphalt—when the particles are smaller than 700 μm—to improve stiffness and abrasion resistance and reduce microcracking [84]. Zhang et al. developed novel supplementary cementitious materials (SCMs) by mixing secondary aluminum dross with dolomite [85]. They investigated the effect of thermal activation on calcined mixes, the hydration properties of pastes containing novel SCMs, and the workability of cemented paste backfill (CPB). High-temperature calcination significantly increased SCM activity, and thermal activation enhanced the activity of SCMs (Figure 8).

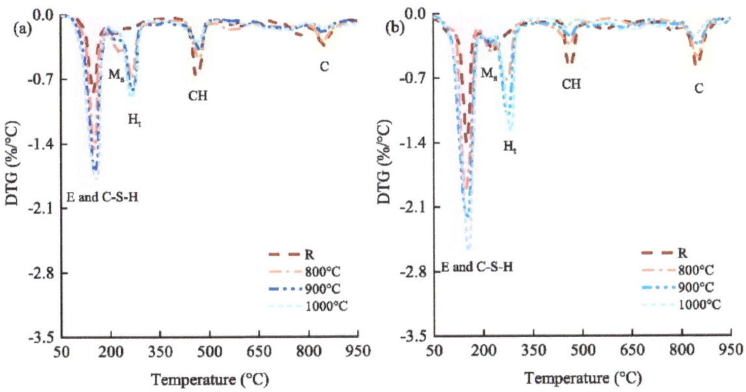

Figure 8. Differential thermogravimetry (DTG) of pastes at curing times of (**a**) 3 days and (**b**) 28 days. E, ettringite ($Ca_6Al_2S_3H_{64}O_{60}$); C-S-H, calcium silicate hydrate; Ms, monosulphate ($Ca_6Al_2SH_{24}O_{22}$); Ht, hydrotalcite ($Mg_6Al_2CH_{24}O_{52}$); CH, portlandite (CaO_2H_2) [85].

The black-dross-leached Al residue, produced during the hydrothermal treatment of Al black dross, can also be used as a raw material for producing Portland cement clinker [85]. Ercoli et al. evaluated the mechanical and thermal properties of geopolymer foams reinforced by carbon fibers, using by-products from the secondary aluminum industry [86]. The flexural and tensile strengths decreased when aluminum-rich by-products were added as additives to foam the geopolymers because of the gas bubbles formed in their structure during the consolidation process (Figure 9 illustrates the gas bubbles in the geopolymer foam, which are not homogeneous in size). However, the Charpy impact strength value was increased because chopped carbon fibers reinforced the geopolymer.

Figure 9. (**a**) Geopolymer foam section and (**b**) magnified image of bubbles generated by the oxidation of the by-product [86]. The gas bubbles were generated during the consolidation process of by-product and alkali activator.

4.2.3. Dross for Engineering Composites

Dirisu et al. used uncarbonized egg shell [87] to reinforce aluminum dross, cement, silicate, and pulverized carbon graphite. They applied a molding process to produce a building ceiling composite and investigated how its composition affected thermal conductivity and microstructure. The sample comprised 30 wt% Al dross, 25 wt% cement, 30 wt% silicate, 5 wt% carbon graphite (CG), and 10% uncarbonized eggshell (UES). The composite could serve as a flame-retardant ceiling because of its lower thermal conductivity relative to existing ceiling tiles. Moreover, the produced composite exhibited a lower heat flux than other previously developed ceiling composites. The production of conventional materials from expensive raw materials is increasingly being replaced by more cost-effective composite materials [88]. Dross, SiC, and TiO_2 nanopowders have been analyzed for use in ceramic composites [88]. SiC is known for its desirable properties, including high hardness, strength, and melting point; good chemical and thermal stability; and oxidation and erosion resistance. TiO_2 can be added to SiC with the aim of improving the sinterability and properties of the composites [89]. TEM images of sintered composites, with and without added SiC and TiO_2 (after being milled for 5 h), show no particle coarsening, and the particle size of the waste Al dross decreases as the weight percent of SiC and TiO_2 (as reinforcement) increases (Figure 10).

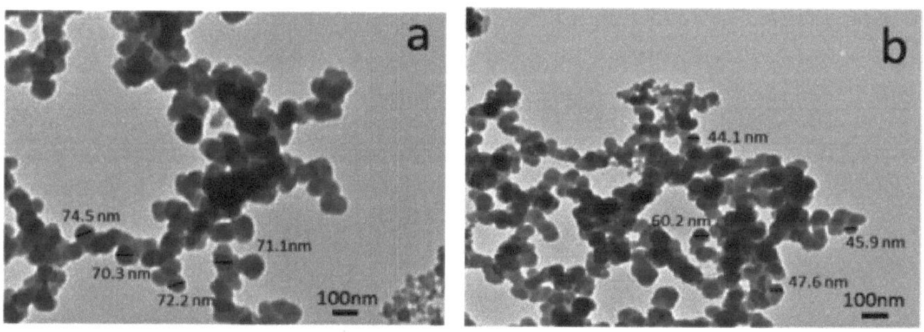

Figure 10. TEM images of (**a**) Al-dross waste powder without reinforcement and (**b**) W15 powder that contains 7.5 wt% SiC and 7.5 wt% TiO_2 [89].

4.2.4. Other Applications

Other uses of aluminum dross include its application in the steel, ceramic, and fertilizer industries. Heo and Park investigated Fe recovery from electric arc furnace (EAF) slag using aluminothermic smelting reduction (ASR) at 1773 K with Al dross as the reductant [90]. During the FeO reduction reaction, solid spinel ($MgO{\cdot}Al_2O_{33}$) and MgO compounds precipitate from the slag. Moreover, the metallic Al in the Al dross reductant can reduce Mn from the EAF slags. The use of black dross to produce a ladle-fluxing agent for the steel industry has been studied [91]. Ewais and Besisa produced magnesium aluminate titanate (MAT)-based ceramics by sintering aluminum dross waste and combining it with rutile ore powders at 1300 °C for 6 h [92]. The obtained samples were thermally stable and did not decompose at high temperatures, presenting a promising method for producing a new advanced ceramic material. Zhang et al. synthesized $MgAl_2O_4$ spinel from secondary aluminum dross [93]. They doped the spinel with rare earth oxides (REO), including Y_2O_3, Eu_2O_3, La_2O_3, and CeO_2, to improve the densification behavior. The densification of the $MgAl_2O_4$ doped with lanthanum oxide was higher when heated at 1473 to 1773 K Dangtungee et al. neutralized aluminum dross and used it as fertilizer [94]. The treated aluminum dross fertilizer enhanced the height and weight of Chinese cabbage relative to untreated plants and controls (Figure 11). Sweet corn, basil, and spring onions also exhibited improved growth with the aluminum treatment. Because no toxic heavy metals are present in the prepared fertilizer, aluminum dross—after neutralization—can be considered an environmentally friendly fertilizer.

Figure 11. Untreated and dross-treated Chinese cabbage after 13 days [94]. The plants showed increased height and weight compared to the control plants.

Aside from metals and oxides, gaseous phases such as NH_3 can be extracted during the synthesis of materials from aluminum dross. Gomes et al. used ammonia (NH_3) released from Al dross as a nitrogen source for sprouting seeds [95]. Vegetable and fruit seeds were planted in soils directly infused with dross-produced NH_3 for 2 h, followed by 8 to 12 h of self-diffusion time to ensure even distribution of the gas throughout the soil. Seeds that prefer acidic soils did not germinate well in soils infused with NH_3, whereas seeds that thrive in alkaline conditions, such as ridge gourd and watermelon seeds, sprouted early and robustly. Shi et al. [96] created a polyaluminum chloride flocculant by treating secondary aluminum dross with hydrochloric acid. The new flocculant effectively removed impurities from wastewater and other waste liquids. Philipson et al. [97] produced a silicon alloy reducing silica in calcia–silica (CaO-SiO_2) slag with aluminum dross as a reductant. The kinetics of reduction were investigated by comparing the Si-alloy and product slag compositions to simulated thermodynamic equilibrium. The Al dross was shown to be as effective a reductant as pure Al, and the results show rapid aluminothermic reduction and good Si-alloy/slag separation for both reductants.

4.3. Salt Slag

Salt slag, a material produced during the melting of aluminum scrap or dross, contains aluminum, aluminum oxide, alkaline chlorides, and impurities such as carbides, nitrides, sulfides, and phosphides [98,99]. The treatment of salt slag is typically carried out in the US, Canada, and Europe, where the landfilling of salt slag is prohibited by law. The treatment of salt cake results in a significant financial return. The salt fraction, including halite (NaCl) and sylvite (KCl) along with residual metallic aluminum, can be recovered, justifying the investment in salt cake recycling facilities by large refining companies [100]. The residues generated post-recycling are non-toxic, and the alumina-containing compounds can be used as new raw materials in other processes and applications, such as in the cement industry and as refractory materials (Figure 12).

Figure 12. Various applications of aluminum salt slag.

4.3.1. Synthesis of Zeolite

Sánchez-Hernández et al. used hydrolyzed salt as a reactant along with Na_2SiO_3 and NaOH solutions to synthesize zeolite [101]. Their single-step method produced zeolite-type NaP ($Na_6Al_6Si_{10}O_{32}\cdot12H_2O$) by adding the corresponding amount of silicate solution. The produced zeolite exhibited characteristics similar to those of conventional chemical reagent–based forms. Jimenez et al. synthesized pollucite and analcime zeolites by extracting aluminum from saline slag [102]. The structures of the synthesized materials are based on $(Al,Si)O_4$ tetrahedra that share corners and feature pores, channels, and/or cavities at the molecular level. The extraction liquor contained aluminum and alkali metal cations. After adding the required amount of Si and undergoing hydrothermal synthesis at 200 °C for 24 h, zeolitic materials were produced (Figure 13 shows SEM images of this material). The crystallinity and water content of analcime were higher than those of pollucite. Morphological analyses revealed the formation of spherical particles, which are larger for analcime solids with pentagonal or polygonal faces. For pollucite, spheres are smaller, and it is possible that they are hollow.

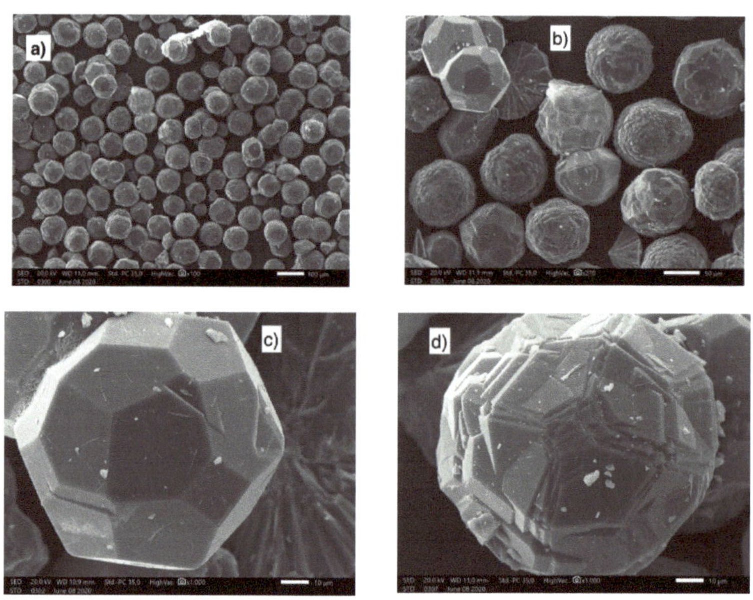

Figure 13. SEM images at different magnifications of ANA zeolites obtained from the aluminum waste [102]. (**a**,**b**) demonstrated the existence of particles with a diameter of less than 100 µm. (**c**,**d**) showed pentagonal faces zeolite.

4.3.2. Application in Construction Materials

Lin et al. explored the use of recycled aluminum salt slag (RASS) as aggregate combined with recycled concrete through the alkali-activation method for stone column materials [103]. Both materials were stabilized by fly ash and ground granulated blast furnace slag to improve soft soils. Increasing the RASS content decreased the unconfined compressive strength [104] of the mixtures; however, the mixture of recycled concrete aggregate with 5% alkali-activated fly ash and slag met the minimum unconfined compressive strength [104] requirement for ground improvement.

4.4. Spent Carbon Cathode

Spent carbon cathode is considered as hazardous solid wastes from aluminum production. Therefore, studies are carried out to investigate cost-effective and environmentally friendly purification techniques for separating carbon and subsequently using it efficiently. Most of the research is on metal recovery (Figure 14), which will be elaborated on in the next subsection. Additionally, the use of spent carbon cathode for other applications will also be discussed.

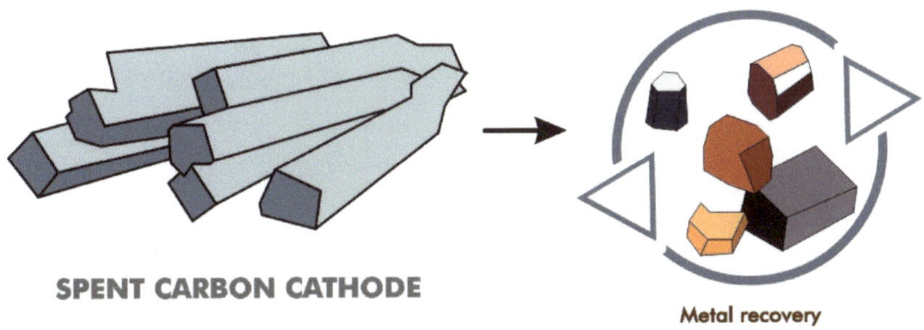

Figure 14. Recovery of various metal from spent carbon cathode.

4.4.1. Metal Recovery

Carbon recovery from spent carbon cathode is studied by several researchers. For example, Yuan et al. [105] synthesized silicon carbide with silicon dioxide at 1600 °C, giving the spent cathode carbon regeneration value, and obtained high-purity carbon powder through ultrasonic leaching. High-purity recovered carbon was also obtained from the research of Lu et al. [106] to be used directly as the anode for Na-ion batteries in order to benefit from these characteristics and reuse the high value-added carbon material. The capacitive nature of Na^+ in recovered carbon material allows the electrode to demonstrate notable electrochemical performance in the electrolyte solvent. However, there is a challenge of physical leaching treatments that produced low carbon extraction [107]. Other metals, such as iron and silicon, have been recovered from spent carbon cathodes for the treatment of red mud [108]. Fluoride can also be recovered from spent carbon cathodes. Robshaw et al. [109] used chelating resin loaded with lanthanum ions. The resin showed great promise in recovering fluoride from aqueous waste streams and demonstrated a large defluoridation capacity.

4.4.2. Other Applications

A hydrothermal acid-leaching technique was employed by Xiao et al. [110] to separate the various components in spent carbon cathode. While carbon and silicon remained in the filter residue, impurity elements like Al, F, and Fe entered the solution. Additionally, they used the carbothermal reduction method to create silicon carbide products. Spent carbon cathode has a highly expanded graphitic structure by nature [111]. This can be advantageous in preparing the anodes of lithium-ion batteries [112]. The anode material used was the purified graphitized spent carbon cathode which has better cycle stability. Carbon anodes for aluminum electrolysis were made by Yao et al. [113] using the recovered carbon material from roasting molten salt. It was performed better compared to carbon anodes made entirely of petroleum coke. This suggested that SCC might be utilized as raw materials to make carbon anodes for aluminum electrolysis. Zhi et al. [114] treated the spent carbon cathode with water washing process, and recovered large amount of sodium fluoride (NaF). Li et al. [115] investigated the thermal behavior of spent carbon cathode block blended with molten salts. It was discovered that NaCl-Na_2CO_3 binary molten salts were efficient in lowering the fluoride content of spent carbon cathode blocks and preserving the energy needed for the thermal treatment procedure.

4.5. Bauxite Residue

The proposed solution to bauxite residue stockpiling involves creating comprehensive use technology or transforming it into a secondary resource. Since the 1950s, multiple research projects have focused on the disposal and use of bauxite, given its distinctive physical and chemical characteristics, such as high pH and the presence of trace amounts of heavy metals, which can pose a serious environmental risk [116]. The production of cleaner bauxite residue is now a primary focus for the aluminum industry. The subsequent subsections will discuss the waste management of bauxite residue, neutralization of impurities, extraction of metal oxide and rare earth elements and the use of this residue in environmental, energy, coating, construction, and geopolymer applications.

4.5.1. Waste Management Methods

Wet processing, dry processing, and semi-dry processing are the three standard methods for disposing of bauxite residue in landfills [117–119]. During wet processing, bauxite residue is first washed in a thickener to facilitate the sedimentation of solids and produce a residue slurry with a solid content of 15% to 30%. The resulting slurry is then pumped into a storage yard comprising earthworks, tailing ponds, and dams to allow the solids to settle. The liquid supernatant in the storage yard is collected and used to generate recovered caustic soda [120]. However, the disposal of the dilute residue slurry after wet processing poses environmental risks to nearby areas/communities because of the potential release of caustic and highly alkaline (pH >12) liquid into the subsoil water [121].

The dry-stacking method and dry cake disposal have been adopted by many aluminum production plants because of their reduced spatial footprint and lower potential for leaks, as landfills are becoming scarcer and space-limited [122]. In this process, the bauxite residue is pre-thickened to produce a residue slurry with a solid content of over 50%, which is then deposited in layers ranging in thickness from 0.4 to 0.7 m. Drum filters or plate and frame filter presses are used to achieve the high solid content of the residue slurry during filtration. Bulldozers turn up the bottom layer to expedite drying and minimize dust emissions. Successive layers of residue slurry are placed on top to expand the spatial storage capacity, thereby reducing the landfill space occupied by bauxite residues [123].

Semi-dry processing refers to the co-disposal of bauxite residue slurry generated by the Bayer and sintering processes. The residue typically exhibits a low permeability coefficient and high shear yield strength because of its hydration and hardening capabilities [124]. Thus, the dried residue from the sintering process can directly replace clay in the construction of dams.

4.5.2. Neutralization of Bauxite Residues

Efforts have been made to partially neutralize bauxite residues to mitigate the environmental impact of their high alkalinity. These approaches include seawater neutralization for plants near the sea, carbonation treatment, acid neutralization, sintering, and bioleaching. The addition of seawater to caustic red mud decreases the pH, leading to the precipitation of minerals such as hydroxycarbonate, carbonate, and hydroxide [125]. Seawater neutralization converts readily soluble, strongly caustic wastes into less soluble, weakly alkaline solids, although it does not completely eliminate hydroxide. The waste's carbonate and bicarbonate alkalinity are primarily removed through reactions of hydroxide with calcium to form aragonite and calcite [126]. Neutralization is considered complete once the liquid from the treated bauxite residue has a pH below 9.0 and a total alkalinity of less than 200 mg/L (measured as calcium carbonate equivalent alkalinity), allowing the decantation of seawater-neutralized bauxite residue to be safely discharged into the marine environment [127].

Red mud neutralization by carbon dioxide gas has been studied [104]. Gaseous CO_2 or flue gas containing CO_3^{2-} is bubbled through aqueous slurries to form carbonic acid in the aqueous phase [128]. This carbonic acid reacts with the basic components of red mud, thereby lowering its pH. However, only a fraction of the alkaline material in red mud is neutralized by gaseous CO_2 within the short contact times necessitated by industrial process rates. Consequently, although the pH of the aqueous phase decreases rapidly upon exposure to CO_2, it quickly returns to unacceptable levels as more alkaline material leaches from the mud. Given that the pH of water exposed to gaseous CO_2 is unlikely to fall below approximately 5.5, the neutralization rate of the solids in the aqueous slurry typically does not meet industrial requirements. Researchers have, therefore, explored the use of high-pressure liquid CO_2 instead of its vapor phase for more effectively reducing the pH of red mud [129], including the approach of neutralizing red mud with CO_2 over multiple cycles [130].

A variety of aqueous acidic solutions, including acidic industrial wastewater, have been considered for neutralizing residual alkalinity. Several studies have assessed the feasibility of treating bauxite residues, such as the Kwinana red mud slurry, with acid [131]. However, the complete neutralization of the residue requires large volumes of reagent, even when using spent (waste) acid, leading to relatively high costs. Additionally, acid treatment introduces significant quantities of impurities into the process's water stream, such as sulfate in the case of sulfuric acid and chloride in the case of hydrochloric acid. Therefore, reintroducing any water from the residue deposits back into the production process is likely to be unacceptable without further treatments to eliminate these added impurities. The treatment of red mud with acidic spent pickling solutions (SPSs) derived from the steelmaking process produces a coagulant—a mixture of aluminum and iron salts—for wastewater treatment. Although sintering the residue can control all leachable soda, this process can be prohibitively expensive given the increased energy consumption required for high-temperature sintering of red mud. Alcoa of Australia has implemented a bioremediation of bauxite residue in Western Australia by adding organic substrate to the red mud to cultivate microorganisms. These microorganisms produce various organic acids and, in some cases, CO_2, which neutralizes the mud [131].

Efforts to find viable solutions for bauxite residue disposal continue, including efforts to improve storage, monitoring, and safety standards to reduce environmental impacts, develop techniques to enhance storage yard reclamation potential, and promote reusability methods to decrease the volume of accumulated bauxite residue [132]. Despite these efforts, red mud still occupies a significant amount of landfill space, and sustainable waste management strategies are needed to use red mud as a secondary resource material.

4.5.3. Extraction of Metal Oxides from Bauxite Residue

Bauxite residue primarily consists of Fe_2O_3, Al_2O_3, SiO_2, TiO_2, Na_2O, and CaO (Table 1). Various methods exist for extracting these oxides. The high Na_2O content limits the potential application of red mud for iron-containing products, as it is not easily soluble in water, resulting in high alkalinity. However, the hydrochemical conversion of goethite to magnetite has been proposed as a solution to this issue [133]. High-iron bauxite residue leaching can occur in the presence of Fe^{2+} (Figure 15), as this ion facilitates the extraction of Al from Al-hematite and Al-goethite via the precipitation of magnetite from the solution. The simultaneous extraction of Al and Na results in a product containing more than 96% Fe_3O_4.

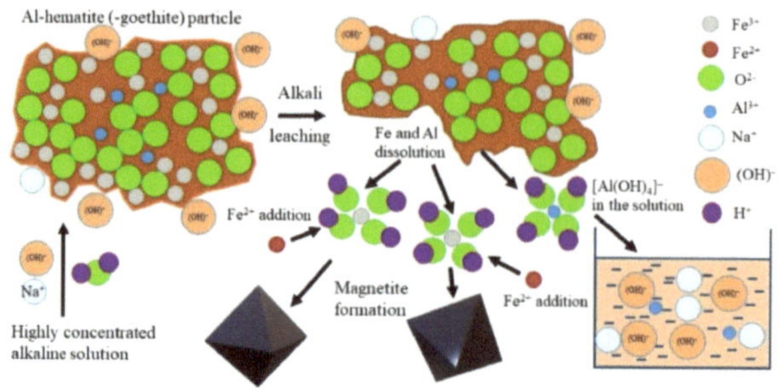

Figure 15. The schematic mechanism of high-iron bauxite residue leaching by a highly concentrated NaOH solution in the presence of Fe^{2+} [133].

Li et al. proposed a method for extracting Al_2O_3 and TiO_2 from red mud smelting separation slag through mineral phase reconstruction conducted under an air atmosphere [134]. Al_2O_3 and SiO_2 are converted into alkaline-soluble $NaAlO_2$ and Ca_2SiO_4 by alkaline hydroxide roasting. The resulting $NaAlO_2$ solution can serve as a source for extracting alumina, with more than 80% of Al_2O_3 selectively dissolved in a 95 °C NaOH leaching solution. Subsequently, SiO_2 is extracted from the residue using an HCl solution, yielding a solution containing SiO_2 and a concentrated residue of undissolved CaO and TiO_2. Hydrochloric acid leaching was used to recover the latter.

A novel process for the coupled treatment of coal gangue and red mud has also been developed [135]. Guo et al. demonstrated that red mud can adjust the Al/Si molar ratio of coal gangue to an appropriate level and that the Na_2O in the red mud can partially substitute for the Na_2CO_3 required for coal gangue activation, thereby reducing the amount of Na_2CO_3 consumed from 100% to 12.1–20.5%.

4.5.4. Extraction of Rare Earth Metals from Red Mud

In addition to the extraction of metal oxides from red mud, several rare earth elements (REEs) may also be extracted [136]. Small amounts of Sc, Y, La, Ce, Pr, Nd, Sm, Eu, Gd, Tb, Dy, Ho, Er, Tm, Yb, and Lu are found in red muds [137]. Red mud has been identified as one of the promising sources of scandium (Sc) [138]. Dai et al. developed a low-cost and environmentally friendly biosorbent for Sc extraction [139]. In their study, H_3PO_4-activated biochar (P40s) produced from pitaya peels was used for Sc adsorption and recovery from red mud. This biochar demonstrated optimal reusability after five adsorption/desorption cycles. The extraction of Al and REEs from red mud via aerobic and anaerobic bi-stage bioleaching has also been explored [140] by Zhang et al., which achieved the highest extraction rates for Al, Ce, Y, and Sc (Figure 16). Thus, the release of valuable elements from bioleached red mud residue can be enhanced by bio-oxidation under anaerobic conditions.

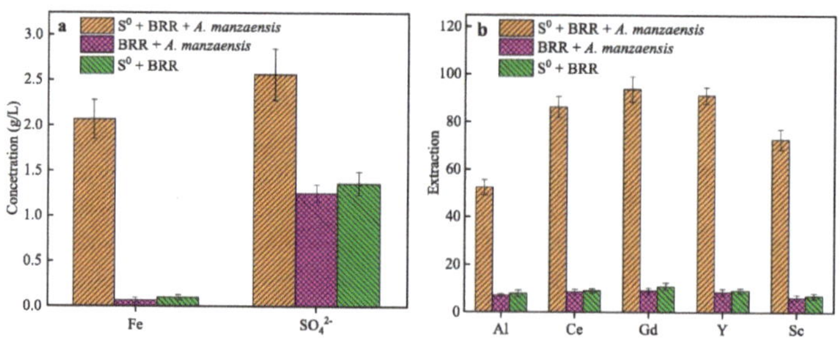

Figure 16. (**a**) [Fe^{2+}] and [SO_4^{2-}] concentrations and (**b**) the extraction rates of Al, Ce, Gd, Y, and Sc during the anaerobic bioleaching of red mud using *Acidianus manzaensis* [140].

Roasting can also remove calcium and obtain rare earth element (REE) oxides from red mud; this process is followed by the selective mixing of sulfuric acid with REE oxides. Calcium, aluminum, and iron components can also be recovered at each stage [141].

4.5.5. Bauxite Residues for Environmental Applications

Carbon capture and storage (CCS) technologies are a significant area of research aimed at reducing greenhouse gas emissions. Direct mineral carbonation is a CCS technology that offers the benefits of permanence, affordability, and promising on-site options, while also enhancing the environmental quality of waste. In recent years, red mud has been explored for gas cleaning treatments.

The use of red mud in reducing concentrations of residual antibiotics in composting has been investigated [142]. Composting has been found to decrease the concentrations of most residual antibiotics, as well as the abundance of antibiotic resistance genes and mobile genetic elements. In another application, Li et al. synthesized zero-valent iron (ZVI) biochar composites from red mud and ginkgo leaves [143]. Ginkgo leaves contain active groups that have a strong adsorption effect on heavy metal ions [144]. Iron phases in red mud were transformed from Fe_2O_3 to Fe^0 after pyrolysis at 800 °C with a red mud:ginkgo leaf mass ratio of 2:1; the iron phase and ketonic functional groups (C=O) were identified as two key active sites (Figure 17). The synthesized composites were effective in removing bisulfate and acid orange 7.

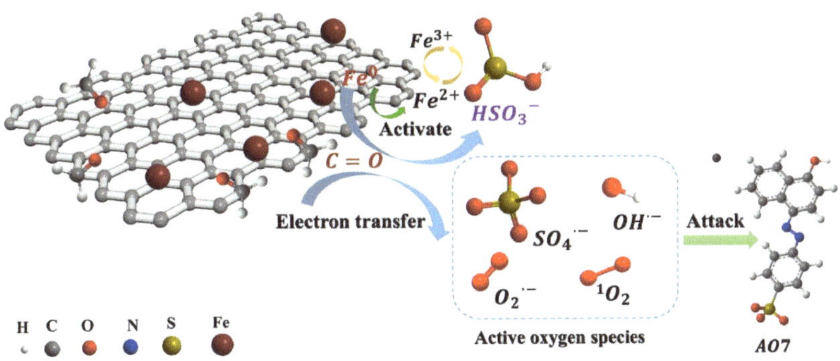

Figure 17. Acid orange 7 (AO7) degradation in the biochar composites [143]; both radical and nonradical processes broke down the adsorbed AO7 molecules. Notably, SO_4^{2-} played a significant part in the degradation process.

Bai et al. proposed the use of red mud for purifying heavy metal pollutants in water [145]. The adsorption strength for Pb, Cd, and Cu can be enhanced by increasing the liquid temperature. Guo et al. synthesized biochar-supported red mud catalysts for the degradation of COVID 19–related drugs such as arbidol and acyclovir [146]. The presence of functional groups in red mud catalysts, which contain oxygen and Fe species (Fe^0 and Fe_3O_4) as well as Ca^{2+} ions, was crucial for the removal of arbidol.

The efficient treatment of industrial oily wastewater before discharge into receiving bodies is critical for preventing environmental pollution and protecting human health. Membranes with superwettable surfaces have been regarded as promising materials for the separation of oil/water emulsions [147]. These surfaces include nitrocellulose membranes coated with red mud (Figure 18).

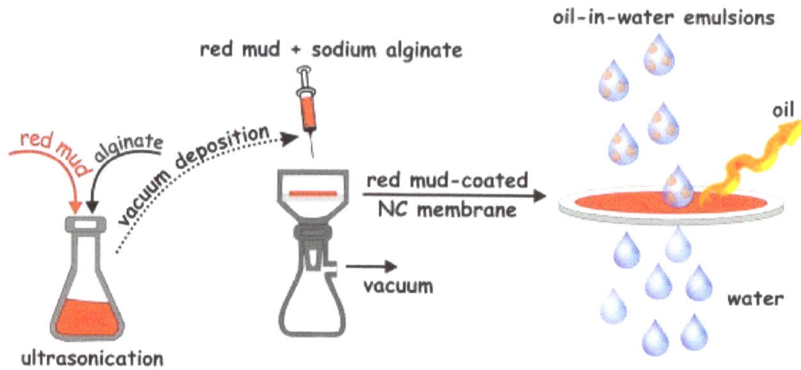

Figure 18. Schematic illustration of the fabrication of red-mud-coated-nitrocellulose (NC) membrane for oil-in-water emulsion separation [147].

4.5.6. Bauxite Residue as Construction Material

Portland cement is the most widely used type of cement for general construction purposes. Most CO_2 emissions from the cement industry occur through the process of decarbonizing limestone during the creation of clinker. Given the rising demand for infrastructure and sanitation, the primary method for reducing cement use (and thus greenhouse gas emissions) in the civil construction sector involves the partial or complete replacement of Portland cement clinker. Several studies have explored producing various types of cement from red mud [148,149]. The moderate amounts of SiO_2 and Al_2O_3 in red mud allow for its potential use as a material for mortar, concrete, bricks, aggregates, and roofing. Zeng et al. discussed the production of a permeable red mud–based brick using a fully automatic production line for ecological RM permeable bricks (Figure 19) [150]. The company's daily output was 3000 m^2, and its production process could digest nearly 100,000 t of RM annually. The brick exhibited a bending strength of 6–8 MPa, exceeding the national standard of 3.2 MPa, with an RM content of up to 80%.

Figure 19. Red mud–based permeable brick produced in Shandong Province, China [150].

Wang et al. incorporated red mud into road cement clinker [151]. For developing Portland cement for road clinker, they first de-alkalized the sintering red mud and then combined it with materials such as clay, fly ash, limestone, and sandstone. Each group of raw materials containing red mud exhibited a lower thermal decomposition temperature than the control materials; this reduction in decomposition temperature related to the accelerated carbonate decomposition process of the raw materials for roads facilitated by the addition of red mud.

The structural characterization of red mud in roof tile manufacturing has also been investigated [152]. Increases in both firing temperature and RM content enhance the compressive strength of fired samples. Ma et al. studied modified Bayer red mud for use as a highway subgrade and assessed its performance in a large-scale application as part of a road trial for the Ji-Qing Highway Reconstruction and Expansion Project in China [153]. The emplaced subgrade spanned approximately 4900 m in length, 11 m in width, and 0.2 m in thickness (Figure 20). The modified red mud replaced the standard 0.2 m of lime-stabilized soil in the upper part of the roadbed. This modified Bayer red mud showed significant improvements in road performance. The pH of the leaching solution decreased, and the compressive strength reached up to 3 MPa within three days. Additionally, the concentration of the leached harmful components decreased by more than 70% relative to the pretreated red mud.

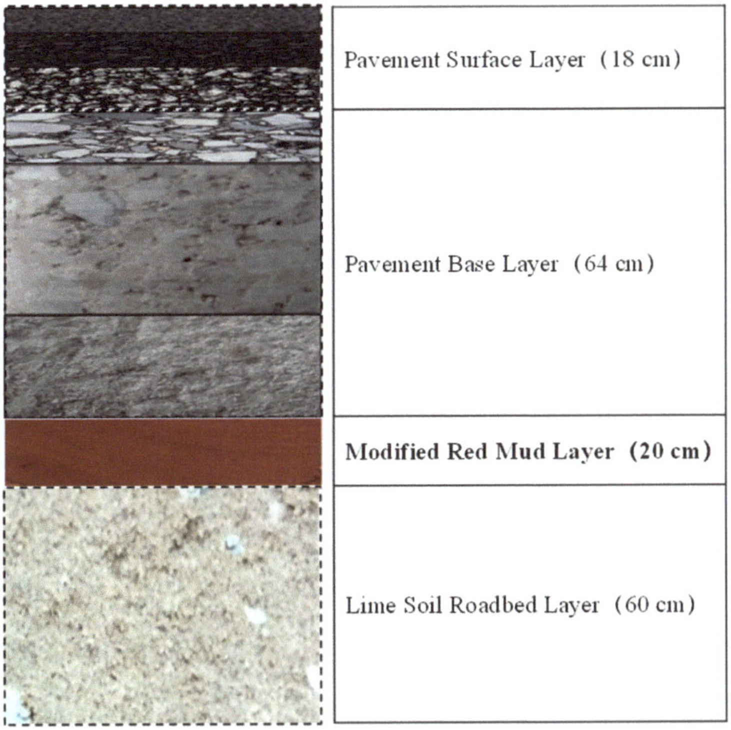

Figure 20. Pavement structural diagram of a demonstration road in Jinan Qingdao Highway, China [153].

Iron and other high-atomic-number (Z) waste materials, such as steel slag and red mud, can be used to create X- and γ-ray shielding concrete [154,155]. Human skin itself blocks alpha particles, whereas beta radiation is stopped by thin sheets of wood or aluminum (Figure 21). The most challenging forms of radiation to attenuate are highly penetrative and dangerous uncharged gamma and neutron radiation.

Aside from concrete, lead is often used to shield against high-energy photons given its high atomic number and density [156]. The challenge with radiation shielding in concrete is predicting radiation attenuation because of variable moisture within the concrete. Although commonly used, lead poses significant health and environmental risks, and its heterogeneous nature complicates predictions of its effectiveness as a radiation barrier [157,158].

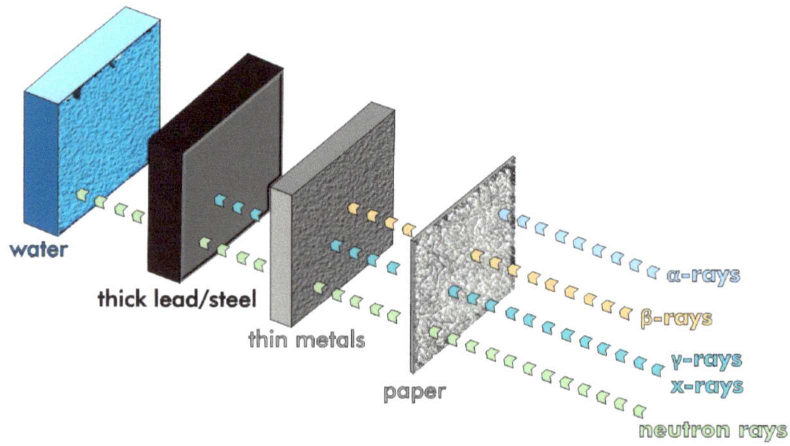

Figure 21. The comparative penetrative power of various radiation sources.

Agrawal et al. fabricated diagnostic X-ray shielding tiles from red mud and barium sulfate ($BaSO_4$) with the addition of sodium hexametaphosphate (SHMP) and kaolin clay to enhance their mechanical strength [159]. The radiation attenuation of the tiles, measuring 10.3 mm and 14.7 mm in thickness, was equivalent to that of 2 mm and 2.3 mm of lead at 125 and 140 kilovoltage peak (kVp), respectively. These tiles are already being used in bone densitometry and catheterization laboratories.

4.5.7. Bauxite Residue for Geopolymers

Geopolymer is an inorganic polymer material composed of aluminosilicate and an alkali activator. The sources of Al_2O_3 and SiO_2 can be derived from natural minerals such as metakaolin and clay, given their abundance in many regions of the world [160]. The drive for more eco-friendly materials has spurred interest in using this metal oxide-rich industrial waste. Such residual materials include fly ash, silica fume, blast furnace slag, and air-cooled slag [161,162]. Gonçalves et al. produced 3D-printed red mud/metakaolin-based geopolymers as methylene blue sorbents from water pollutants (Figure 22) [163]. Lattices with approximately 62% open porosity were achieved, showing no significant adsorption loss after 10 cycles of operation. These geopolymer composites exhibited low leachability with regard to contaminants, especially those with notable compressive strength. The development of geopolymer-based protective coverings has recently garnered increased attention. Fly ash–red mud geopolymers have been explored as coating materials for mild steel [164], with the developed material providing excellent corrosion protection, thereby helping to extend the service life of structures made of mild steel.

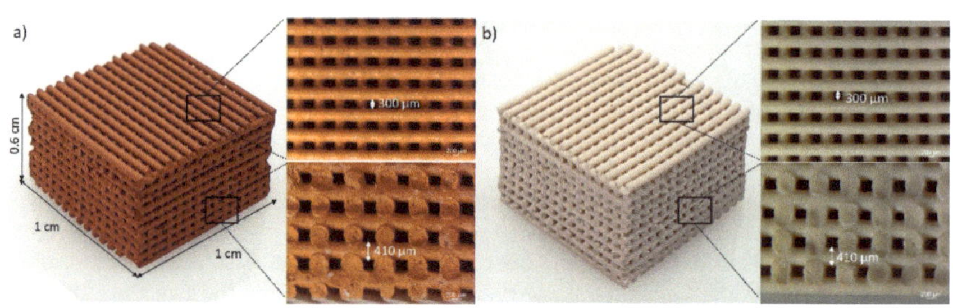

Figure 22. Optical micrographs of 3D printed red mud with (**a**) 50% red mud and (**b**) 100% metakaolin [163].

Xu et al. investigated the effects of nano-SiO_2 (NS) and steel fiber on the properties of red mud–based geopolymer concrete [165]. They found that steel fibers and nano-SiO_2 significantly enhanced the performance of the red mud geopolymer composite, with the most substantial improvements observed when both nano-SiO_2 (NS) and steel fiber were combined. The interface transition zone of the NS-reinforced fiber matrix is depicted in Figure 23.

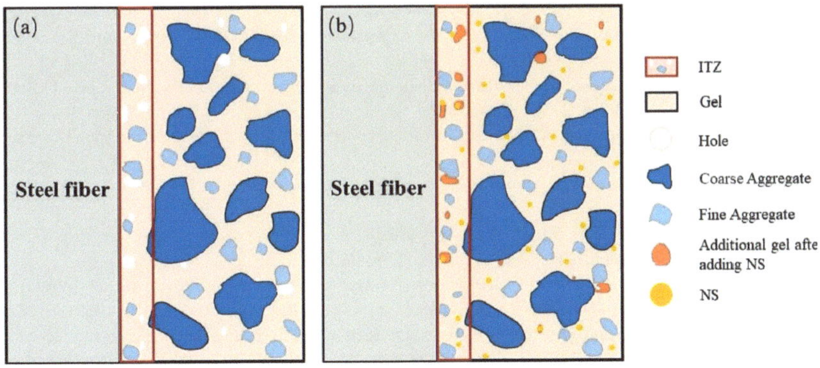

Figure 23. A nano-SiO_2-reinforced fiber matrix interfacial transition zone (ITZ); (**a**) without nano-SiO_2 (NS) and (**b**) with NS [165].

The thermal-related performance of PCM (phase change material) geopolymer composites has also been evaluated [166] by Afolabi et al., which used expanded graphite (EG) to encapsulate PCMs as a composite, then vacuum-impregnated the composite into geopolymer aggregates to create an enhanced thermal wall material for use in buildings and construction. The outdoor thermal storage performance of the EG/PCM composite geopolymer wall material exceeded that of conventional materials such as cement, clay, and gypsum.

5. Conclusions

Despite recent technological advances, the disposal of aluminum dross, salt slag, and bauxite residue remains a challenge for the aluminum industry, with the problems with resource utilization of these residues and critical issues continuing to impact the feasibility of recycling and the economics of the process. Multiple applications must be implemented to reduce the volumes of generated waste. In this review, we discussed the use of these residues in several construction products, including their utility as materials for cement and concrete, as supplemental additives to Portland cement, and as geopolymers to replace cement materials. The valorization of this industrial waste requires controlled processing parameters to ensure that the finished products meet existing application standards. The effectiveness of aluminum industrial waste in new applications must be comparable to alternatives, such as steel and energy plant industrial waste, in terms of quality, price, and environmental risk. The aluminum industry must collaborate with aluminum associations and research institutions to develop and promote the potential use of residues and their large-scale reuse in a variety of applications.

Funding: This work was suppoeted by the Volet Soutien à la diffusion des travaux de recherche et de création (PSDrc 2024–2025) UQAC.

Data Availability Statement: The data presented in this study are available on request from the corresponding author.

Acknowledgments: Authors thank Rio Tinto Saguenay for their support for this study.

Conflicts of Interest: The authors declare no conflicts of interest.

References

1. Davis, J.R. *Alloying: Understanding the Basics*; ASM International: Materials Park, OH, USA, 2001. [CrossRef]
2. Aluminium. Available online: http://www.rsc.org/periodic-table/element/13/aluminium (accessed on 14 March 2023).
3. Kvande, H. The Aluminum Smelting Process. *J. Occup. Environ. Med.* **2014**, *56*, S2–S4. [CrossRef] [PubMed]
4. Schlesinger, M.E. *Aluminum Recycling*, 2nd ed.; CRC Press: Boca Raton, FL, USA, 2013. [CrossRef]
5. Fridlyander, I.N.; Sister, V.G.; Grushko, O.E.; Berstenev, V.V.; Sheveleva, L.M.; Ivanova, L.A. Aluminum Alloys: Promising Materials in the Automotive Industry. *Met. Sci. Heat Treat.* **2002**, *44*, 365–370. [CrossRef]
6. Purchase, C.K.; Al Zulayq, D.M.; O'Brien, B.T.; Kowalewski, M.J.; Berenjian, A.; Tarighaleslami, A.H.; Seifan, M. Circular Economy of Construction and Demolition Waste: A Literature Review on Lessons, Challenges, and Benefits. *Materials* **2022**, *15*, 76. [CrossRef] [PubMed]
7. Rudnick, R.L.; Gao, S. 4.1—Composition of the Continental Crust. In *Treatise on Geochemistry*, 2nd ed.; Holland, H.D., Turekian, K.K., Eds.; Elsevier: Oxford, UK, 2014; pp. 1–51. [CrossRef]
8. Kvande, H. Occurrence and Production of Aluminum. In *Encyclopedia of Inorganic and Bioinorganic Chemistry*; Wiley Online Library: Hoboken, NJ, USA, 2015. [CrossRef]
9. Kamble, P.H.; Bhosale, S.M. Environmental Impact of Bauxite Mining: A Review. *Int. J. Res. Appl. Sci. Eng. Technol.* **2019**, *7*, 86–90. [CrossRef]
10. Kloprogge, J.T.; Duong, L.V.; Wood, B.J.; Frost, R.L. XPS study of the major minerals in bauxite: Gibbsite, bayerite and (pseudo-)boehmite. *J. Colloid Interface Sci.* **2006**, *296*, 572–576. [CrossRef]
11. Hamed, M.Y.; Silver, J.; Wilson, M.T. Studies on the reactions of ferric iron with glutathione and some related thiols. Part III. A study of the iron catalyzed oxidation of glutathione by molecular oxygen. *Inorganica Chim. Acta* **1983**, *80*, 237–244. [CrossRef]
12. Investing News Network. *Top 10 Aluminum-Producing Countries (Updated 2024)*; Investing News Network: Toronto, ON, Canada, 2024.
13. China's Aluminium Industry Now Has a Bauxite Tightness Headache. Available online: https://www.kpler.com/blog/chinas-aluminium-industry-now-has-a-bauxite-tightness-headache (accessed on 17 October 2024).
14. Leading Countries Based on the Production of Alumina Worldwide in 2023. 2024. Available online: https://www.statista.com/statistics/264963/global-alumina-production-by-country/ (accessed on 17 October 2024).
15. Vedanta Aluminium Expands Alumina Refining Capacity To 3.5 Million Tonnes per Annum (MTPA). Available online: https://vedantaaluminium.com/media/press-releases/list/vedanta-aluminium-expands-alumina-refining-capacity-to-3-5-million-tonnes-per-annum-mtpa/ (accessed on 17 October 2024).
16. Zhao, R.; Nowicki, C.; Gosselin, L.; Duchesne, C. Energy and exergy inventory in aluminum smelter from a thermal integration point-of-view. *Int. J. Energy Res.* **2016**, *40*, 1321–1338. [CrossRef]
17. Cullen, J.M.; Allwood, J.M. Mapping the Global Flow of Aluminum: From Liquid Aluminum to End-Use Goods. *Environ. Sci. Technol.* **2013**, *47*, 3057–3064. [CrossRef]
18. Rodionova, I.A. The shifts in the spatial structure of the world bauxite industry and Guinea's position in the industry. *Rev. ESPACIOS* **2020**, *41*, 11–21.
19. Swain, B.; Lee, C.G.; Park, J.R. Assessment of bauxite residue as secondary resource for rare earth metal and valorization challenges: A perspective. *Resour. Conserv. Recycl. Adv.* **2022**, *14*, 200078. [CrossRef]
20. Samantray, J.; Anand, A.; Dash, B.; Ghosh, M.K.; Behera, A.K. *Nepheline Syenite—An Alternative Source for Potassium and Aluminium*; Springer International Publishing: Cham, Switzerland, 2019.
21. Bagani, M.; Balomenos, E.; Panias, D. Nepheline Syenite as an Alternative Source for Aluminum Production. *Minerals* **2021**, *11*, 734. [CrossRef]
22. Terry, B. The acid decomposition of silicate minerals part I. Reactivities and modes of dissolution of silicates. *Hydrometallurgy* **1983**, *10*, 135–150. [CrossRef]
23. Haddad, A.Z.; Pilgrim, C.D.; Sawvel, A.M.; Hohman, J.N.; Gadgil, A.J. On the Conversion of Bauxite Ores to Highly Activated Alumina Media for Water Remediation. *Adv. Sustain. Syst.* **2019**, *3*, 1900005. [CrossRef]
24. BenSaad, L.; Hwi, K.K.; AlSereti, M.; Shahimi, M. Evaluation of the analgesic effects of Libyan fresh pomegranate fruit of Punica granatum. *PharmaNutrition* **2014**, *2*, 108–109. [CrossRef]
25. Kaußen, F.M.; Friedrich, B. Methods for Alkaline Recovery of Aluminum from Bauxite Residue. *J. Sustain. Metall.* **2016**, *2*, 353–364. [CrossRef]
26. Donoghue, A.M.; Frisch, N.; Olney, D. Bauxite mining and alumina refining: Process description and occupational health risks. *J. Occup. Environ. Med.* **2014**, *56* (Suppl. S5), S12–S17. [CrossRef] [PubMed]
27. Farjana, S.H.; Huda, N.; Mahmud, M.A.P. Impacts of aluminum production: A cradle to gate investigation using life-cycle assessment. *Sci. Total Environ.* **2019**, *663*, 958–970. [CrossRef]
28. Crippa, M.; Solazzo, E.; Guizzardi, D.; Monforti-Ferrario, F.; Tubiello, F.N.; Leip, A. Food systems are responsible for a third of global anthropogenic GHG emissions. *Nat. Food* **2021**, *2*, 198–209. [CrossRef]
29. Santamaría, L.; Korili, S.A.; Gil, A. Layered double hydroxides from slags: Closing the loop. *J. Environ. Chem. Eng.* **2022**, *10*, 106948. [CrossRef]
30. Peterson, R.D.; Newton, L. Review of Aluminium Dross Processing. In *Light Metals*; TMS-AIME: Warrendale, PA, USA, 2002; p. 1029.

31. Maung, K.N.; Yoshida, T.; Liu, G.; Lwin, C.M.; Muller, D.B.; Hashimoto, S. Assessment of secondary aluminum reserves of nations. *Resour. Conserv. Recycl.* **2017**, *126*, 34–41. [CrossRef]
32. Mahinroosta, M.; Allahverdi, A. Hazardous aluminum dross characterization and recycling strategies: A critical review. *J. Environ. Manag.* **2018**, *223*, 452–468. [CrossRef] [PubMed]
33. Wibner, S.; Antrekowitsch, H.; Meisel, T.C. Studies on the Formation and Processing of Aluminium Dross with Particular Focus on Special Metals. *Metals* **2021**, *11*, 1108. [CrossRef]
34. Environmental Protection Agency An Gníomhaireacht um Chaomhnú Comhshaoil. European Waste Catalogue and Hazardous Waste List. Available online: https://v4r7y5k5.stackpathcdn.com/wp-content/uploads/2016/08/EWC_HWL.pdf (accessed on 14 March 2023).
35. Zhang, L. *State of the Art in Aluminum Recycling from Aluminum Dross;* Light Metals: Proceedings of Sessions; TMS Annual Meeting: San Antonio, TX, USA, 2006; Volume 931.
36. Das, B.R.; Dash, B.; Tripathy, B.C.; Bhattacharya, I.N.; Das, S.C. Production of η-alumina from waste aluminium dross. *Miner. Eng.* **2007**, *20*, 252–258. [CrossRef]
37. Lazzaro, G.; Eltrudis, M.; Pranovi, F. Recycling of aluminium dross in electrolytic pots. *Resour. Conserv. Recycl.* **1994**, *10*, 153–159. [CrossRef]
38. Stark, T.D.; Martin, J.W.; Gerbasi, G.T.; Thalhamer, T.; Gortner, R.E. Aluminum Waste Reaction Indicators in a Municipal Solid Waste Landfill. *J. Geotech. Geoenvironmental Eng.* **2012**, *138*, 252–261. [CrossRef]
39. Abdulkadir, A.; Ajayi, A.; Hassan, M.I. Evaluating the Chemical Composition and the Molar Heat Capacities of a white Aluminum Dross. *Energy Procedia* **2015**, *75*, 2099–2105. [CrossRef]
40. Hwang, J.; Huang, X.; Xu, Z. Recovery of Metals from Aluminum Dross and Saltcake. *J. Miner. Mater. Charact. Eng.* **2006**, *5*, 47–62. [CrossRef]
41. Zhou, L.; Yao, Z.; Sun, K.; Tian, Z.; Li, J.; Zhong, Q. Methodological Review of Methods and Technology for Utilization of Spent Carbon Cathode in Aluminum Electrolysis. *Energies* **2024**, *17*, 4866. [CrossRef]
42. Gao, L.; Mostaghel, S.; Ray, S.; Chattopadyay, K. Using SPL (spent pot-lining) as an alternative fuel in metallurgical furnaces. *Metall. Mater. Trans. E* **2016**, *3*, 179–188.
43. Yong, Y.; Jianhang, H.; Yongkui, L.; Dapeng, Z.; Hua, W. A new method for simultaneous separation and solidification of arsenic from arsenic-bearing gypsum sludge using waste carbon cathodes. *Sep. Purif. Technol.* **2022**, *291*, 120656. [CrossRef]
44. Lisbona, D.F.; Somerfield, C.; Steel, K.M. Leaching of spent pot-lining with aluminium nitrate and nitric acid: Effect of reaction conditions and thermodynamic modelling of solution speciation. *Hydrometallurgy* **2013**, *134–135*, 132–143. [CrossRef]
45. Andrade-Vieira, L.; Davide, L.; Gedraite, L.; Campos, J.; Azevedo, H. Genotoxicity of SPL (spent pot lining) as measured by Tradescantia bioassays. *Ecotoxicol. Environ. Saf.* **2011**, *74*, 2065–2069. [CrossRef] [PubMed]
46. Xie, M.; Li, R.; Zhao, H.; Liu, W.; Lu, T.; Liu, F. Detoxification of spent cathode carbon blocks from aluminum smelters by joint controlling temperature-vacuum process. *J. Clean. Prod.* **2020**, *249*, 119370. [CrossRef]
47. Yang, K.; Gong, P.; Tian, Z.; Lai, Y.; Li, J. Recycling spent carbon cathode by a roasting method and its application in Li-ion batteries anodes. *J. Clean. Prod.* **2020**, *261*, 121090. [CrossRef]
48. Zhou, H.; Zhang, D.; Jiang, Y.; Zeng, B.; Zhao, C.; Zhang, M.; Zeng, B.; Zhu, X.; Su, X.; Romanovski, V.; et al. Recovery of carbon from spent carbon cathode by alkaline and acid leaching and thermal treatment and exploration of its application in lithium-ion batteries. *Environ. Sci. Pollut. Res.* **2023**, *30*, 114327–114335. [CrossRef]
49. Yang, K.; Gong, P.; Tian, Z.; Peng, K.; Lai, Y. Carbon recovered from spent carbon cathode of aluminum reduction cell towards its valorisation as negative electrodes for lithium ion batteries. *Diam. Relat. Mater.* **2020**, *109*, 108062. [CrossRef]
50. Wang, C.; Mao, S.; Li, L. Study on ultrasonic leaching and recovery of fluoride from spent cathode carbon of aluminum electrolysis. *RSC Adv.* **2023**, *13*, 16300–16310. [CrossRef]
51. Wehr, J.B.; Fulton, I.; Menzies, N.W. Revegetation strategies for bauxite refinery residue: A case study of Alcan Gove in Northern Territory, Australia. *Environ. Manag.* **2006**, *37*, 297–306. [CrossRef] [PubMed]
52. Snars, K.; Gilkes, R.J. Evaluation of bauxite residues (red muds) of different origins for environmental applications. *Appl. Clay Sci.* **2009**, *46*, 13–20. [CrossRef]
53. Cablik, V. Characterization and applications of red mud from bauxite processing. *Gospod. Surowcami Miner. -Miner. Resour. Manag.* **2007**, *23*, 27–38.
54. Vachon, P.; Tyagi, R.D.; Auclair, J.C.; Wilkinson, K.J. Chemical and biological leaching of aluminum from red mud. *Environ. Sci. Technol.* **1994**, *28*, 26–30. [CrossRef] [PubMed]
55. Wang, L.; Chen, L.; Tsang, D.C.W.; Zhou, Y.; Rinklebe, J.; Song, H.; Kwon, E.E.; Baek, K.; Sik Ok, Y. Mechanistic insights into red mud, blast furnace slag, or metakaolin-assisted stabilization/solidification of arsenic-contaminated sediment. *Environ. Int.* **2019**, *133*, 105247. [CrossRef] [PubMed]
56. Rivera, R.M.; Ulenaers, B.; Ounoughene, G.; Binnemans, K.; Van Gerven, T. Extraction of rare earths from bauxite residue (red mud) by dry digestion followed by water leaching. *Miner. Eng.* **2018**, *119*, 82–92. [CrossRef]
57. Mohapatra, B.; Rao, M.; Rao, R.B.; Paul, A. Charactertstics of red mud generated at NALCO refinery, Damanjodi, India. *Light Met.* **2000**, *2000*, 161–165.
58. Borra, C.R.; Blanpain, B.; Pontikes, Y.; Binnemans, K.; Van Gerven, T. Recovery of Rare Earths and Major Metals from Bauxite Residue (Red Mud) by Alkali Roasting, Smelting, and Leaching. *J. Sustain. Metall.* **2017**, *3*, 393–404. [CrossRef]

59. Agrawal, S.; Rayapudi, V.; Dhawan, N. Comparison of microwave and conventional carbothermal reduction of red mud for recovery of iron values. *Miner. Eng.* **2019**, *132*, 202–210. [CrossRef]
60. Harmaji, A.; Sunendar, B. Utilization of Fly Ash, Red Mud, and Electric Arc Furnace Dust Slag for Geopolymer. *Mater. Sci. Forum* **2016**, *841*, 157–161. [CrossRef]
61. Ramdhani, E.P.; Wahyuni, T.; Ni'mah, Y.L.; Suprapto, S.; Prasetyoko, D. Extraction of Alumina from Red Mud for Synthesis of Mesoporous Alumina by Adding CTABr as Mesoporous Directing Agent. *Indones. J. Chem.* **2018**, *18*, 337–343. [CrossRef]
62. Deihimi, N.; Irannajad, M.; Rezai, B. Characterization studies of red mud modification processes as adsorbent for enhancing ferricyanide removal. *J. Environ. Manag.* **2018**, *206*, 266–275. [CrossRef]
63. Xenidis, A.; Harokopou, A.D.; Mylona, E.; Brofas, G. Modifying alumina red mud to support a revegetation cover. *JOM* **2005**, *57*, 42–46. [CrossRef]
64. Sglavo, V.M.; Campostrini, R.; Maurina, S.; Carturan, G.; Monagheddu, M.; Budroni, G.; Cocco, G. Bauxite 'red mud' in the ceramic industry. Part 1: Thermal behaviour. *J. Eur. Ceram. Soc.* **2000**, *20*, 235–244. [CrossRef]
65. Boyarintsev, A.V.; Aung, H.Y.; Stepanov, S.I.; Shoustikov, A.A.; Ivanov, P.I.; Giganov, V.G. Evaluation of Main Factors for Improvement of the Scandium Leaching Process from Russian Bauxite Residue (Red Mud) in Carbonate Media. *ACS Omega* **2022**, *7*, 259–273. [CrossRef] [PubMed]
66. Chaikin, L.; Shoppert, A.; Valeev, D.; Loginova, I.; Napol'skikh, J. Concentration of Rare Earth Elements (Sc, Y, La, Ce, Nd, Sm) in Bauxite Residue (Red Mud) Obtained by Water and Alkali Leaching of Bauxite Sintering Dust. *Minerals* **2020**, *10*, 500. [CrossRef]
67. Valeev, D.; Zinoveev, D.; Kondratiev, A.; Lubyanoi, D.; Pankratov, D. Reductive Smelting of Neutralized Red Mud for Iron Recovery and Produced Pig Iron for Heat-Resistant Castings. *Metals* **2020**, *10*, 32. [CrossRef]
68. Boeva, N.M.; Slukin, A.D.; Makarova, M.A.; Shipilova, E.S.; Melnikov, P.P.; Vnuchkov, D.A.; Zhegallo, E.A.; Zaitseva, L.V.; Bortnikov, N.S. Bauxites of the Tatarka Deposit (Yenisei Ridge, Russia): The First Evidence of a Contact-Karst Origin. *Dokl. Earth Sci.* **2023**, *513*, 1173–1181. [CrossRef]
69. Srikanth, S.; Ray, A.K.; Bandopadhyay, A.; Ravikumar, B.; Jha, A. Phase Constitution During Sintering of Red Mud and Red Mud–Fly Ash Mixtures. *J. Am. Ceram. Soc.* **2005**, *88*, 2396–2401. [CrossRef]
70. Mikulionok, I.O. A state of art and prospects of red mud management. *Energy Technol. Resour. Sav.* **2024**, *79*, 62–87. [CrossRef]
71. Khairul, M.A.; Zanganeh, J.; Moghtaderi, B. The composition, recycling and utilisation of Bayer red mud. *Resour. Conserv. Recycl.* **2019**, *141*, 483–498. [CrossRef]
72. Lemougna, P.N.; Wang, K.-T.; Tang, Q.; Cui, X.-M. Synthesis and characterization of low temperature (<800 °C) ceramics from red mud geopolymer precursor. *Constr. Build. Mater.* **2017**, *131*, 564–573. [CrossRef]
73. Mathisen, A.; Sørensen, H.; Melaaen, M.; Müller, G.-I. Investigation into Optimal CO_2 Concentration for CO_2 Capture from Aluminium Production. *Energy Procedia* **2013**, *37*, 7168–7175. [CrossRef]
74. D'Adamo, I.; Falcone, P.M.; Martin, M.; Rosa, P. A Sustainable Revolution: Let's Go Sustainable to Get Our Globe Cleaner. *Sustainability* **2020**, *12*, 4387. [CrossRef]
75. Falcone, P.M. Tourism-Based Circular Economy in Salento (South Italy): A SWOT-ANP Analysis. *Soc. Sci.* **2019**, *8*, 216. [CrossRef]
76. Nowicki, C.; Gosselin, L. An Overview of Opportunities for Waste Heat Recovery and Thermal Integration in the Primary Aluminum Industry. *JOM* **2012**, *64*, 990–996. [CrossRef]
77. Pocola, A.; Serban, A.; Balan, M. Complex and Efficient Waste Heat Recovery System in Aluminum Foundry. *Energy Procedia* **2017**, *112*, 504–509. [CrossRef]
78. Wang, Z.; Zhou, N.; Jing, G. Performance analysis of ORC power generation system with low-temperature waste heat of aluminum reduction cell. *Physics Procedia* **2012**, *24*, 546–553. [CrossRef]
79. Yang, Y.; Tao, W.; Wang, Z.; Shi, Z. A Numerical Approach on Waste Heat Recovery through Sidewall Heat-Exchanging in an Aluminum Electrolysis Cell. *Adv. Mater. Sci. Eng.* **2021**, *2021*, 3573306. [CrossRef]
80. Jouhara, H.; Nieto, N.; Egilegor, B.; Zuazua, J.; González, E.; Yebra, I.; Igesias, A.; Delpech, B.; Almahmoud, S.; Brough, D.; et al. Waste heat recovery solution based on a heat pipe heat exchanger for the aluminium die casting industry. *Energy* **2023**, *266*, 126459. [CrossRef]
81. Egilegor, B.; Jouhara, H.; Zuazua, J.; Al-Mansour, F.; Plesnik, K.; Montorsi, L.; Manzini, L. ETEKINA: Analysis of the potential for waste heat recovery in three sectors: Aluminium low pressure die casting, steel sector and ceramic tiles manufacturing sector. *Int. J. Thermofluids* **2020**, *1–2*, 100002. [CrossRef]
82. Lv, H.; Xie, M.; Wu, Z.; Li, L.; Yang, R.; Han, J.; Liu, F.; Zhao, H. Effective Extraction of the Al Element from Secondary Aluminum Dross Using a Combined Dry Pressing and Alkaline Roasting Process. *Materials* **2022**, *15*, 5686. [CrossRef]
83. Spoel, H.; Zebedee, W.A. The hot aluminum dross recycling (HDR) system. In Proceedings of the Conference: Annual Meeting and Exhibition of the Minerals, Metals and Materials Society (TMS), Anaheim, CA, USA, 4–8 February 1996.
84. Zuo, Z.; Lv, H.; Li, R.; Liu, F.; Zhao, H. A new approach to recover the valuable elements in black aluminum dross. *Resour. Conserv. Recycl.* **2021**, *174*, 105768. [CrossRef]
85. Zhang, S.; Wang, K.; Li, H.; Zhang, X.; Jiang, Y. Novel SCMs produced by the calcination of secondary aluminium dross with dolomite and their potential usage in cemented paste backfill. *Constr. Build. Mater.* **2023**, *365*, 130119. [CrossRef]
86. Ercoli, R.; Laskowska, D.; Nguyen, V.V.; Le, V.S.; Louda, P.; Łoś, P.; Ciemnicka, J.; Prałat, K.; Renzulli, A.; Paris, E.; et al. Mechanical and Thermal Properties of Geopolymer Foams (GFs) Doped with By-Products of the Secondary Aluminum Industry. *Polymers* **2022**, *14*, 703. [CrossRef]

87. Dirisu, J.O.; Oyedepo, S.O.; Fayomi, O.S.I.; Akinlabi, E.T. Development of Silicate Aluminium Dross Composites for Sustainable Building Ceilings. *Silicon* **2021**, *13*, 1979–1991. [CrossRef]
88. Shehab, E.; Meiirbekov, A.; Amantayeva, A.; Tokbolat, S. Cost Modelling for Recycling Fiber-Reinforced Composites: State-of-the-Art and Future Research. *Polymers* **2023**, *15*, 150. [CrossRef] [PubMed]
89. Taha, M.A.; Nassar, A.H.; Zawrah, M.F. In-situ formation of composite having hard outer layer based on aluminum dross reinforced by SiC and TiO_2. *Constr. Build. Mater.* **2020**, *248*, 118638. [CrossRef]
90. Heo, J.H.; Park, J.H. Thermochemical analysis for the reduction behavior of FeO in EAF slag via Aluminothermic Smelting Reduction (ASR) process: Part II. Effect of aluminum dross and lime fluxing on Fe and Mn recovery. *Calphad* **2017**, *58*, 229–238. [CrossRef]
91. Beheshti, R.; Moosberg-Bustnes, J.; Akhtar, S.; Aune, R.E. Black Dross Processing: Utilization of Black Dross in the Production of a Ladle Fluxing Agent. *J. Sustain. Metall.* **2017**, *3*, 265–273. [CrossRef]
92. Ewais, E.M.M.; Besisa, N.H.A. Tailoring of magnesium aluminum titanate based ceramics from aluminum dross. *Mater. Des.* **2018**, *141*, 110–119. [CrossRef]
93. Zhang, Y.; Guo, Z.; Han, Z.; Xiao, X. Effect of rare earth oxides doping on MgAl2O4 spinel obtained by sintering of secondary aluminium dross. *J. Alloys Compd.* **2018**, *735*, 2597–2603. [CrossRef]
94. Dangtungee, R.; Vatcharakajon, P.; Techawinyutham, L. Aluminium dross neutralization and its application as plant fertilizer. *Mater. Today Proc.* **2022**, *52*, 2420–2426. [CrossRef]
95. Gomes, S.A.; Reghu, V.R.; Ramaswamy, P. Sprouting in Seeds Aided by Nitrogen Sourced from Ammonia Fumes Leached from Aluminum Dross. *Mater. Today Proc.* **2023**, *in press*. [CrossRef]
96. Shi, J.; Huang, Z.; Qin, Q.; Yuan, C.; Chen, J.; Zhou, Y. Experiment study on the preparation of polyaluminum chloride with secondary aluminum dross. *Met. Mine* **2021**, *50*, 206–210.
97. 97. Philipson, H.; Wallin, M.; Einarsrud, K.E.; Tranell, G. Kinetics of Silicon Production by Aluminothermic Reduction of Silica Using Aluminum and Aluminum Dross as Reductants. In Proceedings of the 16th International Ferro-Alloys Congress (INFACON XVI), Trondheim, Norway, 27–29 September 2021.
98. Grande, L.; Vicente, M.Á.; Korili, S.A.; Gil, A. Synthesis strategies of alumina from aluminum saline slags. *Process Saf. Environ. Prot.* **2023**, *172*, 1010–1028. [CrossRef]
99. Feng, H.; Wang, L.; Li, G.; Jin, Q. Removal of impurities for the synthesis of high-property alumina-spinel ceramic from secondary aluminum dross. *Ceram. Int.* **2023**, *49 Pt A*, 34616–34626. [CrossRef]
100. Teixeira, A.B.; Ambrós, W.M.; Sampaio, C.H.; Raposo, F.L.Q.; De Brum, I.A.S.; Moncunill, J.O. Optimization of Water Leaching of Chlorides from Aluminum Salt Slag. *Minerals* **2022**, *12*, 1141. [CrossRef]
101. Pereira, D.A.; de Aguiar, B.; Castro, F.; Almeida, M.F.; Labrincha, J.A. Mechanical behaviour of Portland cement mortars with incorporation of Al-containing salt slags. *Cem. Concr. Res.* **2000**, *30*, 1131–1138. [CrossRef]
102. Jiménez, A.; Misol, A.; Morato, Á.; Rives, V.; Vicente, M.A.; Gil, A. Synthesis of pollucite and analcime zeolites by recovering aluminum from a saline slag. *J. Clean. Prod.* **2021**, *297*, 126667. [CrossRef]
103. Lin, Y.; Maghool, F.; Arulrajah, A.; Horpibulsuk, S. Alkali activation of recycled concrete and aluminum salt slag aggregates for semi-rigid column inclusions. *Constr. Build. Mater.* **2023**, *366*, 130106. [CrossRef]
104. Mucsi, G.; Halyag, N.; Kurusta, T.; Kristály, F. Control of Carbon Dioxide Sequestration by Mechanical Activation of Red Mud. *Waste Biomass Valorization* **2021**, *12*, 6481–6495. [CrossRef]
105. Yuan, J.; Xiao, J.; Li, F.; Wang, B.; Yao, Z.; Yu, B.; Zhang, L. Co-treatment of spent cathode carbon in caustic and acid leaching process under ultrasonic assisted for preparation of SiC. *Ultrason. Sonochemistry* **2018**, *41*, 608–618. [CrossRef]
106. Lu, Y.; Hua, Z.; Chen, C.; Chen, Z.; Li, X.; Yu, H.; Peng, K.; Tian, Z. Stable and fast Na-ion storage of recovered carbon from the spent carbon cathode of aluminum electrolysis. *Sustain. Energy Fuels* **2024**, *8*, 1040–1047. [CrossRef]
107. Li, N.; Xie, G.; Wang, Z.X.; Hou, Y.Q.; Li, R.X. Recycle of spent potlining with low carbon grade by floatation. *Adv. Mater. Res.* **2014**, *881*, 1660–1664. [CrossRef]
108. Xie, W.; Zhou, F.; Liu, J.; Bi, X.; Huang, Z.; Li, Y.; Chen, D.; Zou, H.; Sun, S. Synergistic reutilization of red mud and spent pot lining for recovering valuable components and stabilizing harmful element. *J. Clean. Prod.* **2020**, *243*, 118624. [CrossRef]
109. Robshaw, T.; Tukra, S.; Hammond, D.B.; Leggett, G.J.; Ogden, M.D. Highly efficient fluoride extraction from simulant leachate of spent potlining via La-loaded chelating resin. An equilibrium study. *J. Hazard. Mater.* **2019**, *361*, 200–209. [CrossRef] [PubMed]
110. Xiao, J.; Zhang, L.; Yuan, J.; Yao, Z.; Tang, L.; Wang, Z.; Zhang, Z. Co-utilization of spent pot-lining and coal gangue by hydrothermal acid-leaching method to prepare silicon carbide powder. *J. Clean. Prod.* **2018**, *204*, 848–860. [CrossRef]
111. Zhao, Q.; Wang, Y.; Dong, H.; Wang, J.; Fu, X.; Cui, X.; Li, S.; Li, C. Preparation of anode materials for lithium-ion batteries by spent carbon anode from electrolytic aluminum. *J. Environ. Chem. Eng.* **2021**, *9*, 105932. [CrossRef]
112. Zheng, Z.; Xie, M.; Yu, G.; Wu, Z.; Zhong, J.; Wang, Y.; Zhao, H.; Liu, F. Preparation of lithium-ion battery anode materials from graphitized spent carbon cathode derived from aluminum electrolysis. *Int. J. Miner. Metall. Mater.* **2024**, *31*, 2466–2475. [CrossRef]
113. Yao, Z.; Zhong, Q.; Xiao, J.; Ye, S.; Tang, L.; Zhang, Z. An environmental-friendly process for dissociating toxic substances and recovering valuable components from spent carbon cathode. *J. Hazard. Mater.* **2021**, *404*, 124120. [CrossRef]
114. Zhi, P.; Ou, Y.; Li, C.; Wang, Y.; Zhao, D. Study on harmless and resource utilization of spent cathode carbon washing process. *IOP Conf. Ser. Earth Environ. Sci.* **2020**, *467*, 012177. [CrossRef]

115. Li, B.; Ding, Y.; Jiao, Y.; Xu, F.; Wang, X.; Zheng, F.; Zou, J.; Gao, Q.; Hu, H. A Sustainable Strategy for Spent Cathode Carbon Blocks Hazardous Waste Recycling Using Binary Molten Salt Thermal Treatment. *J. Therm. Sci.* **2024**, *33*, 1082–1093. [CrossRef]
116. Liu, D.Y.; Wu, C.S. Stockpiling and Comprehensive Utilization of Red Mud Research Progress. *Materials* **2012**, *5*, 1232–1246. [CrossRef]
117. Ujaczki, É.; Feigl, V.; Molnár, M.; Cusack, P.; Curtin, T.; Courtney, R.; O'Donoghue, L.; Davris, P.; Hugi, C.; Evangelou, M.W.; et al. Re-using bauxite residues: Benefits beyond (critical raw) material recovery. *J. Chem. Technol. Biotechnol.* **2018**, *93*, 2498–2510. [CrossRef] [PubMed]
118. Zhao, H.; Liu, F.; Xie, M.; Liu, W.; Sohn, H.Y. Recycling and utilization of spent potlining by different high temperature treatments. *J. Clean. Prod.* **2021**, *289*, 125704. [CrossRef]
119. Lyu, F.; Hu, Y.; Wang, L.; Sun, W. Dealkalization processes of bauxite residue: A comprehensive review. *J. Hazard. Mater.* **2021**, *403*, 123671. [CrossRef] [PubMed]
120. Nikraz, H.R.; Bodley, A.J.; Cooling, D.J.; Kong, P.Y.L.; Soomro, M. Comparison of Physical Properties between Treated and Untreated Bauxite Residue Mud. *J. Mater. Civ. Eng.* **2007**, *19*, 2–9. [CrossRef]
121. Mayes, W.M.; Jarvis, A.P.; Burke, I.T.; Walton, M.; Feigl, V.; Klebercz, O.; Gruiz, K. Dispersal and Attenuation of Trace Contaminants Downstream of the Ajka Bauxite Residue (Red Mud) Depository Failure, Hungary. *Environ. Sci. Technol.* **2011**, *45*, 5147–5155. [CrossRef]
122. Xue, S.; Kong, X.; Zhu, F.; Hartley, W.; Li, X.; Li, Y. Proposal for management and alkalinity transformation of bauxite residue in China. *Environ. Sci. Pollut. Res.* **2016**, *23*, 12822–12834. [CrossRef]
123. Rai, S.; Wasewar, K.L.; Lataye, D.H.; Mukhopadhyay, J.; Yoo, C.K. Feasibility of red mud neutralization with seawater using Taguchi's methodology. *Int. J. Environ. Sci. Technol.* **2013**, *10*, 305–314. [CrossRef]
124. McConchie, D.; Clark, M.; Hanahan, C.; Davies-McConchie, F. The Use of Seawater-neutralised Bauxite Refinery Residues in the Management of Acid Sulphate Soils, Sulphidic Mine Tailings and Acid Mine Drainage. In Proceedings of the 3rd Queensland Environment Conference, Brisbane, Australia, 1 January 2000.
125. Dealing with Red Mud—By—Product of the Bayer Process for Refining Aluminium. Available online: https://www.azom.com/article.aspx?ArticleID=2071 (accessed on 14 March 2023).
126. Khaitan, S.; Dzombak, D.A.; Lowry, G.V. Mechanisms of Neutralization of Bauxite Residue by Carbon Dioxide. *J. Environ. Eng.* **2009**, *135*, 433–438. [CrossRef]
127. Kasturi, A.; Gabitto, J.; Tsouris, C.; Custelcean, R. Carbon dioxide capture with aqueous amino acids: Mechanistic study of amino acid regeneration by guanidine crystallization and process intensification. *Sep. Purif. Technol.* **2021**, *271*, 118839. [CrossRef]
128. Sahu, R.C.; Patel, R.K.; Ray, B.C. Neutralization of red mud using CO_2 sequestration cycle. *J. Hazard. Mater.* **2010**, *179*, 28–34. [CrossRef] [PubMed]
129. Rai, S.; Wasewar, K.L.; Lataye, D.H.; Mishra, R.S.; Puttewar, S.P.; Chaddha, M.J.; Mahindiran, P.; Mukhopadhyay, J. Neutralization of red mud with pickling waste liquor using Taguchi's design of experimental methodology. *Waste Manag. Res.* **2012**, *30*, 922–930. [CrossRef]
130. Wang, P.; Liu, D.-Y. Physical and Chemical Properties of Sintering Red Mud and Bayer Red Mud and the Implications for Beneficial Utilization. *Materials* **2012**, *5*, 1800–1810. [CrossRef]
131. Mussel, G.; Summers, J.; Sparling, G. *Bioremediation of Bauxite Residue in Western Australia: An Initial Feasibility Study*; Alcoa of Australia: Perth, Australia, 1993.
132. Aranda-Usón, A.; Portillo-Tarragona, P.; Scarpellini, S.; Llena-Macarulla, F. The progressive adoption of a circular economy by businesses for cleaner production: An approach from a regional study in Spain. *J. Clean. Prod.* **2020**, *247*, 119648. [CrossRef]
133. Shoppert, A.; Valeev, D.; Diallo, M.M.; Loginova, I.; Beavogui, M.C.; Rakhmonov, A.; Ovchenkov, Y.; Pankratov, D. High-Iron Bauxite Residue (Red Mud) Valorization Using Hydrochemical Conversion of Goethite to Magnetite. *Materials* **2022**, *15*, 8423. [CrossRef] [PubMed]
134. Li, S.; Guo, Z.; Pan, J.; Zhu, D.; Dong, T.; Lu, S. Extracting Al_2O_3 and TiO_2 from Red Mud Smelting Separation Slag by Alkali and Acid Leaching Methods. *Metals* **2023**, *13*, 552. [CrossRef]
135. Guo, Y.; Zhao, Q.; Yan, K.; Cheng, F.; Lou, H.H. Novel Process for Alumina Extraction via the Coupling Treatment of Coal Gangue and Bauxite Red Mud. *Ind. Eng. Chem. Res.* **2014**, *53*, 4518–4521. [CrossRef]
136. Borra, C.R.; Pontikes, Y.; Binnemans, K.; Van Gerven, T. Leaching of rare earths from bauxite residue (red mud). *Miner. Eng.* **2015**, *76*, 20–27. [CrossRef]
137. Silva Filho, E.; Alves, M.; Da Motta, M. Lama vermelha da indústria de beneficiamento de alumina: Produção, características, disposição e aplicações alternativas. *Matéria* **2007**, *12*, 322–338. [CrossRef]
138. Ding, W.; Bao, S.; Zhang, Y.; Xiao, J. Efficient Selective Extraction of Scandium from Red Mud. *Miner. Process. Extr. Metall. Rev.* **2022**, *44*, 304–312. [CrossRef]
139. Dai, X.; Thi Hong Nhung, N.; Hamza, M.F.; Guo, Y.; Chen, L.; He, C.; Ning, S.; Wei, Y.; Dodbiba, G.; Fujita, T. Selective adsorption and recovery of scandium from red mud leachate by using phosphoric acid pre-treated pitaya peel biochar. *Sep. Purif. Technol.* **2022**, *292*, 121043. [CrossRef]
140. Zhang, D.-R.; Chen, H.-R.; Nie, Z.-Y.; Xia, J.-L.; Li, E.-P.; Fan, X.-L.; Zheng, L. Extraction of Al and rare earths (Ce, Gd, Sc, Y) from red mud by aerobic and anaerobic bi-stage bioleaching. *Chem. Eng. J.* **2020**, *401*, 125914. [CrossRef]

141. Li, W.; Li, Z.; Wang, N.; Gu, H. Selective extraction of rare earth elements from red mud using oxalic and sulfuric acids. *J. Environ. Chem. Eng.* **2022**, *10*, 108650. [CrossRef]
142. Huang, Y.; Wen, X.; Li, J.; Niu, Q.; Tang, A.; Li, Q. Metagenomic insights into role of red mud in regulating fate of compost antibiotic resistance genes mediated by both direct and indirect ways. *Environ. Pollut.* **2023**, *317*, 120795. [CrossRef]
143. Li, Y.; Wu, Z.; Zhao, C.; Zhang, Y.; Peng, D.; Gong, Z. Facile fabrication of Zero-valent-iron biochar from red mud for bisulfite activation in wastewater treatment: Performance and mechanism. *Environ. Technol. Innov.* **2023**, *30*, 103110. [CrossRef]
144. Cai, Y.-R.; Song, Y.; Chang, C. Adsorption properties and mechanism of ginkgo biloba leaf-based materials for Cd (II) in aqueous solution. *Environ. Sci. Pollut. Res.* **2022**, *29*, 78499–78508. [CrossRef]
145. Bai, B.; Bai, F.; Li, X.; Nie, Q.; Jia, X.; Wu, H. The remediation efficiency of heavy metal pollutants in water by industrial red mud particle waste. *Environ. Technol. Innov.* **2022**, *28*, 102944. [CrossRef]
146. Guo, Z.; Zhang, Y.; Gan, S.; He, H.; Cai, N.; Xu, J.; Guo, P.; Chen, B.; Pan, X. Effective degradation of COVID-19 related drugs by biochar-supported red mud catalyst activated persulfate process: Mechanism and pathway. *J. Clean Prod.* **2022**, *340*, 130753. [CrossRef] [PubMed]
147. Turk, B.; Kazak, O.; Akkaya, G.K.; Tor, A. A simple and green preparation of red mud-coated membrane for efficient separation of oil-in-water emulsions. *J. Environ. Chem. Eng.* **2022**, *10*, 106928. [CrossRef]
148. Viyasun, K.; Anuradha, R.; Thangapandi, K.; Santhosh Kumar, D.; Sivakrishna, A.; Gobinath, R. Investigation on performance of red mud based concrete. *Mater. Today Proc.* **2021**, *39*, 796–799. [CrossRef]
149. Vladić Kancir, I.; Serdar, M. Contribution to Understanding of Synergy between Red Mud and Common Supplementary Cementitious Materials. *Materials* **2022**, *15*, 1968. [CrossRef] [PubMed]
150. Zeng, H.; Lyu, F.; Sun, W.; Zhang, H.; Wang, L.; Wang, Y. Progress on the Industrial Applications of Red Mud with a Focus on China. *Minerals* **2020**, *10*, 773. [CrossRef]
151. Wang, X.; Sun, K.; Li, X.; Ma, J.; Luo, Z. The Effect of Red Mud on Sintering Processes and Minerals of Portland Cement for Roads. *Crystals* **2021**, *11*, 1267. [CrossRef]
152. Pişkin, S.; Kantürk Figen, A.; özkan, E.; özçay, ü. Structural Characterization of Seydişehir Red Mud to Utilization in Roof Tile Manufacturing. *IFAC Proc. Vol.* **2013**, *46*, 484–487. [CrossRef]
153. Ma, S.; Sun, Z.; Wei, J.; Zhang, X.; Zhang, L. Utilization of Modified Red Mud Waste from the Bayer Process as Subgrade and Its Performance Assessment in a Large-Sale Application. *Coatings* **2022**, *12*, 471. [CrossRef]
154. Tyagi, G.; Singhal, A.; Routroy, S.; Bhunia, D.; Lahoti, M. Radiation Shielding Concrete with alternate constituents: An approach to address multiple hazards. *J. Hazard. Mater.* **2021**, *404*, 124201. [CrossRef]
155. Tyagi, G.; Singhal, A.; Routroy, S.; Bhunia, D.; Lahoti, M. A review on sustainable utilization of industrial wastes in radiation shielding concrete. *Mater. Today Proc.* **2020**, *32*, 746–751. [CrossRef]
156. Bijanu, A.; Arya, R.; Agrawal, V.; Tomar, A.S.; Gowri, V.S.; Sanghi, S.K.; Mishra, D.; Salammal, S.T. Metal-polymer composites for radiation protection: A review. *J. Polym. Res.* **2021**, *28*, 392. [CrossRef]
157. Song, R.; Murphy, M.; Li, C.; Ting, K.; Soo, C.; Zheng, Z. Current development of biodegradable polymeric materials for biomedical applications. *Drug Des. Dev. Ther.* **2018**, *12*, 3117–3145. [CrossRef]
158. Tian, X.; Liu, T.; Wang, Q.; Dilmurat, A.; Li, D.; Ziegmann, G. Recycling and remanufacturing of 3D printed continuous carbon fiber reinforced PLA composites. *J. Clean. Prod.* **2017**, *142*, 1609–1618. [CrossRef]
159. Agrawal, V.; Paulose, R.; Arya, R.; Rajak, G.; Giri, A.; Bijanu, A.; Sanghi, S.K.; Mishra, D.; N, P.; Khare, A.K.; et al. Green conversion of hazardous red mud into diagnostic X-ray shielding tiles. *J. Hazard. Mater.* **2022**, *424*, 127507. [CrossRef] [PubMed]
160. Sunendar, B.; Fathina, A.; Harmaji, A.; Mardhian, D.F.; Asri, L.; Widodo, H.B. The effect of CHA-doped Sr addition to the mechanical strength of metakaolin dental implant geopolymer composite. *AIP Conf. Proc.* **2017**, *1887*, 020020. [CrossRef]
161. Harmaji, A.; Hardiansyah, A.; Widianingsih, N.A.; Jannah, R.M.; Soepriyanto, S. The effect of Basic Oxygen Furnace, Blast Furnace, and Kanbara Reactor Slag as Reinforcement to Cement Based Mortar. *JPSE (J. Phys. Sci. Eng.)* **2022**, *7*, 56–61. [CrossRef]
162. Simatupang, P.H.; Harmaji, A.; Anggraini, A.N.; Pane, I.; Sunendar, B.; Imran, I. Utilization of Industrial Waste in Indonesia as Filler for Low and High Calcium Fly Ash based Geopolymer. In Proceedings of the 2nd International Conference on Materials and Metallurgical Technology 2015 (ICOMMET 2015), Surabaya, Indonesia, 4–6 October 2015.
163. Gonçalves, N.P.F.; Olhero, S.M.; Labrincha, J.A.; Novais, R.M. 3D-printed red mud/metakaolin-based geopolymers as water pollutant sorbents of methylene blue. *J. Clean. Prod.* **2023**, *383*, 135315. [CrossRef]
164. Singh Tomar, A.; Gupta, R.; Singh, A.; Thankaraj Salammal, S.; Akram Khan, M.; Mishra, D. Evaluation of corrosion protective properties of fly ash-red mud based geopolymer coating material for mild steel. *Mater. Today Proc.* **2022**, *68*, 181–186. [CrossRef]
165. Xu, Z.; Liu, Q.; Long, H.; Deng, H.; Chen, Z.; Hui, D. Influence of nano-SiO_2 and steel fiber on mechanical and microstructural properties of red mud-based geopolymer concrete. *Constr. Build. Mater.* **2023**, *364*, 129990. [CrossRef]
166. Afolabi, L.O.; Ariff, Z.M.; Megat-Yusoff, P.S.M.; Al-Kayiem, H.H.; Arogundade, A.I.; Afolabi-Owolabi, O.T. Red-mud geopolymer composite encapsulated phase change material for thermal comfort in built-sector. *Solar Energy* **2019**, *181*, 464–474. [CrossRef]

materials

MDPI

Article

Properties of Adhesive Mortars Using Waste Glass

Galyna Kotsay and Wiktor Szewczenko *

Faculty of Civil Engineering, Mechanics and Petrochemistry, Warsaw University of Technology, Łukasiewicza St. 17, 09-400 Płock, Poland; galyna.kotsay@pw.edu.pl
* Correspondence: wiktor.szewczenko@pw.edu.pl

Abstract: This study investigates the use of waste glass as an active aggregate in glass polymers based on water glass, aiming to enhance the sustainability of construction materials by utilizing recyclable waste. Methodologically, the research employs a combination of water glass as a binder with waste glass, analyzing their chemical interaction and the resulting mechanical properties. The primary findings reveal that the inclusion of finely ground waste glass not only promotes the polycondensation and hardening processes of water glass but also significantly influences the adhesive and cohesive strengths of the developed glass polymers. After 7 days of hardening, the tensile strength of these materials exceeds that of standard concrete with values reaching up to 4.11 MPa, indicating strong adhesion capabilities that could pull out fragments of the concrete substrate. Conclusively, the study underscores the potential of waste glass in improving the structural and economic efficiencies of building materials, contributing to a reduction in landfill waste and offering a promising avenue for the innovative use of recyclable materials in construction.

Keywords: waste glass; water glass; adhesive mortar; alkaline activity; polyalkaline effect; tensile strength

1. Introduction

Waste glass is an ideal recyclable secondary raw material [1–3]. It boasts durability and chemical resistance, posing no threat to the environment, yet it burdens landfills as it is not biodegradable [4–7]. As glass production increases annually, so does the volume of cullet [8]. For this reason, the efficient management of glass waste has become a key challenge in waste management [9,10]. The management strategy depends on the type of glass [11]. However, particular difficulties are associated with the unstable chemical composition of the cullet and the challenge of segregating different types of glass during disposal [5,6]. Currently, most recyclable glass is container glass, which is relatively easy to process due to its uniform composition. Using waste glass in construction can significantly contribute to reducing landfill waste and lowering material costs as well as enhance the sustainability of the construction industry by creating innovative materials.

One effective method for using cullet is to produce glass fiber, rock wool, abrasives, and road paints [12,13]. Many publications have proposed using waste glass as an aggregate or a non-clinker component in cement [14–29]. However, the use of glass in cement products is limited due to its high alkali content, which in turn can cause an alkali–silica reaction (ASR) and reduce the strength of the resulting products [30–32]. There is growing interest in using waste glass in alkali-activated materials or geopolymers, where the acceptable glass content ranges from 10 to 50%. It is currently relevant to develop new building materials using waste glass combined with water glass [33–37].

Water glass is a binder with high strength properties and considerable acid resistance except against hydrofluoric acid [36,38]. In construction, sodium and potassium water glass are used for producing fireproof materials, adhesive mortars, wood impregnation, concrete surface hardening, the activation of mineral additives, and a component in geopolymers [39–44]. Its application depends on the modulus (the molar ratio of silicate to sodium or potassium oxide), concentration, and density. In construction, water glass is most commonly used with a modulus ranging from 1.5 to 3.5 and a density from 1.3 to 1.4 kg/dm^3 [36,45,46]. One of the most important properties of this material is its adhesive ability, characterized by adhesive strength (fad), and as water glass hardens and acquires the properties of a solid body, its structural strength, or cohesive strength (fcoh), also becomes crucial [38].

The setting and hardening mechanisms of water glass differ significantly from those of ordinary cement. While the setting process of cement is based on hydration reactions, the setting process of water glass relies on polycondensation reactions. Polycondensation leads to the formation of amorphous or semi-crystalline silicon tetrahedral groups, which are linked by alternating common oxygen atoms. Given its relatively high chemical activity, the interaction of water glass with various chemical reagents offers the possibility of obtaining materials with varying degrees of polymerization. The introduction of finely ground waste glass into water glass initiates a chemical interaction with the alkaline components of the glass powder, resulting in the formation of a new material, glass polymer (GP). The technological properties of this glass polymer depend heavily on the binding agent, water glass, and factors such as the type and surface area of the waste glass.

It is known that adding a filler to water glass promotes the processes of polycondensation and hardening of the water glass [36,47]. However, the potential use of waste glass as an aggregate is limited due to its high alkali content and the low adhesion of glass as an aggregate to cement slurry. Therefore, the purpose of this study was to combine waste glass as an aggregate with water glass as a binder to produce adhesive grouts. This research focuses on the adhesive and cohesive properties, aiming to determine the optimal amount of waste glass in the grouts tested.

2. Materials and Methods

This paper used waste glass produced by REWA [48] and sodium and potassium water glasses produced by the Chemical Plant "Rudniki" S.A. [49]. Their characteristics are presented in Table 1. The density and the Blaine-specific surface waste glass were performed according to the methods described in the standards [50,51].

Table 1. Characteristics of waste glass and sodium and potassium water glasses.

Materials	Oxides (wt %)						Siliceous Module	Density, g/cm^3	Specific Surface, m^2/g
	SiO_2	Na_2O	K_2O	Al_2O_3	CaO + MgO	H_2O			
Waste glass (GW)	72	10	4	1.0	12.0	-	-	2.43	3987
Sodium water glass (SWG)	26.38	8.02	-	-	-	65.6	3.4	1.45	-
Potassium water glass (PWG)	21.41	-	7.59	-	-	71.0	4.4	1.25	-

The adhesive strength of hardened mortars on substrates was determined according to the standard [52]. As a substrate for determining the adhesion strength, a concrete slab measuring 300 × 500 × 50 mm^3 with concrete compressive strength f_c = 70.5 MPa and tensile strength f_t ~4.0 MPa was used. The work investigated the effect of the composition of the glass polymer by changing the water glass to waste glass ratio from 1/1 to 1/3 on its adhesive (f_{ad}) and cohesive (f_{coh}) strength. Glass polymer (GP) was synthesized by mixing glass powder with water glass for 2–3 min until a homogeneous dense mass is obtained. Then, the glass polymer is applied to the surface of a pre-cleaned concrete slab, and the thickness of the applied layer is ~3 mm. After 30 days of hardening, a metal plate measuring 10 × 50 × 50 mm^3 with a hole for a metal rod that is used to attach it to the press was glued to the surface of the glass polymer using epoxy glue. After the epoxy glue has hardened, cuts are made along the side walls of the metal square with an emery disk, separating a glass polymer sample with an area of 50 cm^2. A total of 8 samples are obtained on the surface of the concrete slab.

Possible fracture patterns are given in Figure 1.

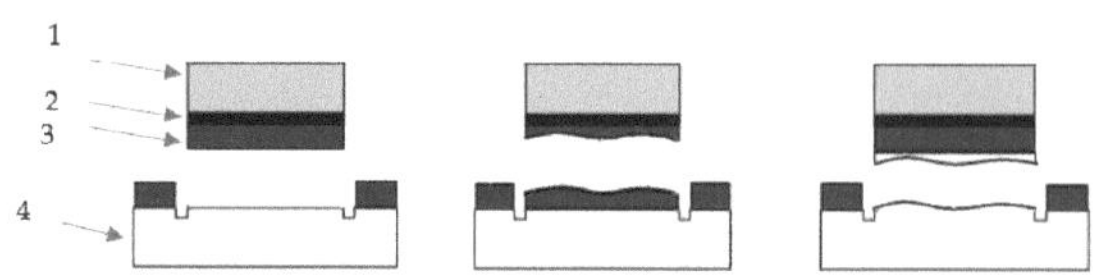

(**a**) (**b**) (**c**)

Figure 1. Fracture patterns (1—pull-head plate; 2—adhesive layer; 3—glass polymer; 4—concrete): (**a**) adhesion fracture—fracture at the interface between glass polymer and concrete; (**b**) cohesion fracture in glass polymer—fracture in the glass polymer itself; (**c**) cohesion fracture in concrete-fracture in the concrete itself.

The Calmetrix I-Cal 2000 HPC calorimeter(Calmetrix Inc., located in Boston, Massachusetts, USA) was utilized to assess the impact of waste glass on the polycondensation reactions of water glass. Heat evolution was recorded for pastes with varying ratios of finely ground waste glass to water glass (1:1 and 1:2). The amount of heat released was monitored every 15 s over 100 h.

Alkaline activity was determined based on a series of six specimens according to [53]. For the analysis of alkali content, a flame photometer FP902 (PG Instruments Limited, Alma Park, Wibtoft, Leicestershire, UK) with an accuracy of ±0.5% was employed. Distilled water was used as an extractant to determine the alkaline activity of individual components and compositions of glass polymers. The results are presented in units of ppm/dm^2 of paste.

Impact and compressive strength tests were measured for glass polymer specimens with dimensions of 10 mm × 10 mm × 80 mm. The strength tests were carried out 5, 9, and 16 days after forming. To research impact strength, a Charpy hammer was employed according to [54]. Compressive strength was determined using a ZD 10/90 static testing machine [55].

3. Results and Discussion

The research aimed to establish the role of the time factor and the glass polymer's (GP) composition on its adhesion to the concrete surface. Table 2 shows the compositions and bond strengths studied. The use of the method described above for determining the adhesion strength of a glass polymer to a concrete surface (f_{ad}) is based on determining the tensile strength of the GP/concrete composition. It is generally accepted that if the separation zone passes through the glass polymer, then the tensile strength will characterize the f_{coh} of the glass polymer. If the separation zone passes through concrete, then the tensile strength will characterize the cohesive strength of concrete ($f_{coh.c}$).

Considering that tensile strength indirectly characterizes the material's cohesive strength, i.e., its structural strength, it can be observed that over 28 days, the cohesive strength represented by f_{coh} for SWG and PWG increases significantly. Specifically, the data show that SWG alone and PWG alone see their tensile strengths increase to 0.65 MPa and 0.70 MPa, respectively, over a period of 28 days. This tenfold increase in the cohesive strength of water glass is associated with the polycondensation process of water glass. Moreover, the combined presence of SWG and PWG in different ratios increases cohesive strength over the first 7 days but shows a negligible change over the duration of 28 days when compared to using SWG or PWG individually.

The introduction of finely ground glass waste dramatically alters this trend. After 7 days of hardening, the tensile strength increases sharply, reaching values that exceed the tensile strength of standard concrete, which in our conditions is approximately 4 MPa. Specifically, a composite with 33.30% SWG and 67.70% glass waste reaches a tensile strength of about 4.11 MPa after 7 days, demonstrating a robust adhesion so powerful that the sample pulls out a fragment of the concrete substrate, as indicated in Figure 2.

Figure 2. Sample of glass polymer glued to concrete surface after a tensile test.

Table 2. The tensile strength of a glass polymer composite (GP) to concrete, depending on the composition.

N	Composition of Glass Polymer (GP) (wt %)			Tensile Strength f_t, (MPa)				Zone Fracture	Note
	Sodium Water Glass (SWG)	Potassium Water Glass (PWG)	Waste Glass (GW)	7 Days	14 Days	21 Days	28 Days		
1	100	-	-	0.06	0.46	0.54	0.65	Fracture in GP	f_{coh} SWG
2	-	100	-	0.07	0.07	0.10	0.70	Fracture in GP	f_{coh} PWG
3	50.0	50.0	-	0.46	0.52	0.71	0.69	Fracture in GP	f_{coh} SWG + PWG
4	80.0	20.0	-	0.47	0.53	0.62	0.67	Fracture in GP	f_{coh} SWG + PWG
5	33.30	-	67.70	4.11	3.79	3.96	4.02	Fracture in concrete	$f_{coh.c}$ *
6	-	33.30	67.70	3.66	3.13	3.21	3.61	Fracture in concrete	$f_{coh.c}$ *

* $f_{coh.c}$—cohesive strength of concrete.

A similar picture is observed for both SWG and PWG. Thus, introducing finely ground glass as an active filler into the composition of a glass polymer sharply increases its cohesive strength (f_{coh}). In order to test this phenomenon, studies were carried out on the adhesive properties of a glass polymer using the example of tensile strength at bending. For this purpose, standard cement beams CEM I 32.5R measuring 40 × 40 × 160 mm^3 were subjected to tensile bending testing. Then, both halves of the same sample were glued with a glass polymer of the appropriate composition and kept in air at room temperature for 28 days. After this, the glued beams were re-tested for tensile strength in bending, while the point of application of the force was located on the bonding line. The results showed that the failure plane was on the bonding line in all cases with glass polymer residues on both halves. This may indicate that the f_{ad} of the glass polymer in all cases exceeds it cohesive strength. At the same time, the compositions of glass polymers (SWG:GW/1:2) and (PWG:GW/1:2) are closest in value to the strength of the standard sample (f_t = 7.5 MPa), respectively, 6.5 and 5.1 MPa, as shown in Table 3. It should be emphasized that the use of dealcalized glass powder reduces the strength of glued samples by more than half, which may indicate the important role of alkalis in the structure of the glass polymer. In addition, it should be noted that potassium water glass is less effective than sodium glass.

As is known, alkalis in the composition of inorganic glasses significantly affect its properties [56,57]. Considering that these glasses contain a fairly high alkalis content, studies were conducted to examine the effect of alkali oxides on the strength characteristics of glass polymers. For this purpose, the dealcalization of finely ground glass waste was carried out by extracting alkalis in distilled water at a temperature of 70 °C, which was followed by drying at 100 °C.

Table 3. Tensile strength in bending of beams glued with various substances.

N	Adhesive Substance	f_t, (MPa)
1	SWG-100%	1.4
2	PWG-100%	0.7
3	GP (SWG:GW/1:2)	6.5
4	GP (PWG:GW/1:2)	5.1
5	GP (SWG:GW */1:2)	2.3
6	GP (PWG:GW */1:2)	1.9

GP—glass polymer. * dealcalized glass powder after alkali extraction.

Comparing the results presented in Tables 2 and 4, one can note a tendency toward a decrease in the tensile strength of samples with dealcalized glass powder.

Table 4. The strength of glass polymer to concrete when using dealcalized glass powder.

N	Composition of Glass Polymer * (wt %)			Tensile Strength in Bending (MPa)				Zone Fracture	Note
	Sodium Water Glass (SWG)	Potassium Water Glass (PWG)	Waste Glass (GW)	7 Days	14 Days	21 Days	28 Days		
1	33.3	-	67.7	3.04	3.39	3.38	3.32	By concrete	$f_{coh \cdot c}$
2	-	33.3	67.7	3.38	3.63	3.38	3.19	By concrete	$f_{coh.c}$

* glass polymer composition (SWG:GW/1:2) and (PWG:GW/1:2).

To determine the influence of the alkaline component on the properties of glass polymer, the method [53,58] was used to determine the alkali content on its surface. First, we determined the alkaline activity (AA) of glass powder before and after dealcalization and the level of alkaline activity of solid glass polymers of various compositions. Alkaline activity was determined on disk samples with a diameter of about 140 mm and a thickness of 3 mm. The exposure time of the extraction process was 5 s. Table 5 presents the research results.

Table 5. Alkaline activity of individual components and compositions of glass polymers of various compositions.

N	Composition of Solid Glass Polymer (wt %)	Alkaline Activity—AA (ppm/dm^2)		$\sum Na^+ + K^+$
		Na^+	K^+	
1	SWG-100	1.940	0.077	2.017
2	PWG-100	0.095	8.060	8.155
3	SWG:PWG/1:1	0.810	2.380	3.190
4	SWG:PWG/4:1	1.840	2.200	4.040
5	GW(powder)	0.080	0.022	0.102
6	GW-dA * (powder)	0.031	0.008	0.039
7	SWG:GW/1:1	1.700	0.190	1.890
8	SWG:GW/1:2	1.460	0.080	1.540
9	PWG:GW/1:1	0.380	3.970	4.350
10	PWG:GW/1:2	0.170	5.560	5.730
11	SWG:GWdA/1:1	1.350	0.055	1.405
12	PWG:GWdA/1:1	0.130	4.400	4.530
13	SWG:PWG:GW/1:1:2	1.460	2.290	3.750
14	SWG:PWG:GWdA/1:1:2	1.020	1.410	2.430

* dA—dealkalized waste glass (GW).

First of all, it should be noted that the dealcalization of finely ground waste glass reduces its alkaline activity by 2.6 times in both sodium and potassium. When glass powder is introduced into the glass polymer at a 1:1 ratio, its alkaline activity decreases by 12.4% in sodium and by 50.8% in potassium compared to SWG-100. Similarly, when glass powder is introduced into the composition of a glass polymer with PWG-100 at the same ratio of water glass to glass powder, an almost twofold decrease in alkaline activity for both potassium and sodium is observed. Thus, by introducing dealcalized glass powder into the composition of the glass polymer, it is possible to reduce the alkaline activity of sodium and potassium by 30% and 35–55%, respectively. The above results demonstrate that by conducting controlled dealcalization, one can regulate the alkaline activity of glass polymers that use waste glass as an aggregate.

The above data highlight the significant role of glass powder in enhancing the adhesion strength and cohesion of the glass polymer. It can be postulated that a chemical interaction occurs between the water glass and the finely ground fraction of glass waste, resulting in the formation of a solid-phase substance—a glass polymer with distinct structural and property differences from solidified water glass. To test this hypothesis, microcalorimetric studies were conducted using a microcalorimeter. Figure 3 illustrates the results of the heat release upon the contact of water glass with waste glass.

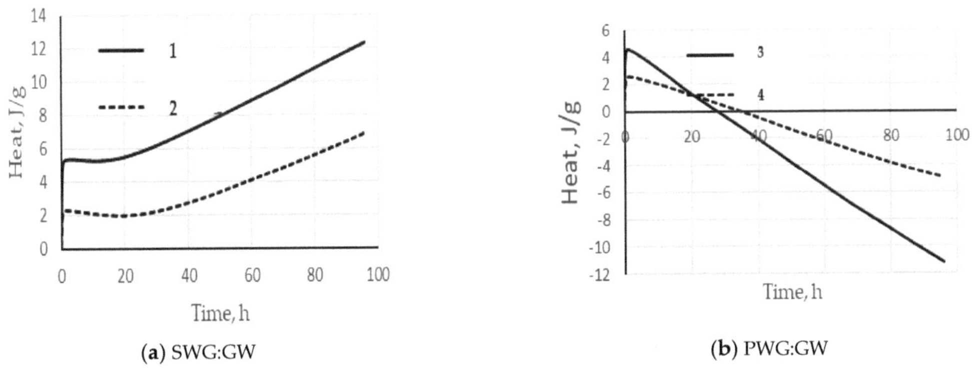

Figure 3. The process of heat release over time when adding waste glass (GW) to water glass (SWG or PWG) at different ratios: (**a**) 1—SWG:GW/1:1; 2—SWG:GW/1:2; (**b**) 3—PWG:GW/1:1; 4—PWG:GW/1:2.

Analyzing the results presented in Figure 3, it should be noted that sodium and potassium water glasses react differently with finely ground glass. In the case of SWG, heat release is observed after 20 h after contact of SWG with GW for both the 1:1 and 1:2 ratios. In the second case, a decrease in the amount of heat by 2.7 times is observed. This may indicate that the exotherm of the process depends on the mass ratio of the reacting components with the main role played by the aggregate—finely ground waste glass.

As is known [37,59], finely ground glass waste belongs to the group of chemically stable glasses, the weak link of which is the alkaline component. In this case, alkaline cations modifiers are connected in the structural network by an ionic bond with a non-bridging oxygen anion. The value of this bond is about 600–1100 kJ/mol [60], which allows the alkaline cation, under certain conditions, to easily leave its place and take part in chemical reactions of glass with various chemical reagents present in the solution. Therefore, one would expect an increase in the alkaline activity of the glass polymer; however, as studies have shown, upon contact with glass powder with water glass, which is known to have an alkaline environment with a pH of 11–12, when glass powder is added, a decrease in alkaline activity is observed from 1.94 to 1.70 ppm/dm^2. At the same time, the contribution of glass alkalis is almost 25 times lower than that of water glass; see Table 5. For all compositions given in Table 5, a decrease in the alkaline activity of glass polymers is observed with the introduction of finely ground glass. This contradiction can be explained by reducing the proportion of aqueous glass with a higher alkaline activity per unit surface when it is replaced with glass powder with a significantly lower alkaline activity. The same pattern is confirmed when the SWG:GW ratio increases from 1:1 to 1:2, as shown in Table 5.

As is known, alkalis belong to the reagents of the second group [56], which dissolve glass. In this case, the entire chemical component of the glass goes into solution. Thus, the chemical interaction of water glass with finely ground glass waste can be considered a process of slow dissolution of the latter in an alkaline glass polymer environment, which is accompanied by heat release, as shown in Figure 3.

In the case of potassium water glass, a completely different picture takes place; the interaction process is accompanied by heat absorption, as shown in Figure 3. This absorption may be associated with the replacement of potassium cations in water glass with sodium cations released by the glass powder with the formation of a phase of the Na_2O composition, SiO_2 nH_2O. Thus, two phases may be present in an ionic solution of potassium water glass: $((Na_2O \cdot mSiO_2) + (K_2O \cdot mSiO_2)) \cdot nH_2O$. The combined presence of two types of alkalis in a water glass solution leads to a polyalkaline effect [61,62], which is expressed as a decrease in the alkaline activity of the mixture, Table 4.

The results presented in this article indicate that finely ground waste glass is an active aggregate in glass polymers based on water glass. The mechanism of interaction of the polymer with the alkaline component of the glass depends on the type of water glass. The result of this interaction is a change in the structure of the glass polymer, which is reflected in the strength cohesion, as shown in Table 3.

Since cohesive strength characterizes the mechanical strength of a material, the effect of glass powder on the cohesive strength of a glass polymer was evaluated by determining its mechanical strength. Table 6 presents the results of determining mechanical strength depending on the composition of the glass polymer and hardening time.

Table 6. Dependence of impact strength and compressive strength of glass polymer samples on composition and hardening time.

N	Composition of Glass Polymer	Impact Strength (J)			Compressive Strength (MPa)		
		by 5 Days	by 9 Days	by 16 Days	by 5 Days	by 9 Days	by 16 Days
1	SWG:GW/1:1	0.213	0.220	0.225	2.86	4.67	5.83
2	SWG:GW/1:2	0.222	0.230	0.234	2.86	4.70	5.85
3	SWG:GW/1:3	0.168	0.171	0.201	2.44	3.82	4.80

The results presented in Table 6 show that for composition N1, the increase in impact strength over 11 days is 5.2%; for N2, it is 5.4%; and for position 3, it is16.4%. The compressive strength increases by 2 times, 1.7 times and 2 times, respectively. Thus, the mechanical strength of the glass polymer increases as the curing time increases, which may be due to the compaction of the structure due to dehydration. In addition, it should be noted that the strength characteristics decrease with an increase in the amount of glass powder in the glass polymer composition.

It is known that any adhesive's ability can be assessed by the adhesion value, i.e., the bond that occurs upon contact of the surface layers of two dissimilar bodies, in our case, liquid glass and glass powder substrate. As a substrate, finely ground glass waste is used on one side, and the surface of cement concrete is on the other side. Analysis of existing theories of adhesion suggests that in the case of water glass as an adhesive and fine glass as an aggregate, adhesion is due to their chemical interaction at the phase boundary. On the one hand, we have an ionic solution of aqueous glass, and on the other, we have solid glass particles, on the surface of which there are structural fragments of $Si\text{-}O^{-}Na^{+}$ with an ionic bond with a binding energy from 600 to 1100 kJ/mol and Si-OH with a hydrogen bond with a binding energy from 1 to 25 kJ/mol [60]. Comparing these values with the $Si\text{-}O^{-}$ covalent bond energy (60–700 kJ/mol), we can conclude that the chemical bond between water glass and glass powder particles is provided to the greatest extent by the hydrated surface of the glass powder according to the following scheme:

$$\text{Si-O-H(GW)} + \text{OH}^{-}\text{(SWG)} \rightarrow \text{Si-O-Si (phase boundary)} + \text{H}_2\text{O} \quad (1)$$

To a small extent, alkali metal cations will play a role in the formation of interfacial bonds according to the following scheme:

$$\text{Si-O}^- + \text{Na(GW)} + \text{OH}^-\text{(SWG)} \rightarrow \text{Si-O-Si(phase boundary)} + \text{NaOH} \quad (2)$$

In the latter case, only cations with the smallest ionic bond values will take part in the formation of interfacial bonds.

In the case of the dealcalization of glass powder, the extraction of alkali cations from the surface of glass grains takes place, and their replacement is with hydrogen cations of water glass. Unfortunately, it is impossible to determine the degree of increase in the cohesive strength of the glass polymer, because tensile strength allows us to determine only the cohesive strength of concrete, which is significantly lower than the cohesive strength of the glass polymer.

When water glass or a glass polymer is applied to the surface of concrete, a mechanical interaction occurs in contrast to the contact of water glass with finely ground glass particles. The interaction is based on the penetration of water glass into the surface pores of the concrete and the formation of a so-called root layer, which ensures strong adhesion between the water glass and the concrete surface. In this case, the penetration depth depends on the pore size, the viscosity of the water glass, the surface tension value, etc. It can also be assumed that the adsorption factor of water glass influences the mechanical interaction between the internal walls of the pores and the surface of the concrete. Thus, in a glass polymer used as an adhesive, three adhesion mechanisms occur: mechanical, chemical, and adsorptive.

It should be noted that the structure of a glass polymer filled with waste glass is very similar to the structure of glass–ceramic materials, in which the solid crystalline phase makes up 80–90%, and the rest is a glassy layer. Moreover, the mechanical strength of such materials significantly exceeds the strength of the glassy layer; for example, the bending strength of glass is about 5–7 kg/mm^2, and that of glass ceramic is 7–35 kg/mm^2. A similar picture is observed in the case of glass polymer. The cohesive strength of sodium water glass as an interlayer is ~0.6 MPa, and the cohesive strength of glass polymer is >5 MPa, which exceeds the tensile strength of cement concrete.

4. Conclusions

This study has been successfully demonstrated by the potential of waste glass as an active aggregate in the production of glass polymers utilizing water glass as a binder. The research findings indicate the following.

The addition of finely ground waste glass to water glass significantly enhances the mechanical properties of the resultant glass polymer. This includes increases in both adhesive and cohesive strengths, which surpass those of traditional concrete, particularly after 7 days of curing where the tensile strength reaches up to 4.11 MPa.

The study identified optimal proportions of waste glass in adhesive grouts, balancing the chemical interactions and mechanical strengths. This optimization contributes to the efficiency of material use and sustainability in construction.

The modified glass polymers exhibit strong adhesion capabilities and are able to pull out fragments from the concrete substrate, suggesting significant potential for commercial applications in construction, especially in contexts requiring enhanced durability and strength.

The work established that the adhesion strength is greatly influenced by the alkaline activity of powdered glass waste, which, as is known, can fluctuate significantly due to poor-quality segregation. In this regard, to improve the quality of the glass polymer, it is proposed first to determine the alkaline activity of individual batches of waste glass and, if necessary, to adjust their alkaline activity by dealkalization from two components—sodium water glass and powdered silicate industrial waste glass.

By incorporating waste glass, the study contributes to waste reduction and promotes sustainability in the construction industry. The approach not only helps to manage glass waste but also reduces the environmental footprint of building materials.

5. Patents

PL237507B1 Method of determining the alkaline activity of cement products.

Author Contributions: Conceptualization, W.S.; Methodology, W.S. and G.K.; Validation, W.S. and G.K.; Formal analysis, W.S. and G.K.; Investigation, W.S. and G.K.; Writing—original draft preparation, W.S. and G.K.; Writing—review and editing, G.K. All authors have read and agreed to the published version of the manuscript.

Funding: This paper was co-financed by the Faculty of Civil Engineering, Mechanics and Petrochemistry research grant, Warsaw University of Technology.

Institutional Review Board Statement: Not applicable.

Informed Consent Statement: Not applicable.

Data Availability Statement: The original contributions presented in the study are included in the article, further inquiries can be directed to the corresponding author.

Conflicts of Interest: The authors declare no conflicts of interest.

References

1. Blengini, G.A.; Busto, M.; Fantoni, M.; Fino, D. Eco-Efficient Waste Glass Recycling: Integrated Waste Management and Green Product Development through LCA. *Waste Manag.* **2012**, *32*, 1000–1008. [CrossRef] [PubMed]
2. Almendro-Candel, M.B.; Jordán Vidal, M.M. Glasses, Frits and Glass-Ceramics: Processes and Uses in the Context of Circular Economy and Waste Vitrification. *Coatings* **2024**, *14*, 346. [CrossRef]
3. Butler, J.H.; Hooper, P.D. Glass Waste. In *Waste*; Elsevier: Amsterdam, The Netherlands, 2019; pp. 307–322.
4. Delbari, S.A.; Hof, L.A. Glass Waste Circular Economy—Advancing to High-Value Glass Sheets Recovery Using Industry 4.0 and 5.0 Technologies. *J. Clean. Prod.* **2024**, *462*, 142629. [CrossRef]
5. Bisikirske, D.; Blumberga, D.; Vasarevicius, S.; Skripkiunas, G. Multicriteria Analysis of Glass Waste Application. *Environ. Clim. Technol.* **2019**, *23*, 152–167. [CrossRef]
6. Dyer, T.D. Glass Recycling. In *Handbook of Recycling*; Elsevier: Amsterdam, Netherlands, 2014; pp. 191–209.
7. Kazmi, D.; Williams, D.J.; Serati, M. Waste Glass in Civil Engineering Applications—A Review. *Int. J. Appl. Ceram. Technol.* **2020**, *17*, 529–554. [CrossRef]
8. Beerkens, R.G.C.; Santen, E.V. Recycling in Container Glass Production: Present Problems in European Glass Industry. In Proceedings of the 66th Conference on Glass Problems: Ceramic Engineering and Science Proceedings, Urbana, IL, USA, 24–26 October 2005; pp. 181–202.
9. Bristogianni, T.; Oikonomopoulou, F. Glass Up-Casting: A Review on the Current Challenges in Glass Recycling and a Novel Approach for Recycling "as-Is" Glass Waste into Volumetric Glass Components. *Glass Struct. Eng.* **2023**, *8*, 255–302. [CrossRef]
10. Moser, H. Glass Recycling Technology of Today. In Proceedings of the 72nd Conference on Glass Problems: A Collection of Papers Presented at the 72nd Conference on Glass Problems, Columbus, OH, USA, 18–19 October 2011; pp. 217–220.
11. Sordoń-Kulibaba, B. Zagospodarowanie Odpadów Szklanych. *Świat szkła* **2008**, *7–8*, 46–49.
12. Duda, J.; Szamałek, K. *Energia i Środowisko w Technologiach Materiałów Budowlanych, Ceramicznych, Szklarskich i Ogniotrwałych: Praca Zbiorowa*; Energia: Warsaw, Poland, 2010; ISBN 9788362105311.
13. Thomason, J.L. Glass Fibre Sizing: A Review. *Compos. Part A Appl. Sci. Manuf.* **2019**, *127*, 105619. [CrossRef]
14. Bignozzi, M.C.; Saccani, A.; Barbieri, L.; Lancellotti, I. Glass Waste as Supplementary Cementing Materials: The Effects of Glass Chemical Composition. *Cem. Concr. Compos.* **2015**, *55*, 45–52. [CrossRef]
15. Khan, M.N.N.; Saha, A.K.; Sarker, P.K. Reuse of Waste Glass as a Supplementary Binder and Aggregate for Sustainable Cement-Based Construction Materials: A Review. *J. Build. Eng.* **2020**, *28*, 101052. [CrossRef]
16. Tamanna, N.; Tuladhar, R. Sustainable Use of Recycled Glass Powder as Cement Replacement in Concrete. *Open Waste Manag. J.* **2020**, *13*, 1–13. [CrossRef]
17. Ibrahim, K.I.M. Recycled Waste Glass Powder as a Partial Replacement of Cement in Concrete Containing Silica Fume and Fly Ash. *Case Stud. Constr. Mater.* **2021**, *15*, e00630. [CrossRef]
18. Ibrahim, S.; Meawad, A. Towards Green Concrete: Study the Role of Waste Glass Powder on Cement/Superplasticizer Compatibility. *J. Build. Eng.* **2022**, *47*, 103751. [CrossRef]
19. Rutkowska, G.; Lipiński, R.; Wichowski, P. Wpływ Rozdrobnionych Odpadów Szklanych Na Wybrane Właściwości Betonów Sporządzonych z Ich Udziałem. *Przegląd Nauk. Inżynieria I Kształtowanie Sr.* **2019**, *27*, 463–475. [CrossRef]

20. Zanwar, A.B.; Patil, Y.D. Utilization of Waste Glass Powder as a Cementitious Material in Concrete. *IOP Conf. Ser. Mater. Sci. Eng.* **2021**, *1070*, 012040. [CrossRef]
21. He, Z.H.; Zhan, P.M.; Du, S.G.; Liu, B.J.; Yuan, W.B. Creep Behavior of Concrete Containing Glass Powder. *Compos. Part B Eng.* **2019**, *166*, 13–20. [CrossRef]
22. Rashad, A.M. Recycled Waste Glass as Fine Aggregate Replacement in Cementitious Materials Based on Portland Cement. *Constr. Build. Mater.* **2014**, *72*, 340–357. [CrossRef]
23. Vaitkevičius, V.; Šerelis, E.; Hilbig, H. The Effect of Glass Powder on the Microstructure of Ultra High Performance Concrete. *Constr. Build. Mater.* **2014**, *68*, 102–109. [CrossRef]
24. Kamali, M.; Ghahremaninezhad, A. An Investigation into the Hydration and Microstructure of Cement Pastes Modified with Glass Powders. *Constr. Build. Mater.* **2016**, *112*, 915–924. [CrossRef]
25. Patel, D.; Tiwari, R.P.; Shrivastava, R.; Yadav, R.K. Effective Utilization of Waste Glass Powder as the Substitution of Cement in Making Paste and Mortar. *Constr. Build. Mater.* **2019**, *199*, 406–415. [CrossRef]
26. Aliabdo, A.A.; Abd Elmoaty, A.E.M.; Aboshama, A.Y. Utilization of Waste Glass Powder in the Production of Cement and Concrete. *Constr. Build. Mater.* **2016**, *124*, 866–877. [CrossRef]
27. Chandra Paul, S.; Šavija, B.; Babafemi, A.J. A Comprehensive Review on Mechanical and Durability Properties of Cement-Based Materials Containing Waste Recycled Glass. *J. Clean. Prod.* **2018**, *198*, 891–906. [CrossRef]
28. Ghamari, A.; Powęzka, A.; Kytinou, V.K.; Amini, A. An Innovative Fire-Resistant Lightweight Concrete Infill Wall Reinforced with Waste Glass. *Buildings* **2024**, *14*, 626. [CrossRef]
29. Abderraouf Belkadi, A.; Kessal, O.; Berkouche, A.; Noui, A.; Eddine Daguiani, S.; Dridi, M.; Benaniba, S.; Tayebi, T. Experimental Investigation into the Potential of Recycled Concrete and Waste Glass Powders for Improving the Sustainability and Performance of Cement Mortars Properties. *Sustain. Energy Technol. Assess.* **2024**, *64*, 103710. [CrossRef]
30. Serpa, D.; Santos Silva, A.; de Brito, J.; Pontes, J.; Soares, D. ASR of Mortars Containing Glass. *Constr. Build. Mater.* **2013**, *47*, 489–495. [CrossRef]
31. Shevchenko, V.V. ASR Effect in Glasses Used as Additives to Portland Cement. *Glass Phys. Chem.* **2012**, *38*, 466–471. [CrossRef]
32. Jiang, Y.; Ling, T.-C.; Mo, K.H.; Shi, C. A Critical Review of Waste Glass Powder—Multiple Roles of Utilization in Cement-Based Materials and Construction Products. *J. Environ. Manag.* **2019**, *242*, 440–449. [CrossRef] [PubMed]
33. Musgraves, J.; Hu, J.; Calves, L. *Handbook of Glass*; Musgraves, J.D., Hu, J., Calvez, L., Eds.; Springer Handbooks; Springer International Publishing: Cham, Switzerland, 2019; ISBN 978-3-319-93726-7.
34. Szewczenko, W. *Glass in Building*; DULP: Lviv, Ukraine, 1996.
35. Falcone, J. Silicon Compounds: Anthropogenic Silicas and Silicates. In *Kirk-Othmer Encyclopedia of Chemical Technology*; Wiley: Hoboken, NJ, USA, 2005.
36. Roggendorf, H. Water Glass. In *Encyclopedia of Glass Science, Technology, History, and Culture*; Wiley: Hoboken, NJ, USA, 2021; pp. 857–865.
37. Scholze, H. Glass Nature, Structure, and Properties. In *Glass*; Springer New York: New York, NY, USA, 1991; pp. 156–364.
38. Baliński, A. *O Strukturze Uwodnionego Krzemianu Sodu Jako Spoiwa Mas Formierskich*; Instytut Odlewnictwa: Kraków, Poland, 2009.
39. Czarnecki, L.; Broniewski, T.; Henning, O. *Chemia w Budownictwie*; Wydawnictwo Arkady: Warszawa, Poland, 2010.
40. Łukowski, P. *Modyfikacja Materiałowa Betonu*; Stowarzyszenie Producentów Cementu: Kraków, Poland, 2016.
41. Shi, C.; Roy, D.; Krivenko, P. *Alkali-Activated Cements and Concretes*; CRC Press: Boca Raton, FL, USA, 2003; ISBN 9780429180712.
42. Provis, J.L.; Van Deventer, J.S.J. *Alkali Activated Materials: State-of-the-Art Report, RILEM TC 224-AAM.*; Springer Science & Business Media: Berlin, Germany, 2013.
43. Provis, J.L. Alkali-Activated Materials. *Cem. Concr. Res.* **2018**, *114*, 40–48. [CrossRef]
44. Davidovits, J. *Geopolymer Chemistry&Applications*, 5th ed.; Institut Geopolymere: Saint-Quentin, France, 2020.
45. Ogłaza, L. Sodium and Potassium Water Glasses: Present and New Challenges. *Chemik* **2010**, *64*, 133–136.
46. *The OxyChem Sodium Silicates Handbook*; Occidental Petroleum Corporation: Houston, TX, USA, 2018. Available online: https://www.oxy.com/siteassets/documents/chemicals/products/other-essentials/silicate.pdf (accessed on 1 March 2024).
47. Rabek, J.F. Contemporary Knowledge of Polymers; 2017.
48. Available online: https://pph-Rewa.Pl/Produkt/Kulki-Szklane/ (accessed on 30 July 2023).
49. Available online: http://www.Zchrudniki.Com.Pl (accessed on 30 July 2023).
50. *PN-EN, 1936: 2010*; Natural Stone Test Methods—Determination of Real Density and Apparent Density, and of Total and Open Porosity. Polish Committee for Standardization: Warsaw, Poland, 2013.
51. *PN-EN, 196–6*; Cement Testing Methods—Part 6: Determination of Grinding Degree. Polish Committee for Standardization: Warsaw, Poland, 2019; p. 17.
52. *PN-EN 1015-12:2002*; Methods of Test for Mortars for Masonry—Part 12: Determination of Adhesion Strenngth of Hardened Rendering and Plastering Mortars on Substrates. Polish Committee for Standardization: Warsaw, Poland, 2002; p. 12.
53. PL237507B1; Method of Determining the Alkaline Activity of Cement Products. 2018; Warsaw, Poland, p. 5.
54. *PN-EN ISO 148-1:2017-02*; Metallic Materials—Charpy Pendulum Impact Test—Part 1: Test Method. Polish Committee for Standardization: Warsaw, Poland, 2017.
55. *PN-EN, 196–1:2016*; Methods of Testing Cement. Determination of Strength. Polish Committee for Standardization: Warsaw, Poland, 2016; p. 34.

56. Yashchyshyn, Y. *Glass Technology.Physics and Chemistry of Glass. Part 1*; Beskyd Bit Publishing House: Lviv, Ukraine, 2008.
57. Yashchyshyn, Y.; Vakhula, Y.; Zheplinsky, T.; Koziy, O. *Glass Technology. Part 3*; Publishing House: Lviv, Ukraine, 2011.
58. Kotsay, G.; Szewczenko, W. Extraction of Alkalis from Silicate Materials PART 2—Crystalline Silicate Materials. *Materials* **2022**, *15*, 6059. [CrossRef]
59. Handke, M. *Krystalochemia Krzemianów*; AGH. Uczelniane Wydawnictwa Naukowo-Dydaktyczne: Kraków, Poland, 2008.
60. Sedliak, J.; Mamiński, M. *Adhesives and Adhesive Processes*; SGGW Publishing House: Warsaw, Poland, 2016.
61. Szewczenko, W.; Kotsay, G. Influence of Water Glass Introduction Methods on Selected Properties of Portland Cement. *Materials* **2021**, *14*, 3257. [CrossRef] [PubMed]
62. Shevchenko, V.V.; Kotsay, G.N. The Effect of Additives of Water-Soluble Glass on the Extraction of Alkali from Portland Cement. *Glass Phys. Chem.* **2019**, *45*, 596–598. [CrossRef]

materials

MDPI

Article

Effective Carbon Dioxide Mitigation and Improvement of Compost Nutrients with the Use of Composts' Biochar

Sylwia Stegenta-Dąbrowska [1], Ewa Syguła [1], Magdalena Bednik [2] and Joanna Rosik [1,*]

[1] Department of Applied Bioeconomy, Wrocław University of Environmental and Life Sciences, Chełmońskiego Street 37a, 51-630 Wrocław, Poland; sylwia.stegenta-dabrowska@upwr.edu.pl (S.S.-D.); ewa.sygula@upwr.edu.pl (E.S.)

[2] Institute of Soil Science, Plant Nutrition and Environmental Protection, Wrocław University of Environmental and Life Sciences, Grunwaldzka Street 53, 50-375 Wrocław, Poland; magdalena.bednik@upwr.edu.pl

* Correspondence: 118318@student.upwr.edu.pl

Abstract: Composting is a process that emits environmentally harmful gases: CO_2, CO, H_2S, and NH_3, negatively affecting the quality of mature compost. The addition of biochar to the compost can significantly reduce emissions. For effective CO_2 removal, high doses of biochar (up to 20%) are often recommended. Nevertheless, as the production efficiency of biochar is low—up to 90% mass loss—there is a need for research into the effectiveness of lower doses. In this study, laboratory experiments were conducted to observe the gaseous emissions during the first 10 days of composting with biochars obtained from mature composts. Biochars were produced at 550, 600, and 650 °C, and tested with different doses of 0, 3, 6, 9, 12, and 15% per dry matter (d.m.) in composting mixtures, at three incubation temperatures (50, 60, and 70 °C). CO_2, CO, H_2S, and NH_3 emissions were measured daily. The results showed that the biochars effectively mitigate CO_2 emissions during the intensive phase of composting. Even 3–6% d.m. of compost biochars can reduce up to 50% of the total measured gas emissions (the best treatment was B650 at 60 °C) and significantly increase the content of macronutrients. This study confirmed that even low doses of compost biochars have the potential for enhancing the composting process and improving the quality of the material quality.

Keywords: composting; biochar; GHG emissions; process intensification; carbon monoxide

1. Introduction

Industrial composting is one of the most environmentally friendly ways to manage green waste. Due to the biological nature of the composting process, it leads to the emission of harmful gaseous compounds such as greenhouse gases (GHGs). During composting, oxygen (O_2) is consumed and carbon dioxide (CO_2) is released, as well as other volatile substances produced by microorganisms such as methane (CH_4) and nitrogen oxide (N_2O) [1]. CO_2 is the main gas generated during the microbial decomposition of organic matter, which plays an essential role in the balance of heat, and the rise in atmospheric CO_2 levels will have an impact on climate change [2]. CH_4 is produced under anaerobic conditions by methanogens. It has 26 times the capacity of CO_2 to absorb infrared light and exacerbate the greenhouse effect, making it a powerful greenhouse gas [3]. The composting process also generates ammonia (NH_3), which is a critical precursor to atmospheric particulate matter [4]. In the nitrogen conversion process, it is unavoidable to produce N_2O and NH_3. N_2O is a significant contributor to the depletion of the ozone layer. The warming potential of a single molecule of N_2O is 296 times higher than that of CO_2 [5]. Greenhouse gas emissions from composting are not only harmful to the environment but also negatively affect the final parameters of the stabilized material [6]. Several studies have revealed that organic waste from pesticide use contains a significant amount of mobile heavy metal fractions. This poses a risk of soil contamination and limits the possibility of their cyclic use since heavy metals are densified during the composting process [7].

In recent years, biochar has gained popularity as an effective tool for reducing gaseous emissions due to its special physicochemical properties. Biochar is a carbon-rich product from biomass burning in the absence of oxygen, called pyrolysis. This pyrolysis product is gaining attention due to its polyaromatic and microporous structures, large specific surface area, varied surface functional groups, and high cation exchange capacity, as described in the recent literature [8–11]. Biochar is widely recognized as a promising material for improving soil fertility and sequestering carbon on the centenary time scale, thereby reducing climate change [12–14]. In addition, biochar increases aggregation and the soil's ability to retain water, which is particularly positive for plants growing under water stress, increasing grain yield [15]. In the case of composting, biochar can be a particularly useful tool that positively influences the course of the aerobic stabilization process [16]. The application of biochar has a significant impact on the moisture content, oxygen availability, temperature, pH, and C/N ratio during the composting of organic waste. Additionally, biochar promotes the aeration rate, enhances gas exchange, and prevents the formation of anaerobic zones, which lead to a significant reduction in greenhouse gases and odor emissions during the composting process [17]. Biochar has a high potential for CO_2 mitigation, is durable, and has low reversal risk. It may also benefit food security and soil fertility when applied to soil [18,19]. In addition, the remarkable properties of biochar make it a particularly useful tool for the sorption of heavy metals, which reduces soil ecotoxicity [20–22].

So far, the most popular substrates for biochar production have been waste biomass with a high proportion of organic matter. However, the amounts of biochar added to compost to effectively remove CO_2 could even be 19% [23], while the production efficiency of such wood biochars is extremely low—up to 90% weight loss during production [24]. This has made the addition of such biochars less effective. There is still a lack of studies outlining the use of biochars made from biomass with a higher proportion of the mineral fraction-like compost and their effect on the composting process. The first studies are emerging on the promising effect of home compost biochars on the adsorption of volatile organic compounds (VOCs) emitted during food waste storage. The compost used in this study was primarily characterized by a low organic content (<30%), which caused a change in the specific surface area of the biochar and improved the adsorption process capabilities [25].

This study aims for the first time to test the compost biochars' capabilities to reduce greenhouse gas emissions (GHG), including CO_2 mitigation, and to improve the quality of compost by increasing the concentration of nutrients in the composting matrix. The authors provide recent findings on the GHG mitigation from the composting process by composts' biochar. Furthermore, this study provides recommendations for use of compost biochars for lowering the specific gaseous compounds emitted during the intensive phase of composting. Overwhelming evidence and benefits of using compost biochars are presented to improve the composting process and its ability to retain heavy metal contaminants.

2. Materials and Methods

2.1. Feedstock and Biochar Characteristics

The compost biochars were prepared using a laboratory muffle furnace (SNOL, model 8.1/1100, Utena, Lithuania) in retention time 1 h, temperature 550, 600, and 650 °C, heating rate of 10 $°C \cdot min^{-1}$; biochars were labeled B550, B600, and B650. The material used for compost biochars' production was mature, certified compost originating from a composting plant (Best-Eko; Rybnik, Poland). CO_2 was supplied into the chamber during the entire pyrolysis process to maintain an inert atmosphere. After carbonization, the furnace was turned off and left to cool. For this study, the compost biochars' samples were sifted through a 1 mm sieve. As a feedstock for the composting process in the experiment, a mixture of green waste and sewage sludge in a ratio of 9:1 was used originating from the composting company (Best-Eko, Rybnik, Poland).

For feedstock and biochar characteristics, some basic analyses were performed: pH in water solution in 1:10 mass ratio (Elmetron, CPC-411, Zabrze, Poland) and moisture content (MC) according to PN-EN 14346:2011 (laboratory dryer, WAMED, model KBC-65W, Warszawa, Poland) [26,27]. Ultimate analysis of elemental composition C, H, N, S was conducted according to PN-EN ISO 16948:2015-07 (Perkin Elmer, model 2400 Series, Waltham, MA, USA) [28]. The O content of compost biochars and feedstocks was determined by mass balance (O% = 100 − C% − H% − N% − ash%) [29,30].

Near-total concentrations of macronutrients (Ca, Na, Mg, K, P) and selected metals (Zn, Cd, Pb, Ni, Cr, Cu, Mn, Hg) were determined after digestion with aqua regia. Briefly, 1 g of the sample was placed in a glass falcon and 2 mL of 30% hydrogen peroxide was added, to remove the excess of organic matter. Then, samples were treated with 7.5 mL of concentrated HCl and 2.5 mL of concentrated HNO_3, left overnight and boiled for 5–6 h [31]. After the procedure has finished, the obtained suspension was filtered and filled up with distilled water to the volume of 50 mL. In the extracts, the content of Ca, Na, Mg, and K was determined on MP-AES 4200 Spectrometer (Agilent Technologies, Santa Clara, CA, USA) [32]. The other 8 elements: Zn, Pb, Cd, Cu, Cr, Ni, Mn, and P were examined on an ICP-OES (iCAP 7400, Thermo Scientific, Waltham, MA, USA) (PN-EN ISO 11885:2009 [33]). A separate procedure was applied for Hg, whose concentration was determined in the solid sample, using atomic absorption spectrometry method with the amalgamation technique (MA-2 analyzer; Nippon Instruments Corporation, Osaka, Japan).

For compost biochars' samples, the specific surface area (SSA) and FTIR-ATR spectroscopy was also measured. SSA were performed with sorption of N_2 in 77K (Micromeritics ASAP 2020; Norcross, GA, USA). Augmented total reflection–Fourier-transform infrared (ATR)-FTIR measurements were performed with a Nicolet iN10 integrated infrared microscope with Nicolet iZ10 external FT-IR module (Thermo Fischer Scientific, Waltham, MA, USA) equipped with a deuterated-triglycine sulfate (DTGS) detector and a diamond ATR module. For each spectrum, 32 scans were averaged in the mid IR range of 400–4000 cm^{-1} at a spectral resolution of 4 cm^{-1}.

2.2. Experiments Configuration

The 15 g (±0.5 g) mixture of feedstock (green waste and sewage sludge in a ratio of 9:1), was placed in the 1 dm^3 reactors (Figure 1), with the intended dose of compost biochars to individual reactors. Then, the reactors were placed in a thermostatic cabinet for incubation at the following temperatures: 50 °C, 60 °C, 70 °C (Figure 1). Gas concentration measurements were made using a portable gas analyzer in each 24 h. The measured concentrations were later converted into emissions according to the equations in Section 2.3. Biochars were added to the reactors in the following doses: 0%, 3%, 6%, 9%, 12%, and 15% per weight on a dry basis (Figure 1). The compost biochars made from compost were obtained at different temperatures and different rates of temperature increase: 550 °C 10 °C·min^{-1}, 600 °C 10 °C·min^{-1}, 650 °C 10 °C·min^{-1}, with the 1 h retention time. The decisive factor in choosing these variants was their specific surface area (research on the selection of specific types of biochars for testing was not an element of this engineering work). At the same time, during gas emission tests in the reactors, the larger sample of substrate was incubated in the same conditions in the amount of approximately 350 g ± 5 g, which was used to perform post-process analyses—pH, loss of ignition (LOI), moisture content (MC), ultimate analysis CHNS, elemental analysis: Zn, Cd, Pb, Ni, Cr, Cu, K, P, Mn, Ca, Mg, Na, and specific surface area (SSA) (only for compost biochars' samples).

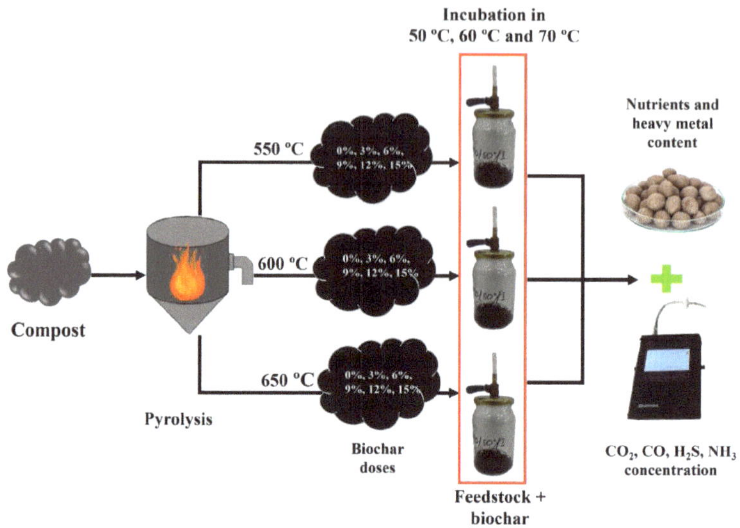

Figure 1. Experiment configuration.

2.3. Analyses of Process Gas Emissions

Measurements of selected gases (CO_2, CO, H_2S and NH_3) concentrations during the aerobic digestion process were carried out for 10 days, once a day, using a portable electrochemical gas analyzer (Nanosens DP-28 BIO; Wysogotowo, Poland). Concentrations of CO, H_2S, and NH_3 were determined in ppm in the following ranges: CO 0–2000 ppm (±20 ppm), H_2S, NH_3 0–1000 ppm (±10 ppm), and CO_2 0–100% (±2%). The measurements were made on a reactor previously removed from the thermostatic cabinet. Each measurement lasted 45 s, followed by automatic cleaning of the analyzer. After the measurements, the cover of the reactor was opened for 30 min to aerate the material. The following formulas were used to convert measured concentration of gases for emissions:

- Conversion to volume, for H_2S, CO and NH_3 (ppm):

$$V = \frac{1000 \times M \times \left(1.66 \times 10^{-24}\right) \times \left(\left(2.68839 \times 10^{22}\right) \times a\right)}{1,000,000} \quad (1)$$

where

V—gas volume;
M—molar mas;
a—accumulated concentration, ppm.

- Conversion to volume, applies to CO_2 (%) only:

$$V = \frac{1000 \times M \times \left(1.66 \times 10^{-24}\right) \times \left(\left(2.68839 \times 10^{22}\right) \times a\right)}{100} \quad (2)$$

where

V—gas volume;
M—molar mas;
a—accumulated concentration, ppm.

- Specific emissions, applies to H_2S, CO, NH_3 (ppm):

$$E = \frac{V}{DTS \times 1000} \quad (3)$$

where

E—emissions, μg·g d.m.;
V—gas volume;
DTS—dry total solids.

- Specific emissions, applies to CO_2 (%) only:

$$E = \frac{V}{DTS} \quad (4)$$

where

E—emissions, μg·g d.m.;
V—gas volume;
DTS—dry total solids.

2.4. Statistical Analysis

For statistical analysis, 13.3 Statistica software (TIBCO Software Inc., Palo Alto, CA, USA) was used. For the statistical differences of the contribution of VOCs, one-way ANOVA was applied according to Tukey's test at a significance level $p < 0.05$, including previous verification of normality and homogeneous variance using the Levene test. For all relevant cases, standard deviation (SD) was applied.

3. Results and Discussion

3.1. Compost, Feedstock and Compost Biochars' Characteristics

For biochar production, commercial certified compost from the Best-Eko composting company (Rybnik, Poland) was used, as a potential source of nutrients and a good material for gas emission mitigation, with beneficial properties enhanced in the pyrolysis process. Used compost has a typical pH value—near to neutral (Table 1)—and it is observed that with the temperature of pyrolysis, the pH increased. Biochars have a strongly alkaline pH, from 8.61 in variant B550 to 9.50 in B650. The increase in pH with the temperature of pyrolysis is typical, and similar results were observed with the pyrolysis of Japanese larch (*Larix kaempferi Sarg.*) and dairy manure (up to 11 pH) [34]. The pH values of biochars are positively correlated with the formation of carbonates and the contents of inorganic alkalis. These groups are the main cause of the alkaline reaction [35]. Higher pH has been associated with the increase in ash content and formation of oxygen functional groups that occur during the pyrolysis process. The feedstock (mix of green waste and sewage sludge) has an acidic reaction with pH 4.77, which is not optimal in composting, but is mostly observed for green waste [36]. This indicates a potentially promising use of the addition of compost biochars in neutralizing the acidic reaction of the substrate.

The compost was characterized with a huge content of mineral fraction (almost 75%) and this content increased in biochars with the pyrolysis temperature. Biochars derived from livestock manure also have more inorganic fraction—e.g., chicken manure, 40.6% d.m. [37]. In other studies, mixes of chicken manure, clay, minerals, and specific organic compounds were used to produce the organo–mineral biochar [38]. This type of biochar studied by other authors was shown to increase soil mineral content, improve soil cation exchange capacity (CEC), decrease the mineralization rate of organic matter, as well as having potential to maintain nutrients longer in the soil compared to a pure biochar.

The compost used in this study was rich in macronutrients: Ca (17,760 mg·kg^{-1}), K (6155 mg·kg^{-1}), Mg (2697 mg·kg^{-1}), Na (1793 mg·kg^{-1}), and P (5835 mg·kg^{-1}), which is typical for high inorganic content biochars [37]. In produced compost biochars, the content of nutrients become concentrated and follow the order: B650 < B600 < B550. The same trend was observed for metal content—all of them increased, except the Zn, which decreased in biochars below the level measured in compost. The content of C, H, N, S, and O decreased with the pyrolysis process, which is typical for biochars produced from substances with a low organic content and was previously observed, for example, in sewage sludge [39].

More importantly, the produced biochars were rich in P, which is a crucial element for living organisms, and there is a decreasing supply of phosphate on the market [40]. Due to the high content of P in biochars, future compost is potentially valuable for use as a fertilizer. At the same time, the metal content in all analyzed materials was significantly below the standard for fertilizers in Poland and in the EU regulations [41].

Table 1. Characteristics of compost, compost biochars' variants, and feedstock (mix of green waste and sewage sludge).

Parameters	Compost	Feedstock	Biochars' Variant		
			B550	B600	B650
pH	7.32 ± 0.04	4.77 ± 0.04	8.61 ± 0.03	9.10 ± 0.03	9.50 ± 0.04
LOI, % d.m.	24.80 ± 0.00	82.01 ± 3.70	11.0 ± 0.00	11.9 ± 0.00	9.30 ± 0.00
Ca, mg·kg^{-1}	17,760 ± 109	14,878 ± 92.5	19,795 ± 175	18,460 ± 105	15,975 ± 80
K, mg·kg^{-1}	6155 ± 65	8608 ± 30	7865 ± 65	7665 ± 60	7390 ± 65
Mg mg·kg^{-1}	2697 ± 29	2405 ± 22	3417 ± 35	3382 ± 35	2967 ± 10
Na, mg·kg^{-1}	1793 ± 1.1	773 ± 5.8	2263 ± 1.5	2168 ± 1.0	2143 ± 1.0
P, mg·kg^{-1}	5835 ± 1167	6300 ± 1260	6865 ± 1373	6953 ± 1391	6749 ± 1350
Pb, mg·kg^{-1}	35.52 ± 7.11	25.57 ± 0.70	45.81 ± 9.16	39.59 ± 7.92	31.01 ± 6.20
Cd, mg·kg^{-1}	1.66 ± 0.33	0.800 ± 0.00	1.98 ± 0.40	1.91 ± 0.38	1.15 ± 0.23
Zn, mg·kg^{-1}	553.8 ± 110.8	253.3 ± 1.5	585.3 ± 117.3	511.8 ± 102.3	426.8 ± 85.4
Cr, mg·kg^{-1}	60.7 ± 12.2	33.0 ± 1.6	141.3 ± 28.3	96.3 ± 19.3	56.6 ± 11.3
Ni, mg·kg^{-1}	11.9 ± 2.4	20.7 ± 0.05	43.2 ± 8.6	32.1 ± 6.4	10.6 ± 2.1
Cu, mg·kg^{-1}	63.7 ± 12	44.03 ± 0.2	72.7 ± 13.8	62.1 ± 11.7	45.2 ± 8.3
Mn, mg·kg^{-1}	362.3 ± 72.5	253.3 ± 1.5	416.8 ± 83.4	391.5 ± 78.3	341 ± 68.2
Hg, mg·kg^{-1}	0.106 ± 0.03	0.08 ± 0.02	<0.001	<0.001	<0.001
C, %	13 ± 3	35.89 ± 4.1	8.3 ± 1.7	9.3 ± 1.8	10 ± 3
H, %	1.2 ± 0.2	4.82 ± 0.9	0.21 ± 0.04	0.23 ± 0.05	0.05 ± 0.2
N, %	1.1 ± 0.2	2.37 ± 0.2	0.48 ± 0.1	0.62 ± 0.12	0.44 ± 0.2
S, %	0.38 ± 0.08	1.64 ± 0.05	0.31 ± 0.06	0.33 ± 0.07	0.31 ± 0.08
O, %	8.7 ± 0.4	37.29 ± 1.9	1.7 ± 0.3	0.82 ± 0.2	0.4 ± 0.4

The important parameter in the assessment of biochars' absorption properties is their specific surface area (SSA). In the present study, values of SSA (Figure 2) sharply increased along with the temperature of pyrolysis, from 6.06 (B550) to 39.2 m^2·g^{-1} (B650). Kumar et al. [16] showed that, in general, municipal solid biochars and livestock manure biochars have a lower SSA and porosity than plant biochars, e.g., the authors of [42] observed an increase in SSA parameters for bamboo biochars to 181.05 m^2·g^{-1} at 600 °C, in comparison to chicken manure biochars—11.84 m^2·g^{-1} at 500 °C [37].

FTIR-ATR Spectroscopy

The spectra of the compost sample shows (Figure 3): a band at 3320 cm^{-1} that can be ascribed to C–H bonds and to O–H of alcohols, phenols, or O–H carboxyl and also to N–H vibrations in amide functions; two characteristic bands at 2925 and 2845 cm^{-1} that may be attributed to asymmetric and symmetric vibrations of C–H stretching of CH_3 and CH_2 groups [43]. The peaks at 1700 and 1600 cm^{-1} were assigned to C=O stretching for carboxyl groups and aromatic C=C bonds, respectively [44]. The peak at 1270 cm^{-1} was attributed to phenolic–OH stretching [45]. Typical for the pyrolysis, all of the bands listed above that were characteristic of compost disappeared, when no new bands typical of biochar samples appeared, as was previously observed in wooden biochars [34]. This could be an effect of changes in the proportion between mineral and organic fraction—at the end of pyrolysis, 90% of mineral fraction was observed. This indicates that some of the chemical bonds were broken. In compost biochars, when compared to raw compost, the peak at 1270 cm^{-1} attributed to phenolic–OH stretching was invisible, when strong absorption peaks at 750–870 cm^{-1} belonging to stretching vibrations of aromatic ring C–H or C–N, R–O–C, or R–O–CH_3 groups appear [37]. The strong peak at around 1020 cm^{-1} indicates the combination of C–O stretching of polysaccharides, in addition to Si–O–Si bonds of silica, and to the group Si–O–C that was visible in all samples [44].

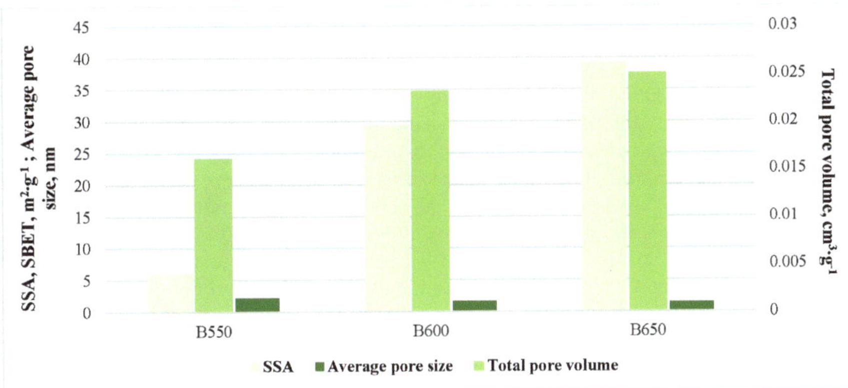

Figure 2. BET results for compost biochars.

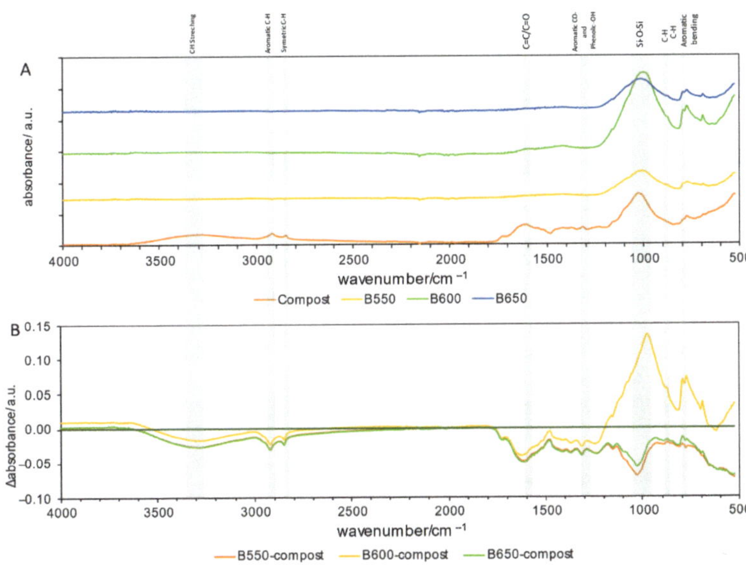

Figure 3. FTIR results of compost and compost biochars; (**A**) shows the FTIR spectra of compost and compost biochars (B550, B600, B650), and (**B**) shows the subtraction spectra of samples. The line indicates zero absorbance change for each subtraction spectrum.

3.2. *Effect of Biochars on Composting*

3.2.1. Effect of Biochars on pH and LOI during the Initial Stage of Composting

The material incubated for 10 days was subjected to pH and loss on ignition (LOI) testing (Supplementary Materials Table S1). In the case of pH, values of a mildly alkaline reaction were observed, especially in the reactors incubated at 50 °C. In the case of biochar doses, an increase in the amount of hydrogen ions relative to the control sample was observed regardless of the production temperature. In the material incubated at 60 °C, it was observed that the pH values for the biochar produced at 650 °C differed significantly from the biochar produced at 550 °C and 600 °C. In the material with the addition of B650, a slightly more acidic environment was present than in the case of the other biochar; only for the dose of 15% was there an increase in pH, which finally amounted to 8.54. In contrast, for the materials incubated at 70 °C, the pH decreased, relative to the other temperatures, which was most likely due to the intensification of the composting process [46]. No significant correlations were observed between the biochar dose and the amount of hydrogen ions. The LOI values in the vast majority of cases ranged from about 70% d.m. to as high as 87% d.m., and a decrease in the value of the parameter was observed with the increasing biochar dose. In addition, this study showed that the initial LOI for biochar produced at 650 °C tended to be higher than for other production temperatures. High parameter values indicate a high content of organic matter in the material.

Sundberg et al. studied the effect of pH and microbial content on the odors of composted food waste. The researchers examined pH values for materials composted on days 3, 8, and 16 of the process, and incubated the material at 50 °C and 70 °C. In each case, the results indicated an increase in pH over time, while samples incubated at 70 °C reached higher values much faster [47]. On the other hand, Wasaq et al. optimized the composting of food waste using biochar produced at temperatures of 350 °C and 450 °C. Doses of 10% and 15% biochar were applied to the compost. The study showed that both the dosage and the type of biochar used had a significant effect on the final pH of the material after the process. The hydrogen ion content of the 10% dose was lower than that of the 15% dose, and a similar relationship was observed for the temperature of biochar production. The addition of biochar had a beneficial effect on the ammonification and denitrification process, which also correlated with an increase in the pH value of the composted materials [48]. Czekała et al., in their study on the co-composting of pig manure with the addition of biochar, also observed an increase in pH depending on the biochar dose [49]. Jia et al. [50], Jindo et al. [51] and Zhang [52] linked a rapid increase in pH to accelerated fatty acid degradation and NH_3 emissions. The high pH value in the compost can also be linked to the high content of hydrogen cations in the biochar studied [44].

3.2.2. The Emission Change

During the initial phase of composting, some potentially harmful gases are released. In this study, the CO_2, H_2S, NH_3, and CO emissions, and the effect of different type of compost biochars and the doses were measured (Supplementary Material Table S1, Figure 4). During the thermophilic phase, the composting temperature can increase from 50 to 80 °C. The thermophilic phase T > 55 °C usually lasts 7–10 days, so 50, 60, and 70 °C temperature variants were tested to simulate the intensive phase of composting [53]. The observed emissions at variant 0—without the addition of biochar was:

- CO_2
    - 429.1–565.0 mg·g^{-1} d.m. for incubation at 50 °C;
    - 410.8–522.5 mg·g^{-1} d.m. for incubation at 60 °C;
    - 153.4–270.9 mg·g^{-1} d.m. for incubation at 70 °C.
- CO
    - 477.7–653.5 µg·g^{-1} d.m. for incubation at 50 °C;
    - 503.5–861.9 µg·g^{-1} d.m. for incubation at 60 °C;
    - 508.8–784.1 µg·g^{-1} d.m. for incubation at 70°C.
- H_2S
    - 60.4–85.5 µg·g^{-1} d.m. for incubation at 50 °C;
    - 129.4–191.5 µg·g^{-1} d.m. for incubation at 60 °C;
    - 52.8–85.3 µg·g^{-1} d.m. for incubation at 70 °C.
- NH_3
    - 0.0–34.1 µg·g^{-1} d.m. for incubation at 50 °C;
    - 59.0–188.5 µg·g^{-1} d.m. for incubation at 60 °C;
    - 0.0–119.1 µg·g^{-1} d.m. for incubation at 70 °C.

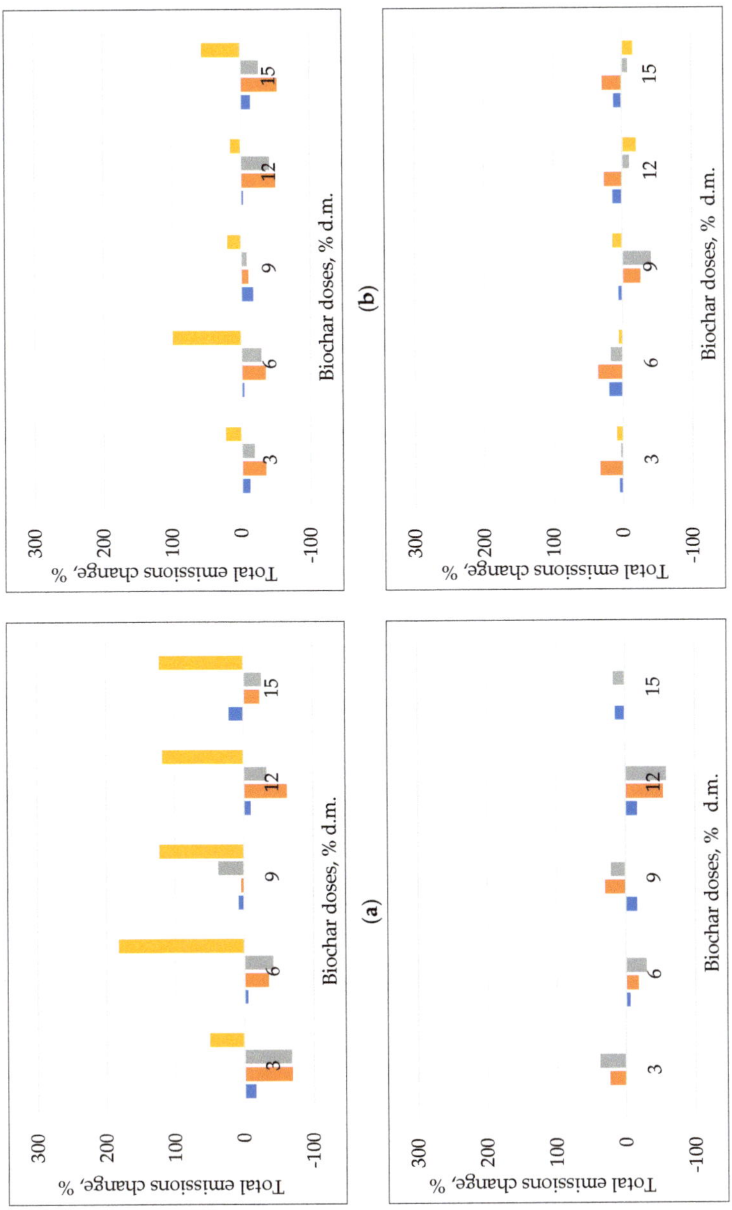

Figure 4. *Cont.*

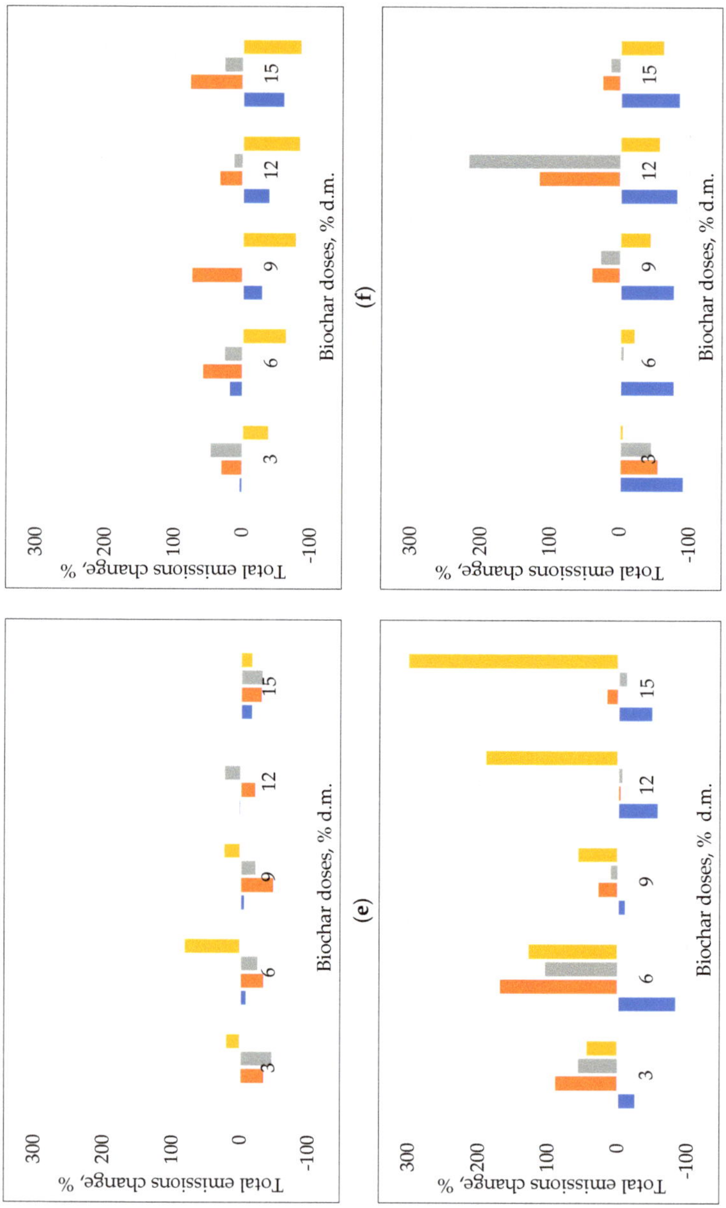

Figure 4. *Cont.*

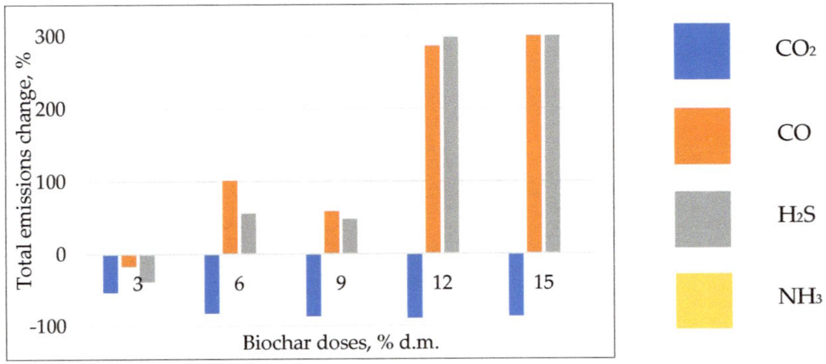

(i)

Figure 4. Total gas emission change, during the first 10 days of composting process at laboratory scale, (**a**) B550, incubation at 50 °C, (**b**) B600, incubation at 50 °C, (**c**) B650, incubation at 50 °C, (**d**) B550, incubation at 60 °C, (**e**) B600, incubation at 60 °C, (**f**) B650, incubation at 60 °C, (**g**) B550, incubation at 70 °C, (**h**) B600, incubation at 70 °C, (**i**) B650, incubation at 70 °C.

CO_2 Emissions

CO_2 was the main gas generated during the bio-oxidative phase of composting. Additionally, CO_2 is rarely produced as a pure component [54]. The adsorption of CO_2 from gas mixtures requires knowledge of the impact of the other gases on the adsorption capacity. Gases can compete for adsorbent sites, form complexes with adsorbed compounds, and interact in the gas phase which will impact adsorption efficiency [55]. There are multiple articles available discussing biochar adsorption of pure CO_2 from gas mixtures [55,56], although the use of biochar for CO_2 capture in a composting matrix is described less [57]. The observed trend for CO_2 emissions was similar at 50 and 60 °C, and dynamic from the beginning of the observation (Supplementary Materials Figure S1). In contrast, at 70 °C, the CO_2 emissions became more static. The CO_2 cumulative emissions were ~500 CO_2 mg·g d.m. at 50 and 60 °C, and <200 CO_2 mg·g d.m. at 70 °C (Supplementary Materials Table S2). In other observations, after 10 days of composting, those emissions were similar (120–170 mg·g d.m.), but the authors showed that the form of the applied biochars is important, especially after 10 days of composting—granulation of biochars result in CO_2 mitigation [1]. On the other hand, the granular biochar resulted in the faster cooling of the composted material, which could mean that the smaller amount of CO_2 was simply due to the slower decomposition process of the organic matter. In other research, the authors proved that the addition of bamboo biochar is effective in reducing CO_2, but during the cooling and maturity period of the composting. Yan et al. proved that biochar can reduce the abundance of key CO_2 emitting enzymes in the TCA cycle at the later stage of composting, thereby reducing carbon losses [58]. The addition of biochar usually decreased CO_2 emissions especially in doses of 3 and 6% d.m. (Supplementary Materials Table S2). The most effective treatment for the mitigation of CO_2 emissions was the addition of biochars at a temperature of 70 °C, when the total reduction was above 80% compared to variants zero (without the addition of biochar). A much lower reduction was observed at 50 and 70 °C, usually <20%. At those temperatures, the most effective biochar was B650. Those observed reductions in CO_2 were much higher than previously observed by Czekała et al. [49], which was 6.9 and 7.4% with a dose of 5 and 10%, respectively. In other research, emitted CO_2 was 2 and 3% higher with the addition of biochars than in the control [59]. In other experiments, bamboo biochar amendments reduced CO_2 emission by 16.77% [58].

Unfortunately, in some cases, increases in CO_2 emissions were observed, especially at 70 °C, and with biochar doses above 10% d.m. (Supplementary Table S1). The most effective CO_2 reduction with the addition of biochar was at a temperature of 70 °C—the total reductions were up to 80% (Figure 3). Ottani et al. observed that the addition of 3–5% d.m. biomass biochar emitted 5% more CO_2 than the control, with the p-values of the Tukey test being not significant. Ottani et al. explained that the addition of biochars facilitated the aerobic degradation processes of the organic substance, causing an increase in CO_2 in the first 10 days of composting [60].

CO Emissions

It was observed that emissions of CO increased with time and temperature (Supplementary Materials Figure S2), which was previously indicated with our own study [61]. This is an effect of the thermochemical reactions and biological activities of microorganism. The highest emissions of CO were registered at the beginning of the composting process, as found in other composting matrices like green waste, manure, and municipal solid waste [62,63]. The biggest emissions of CO were observed in variants with the addition of B650 ~2000 μg·g d.m. All of the tested biochars show the highest emissions of CO at 70 °C (up to 1200 $\mu g \cdot g^{-1}$ d.m.), in comparison to other variants where those cumulative emissions were ~500–800 $\mu g \cdot g^{-1}$ d.m.

The best reduction in CO emissions was observed at a temperature of 50 °C, with a 30–60% reduction for B550, and 20–30% for B600. Variant B650 showed no clear effect, and this biochar was less effective, especially at 70 °C, when massive emissions of CO occurred (>300% increase in observed emissions compared to variant 0). Increasing the dose of B550 and B600 usually did not have a specific effect for the decrease in CO, only a slight reduction was observed with dose 3, and 12% on dry matter. B550 was the most effective in CO mitigation at all incubation temperatures. Although CO production through composting is well documented, observations of the addition of biochars for CO emissions is very limited. In this study, significant changes in CO emissions were observed; meanwhile, in other research, average CO emissions did not statistically differ between 3% biochar and control piles [62]. Hellebrand and Schade demonstrated that the CO produced during composting is generated by abiotic reactions and not by microorganisms [64], but our own research shows that CO emissions differ in biotic and abiotic conditions [63], and there are a lot of different pathways when CO can be generated during the composting by microorganisms [65]. As biochar can positively affect the microbial community—the population of actinomycetes, cellulolytic and proteolytic bacteria showed increases over the period of composting [66]—this effect should be examined in future studies.

H_2S Emissions

H_2S has the highest odor potential generated during the composting process and is produced by the decomposition of sulfur-containing organic components by sulfate-reducing bacteria under anaerobic conditions. Most biochars are alkaline [67], and have a relatively high catalytic center dispersion in the pore system, thus, making them conducive for the effective oxidation of H_2S. The emissions decreased with temperature, which was consistent with previous studies, when H_2S was released mainly in the thermophilic phase (Supplementary Materials Figure S3). The H_2S emission rate increased rapidly in all treatment groups, which was mainly attributed to the reduction in O_2 content in the piles due to the rapid decomposition of OM by aerobic microorganisms. H_2S was generated from two main pathways: the anaerobic decomposition of proteins and other sulfur-containing compounds coupled with the anoxic synthesis of sulfate-reducing bacteria [68]. The observed cumulative emissions were similar to those observed before by Ouyang, when the maximum emissions of all treatments were 200 $\mu g \cdot g^{-1}$ d.m. [69], during sludge aerobic composting. In this study, highest emissions were observed with the addition of B650, >200 $\mu g \cdot g^{-1}$ d.m., and the lowest with the B550 variant, ~50 H_2S $\mu g \cdot g^{-1}$ d.m. at a temperature of 50 °C and 100 H_2S $\mu g \cdot g^{-1}$ d.m. at 60 °C and 70 °C. Ouyang et. el. observed that the addition of biochar reduced from 12.91% to 50.47% of H_2S. The biggest H_2S reduction was observed at 50 °C with the addition of B550, but there was no direct effect of the biochar dose. Observed emissions mostly increased with temperature, and the effect of compost biochars became unclear—both the reduction and increases in H_2S emissions with the addition of biochar were observed (Figure 4). Ky Nguyen et al. suggested that biochar could reduce the H_2S emission, increasing with a proportion of biochar proportion of up to 20% [70]. On the other hand, ventilation of the piles and enhancing the microbial activity, along with effective reduction in the local anaerobic zone in the piles can reduce the sources of H_2S production. Awasthi et al. showed that the pore diameter of biochar in the range of 25~45 μm was favorable for H_2S adsorption [71]. According to the BET results, the pore diameter of compost biochars was 2.2, 1.6, and 1.4 nm for B550, B600, and B650, respectively, which is below the favorable level for H_2S mitigation.

NH_3 Emissions

Biochars influence the productivity of the composting process by enhancing microbial performance, altering physicochemical properties, enhancing substrate degradation, and resulting in rapid humification and gas emissions [72,73]. Biochars can be also used for effectively reducing ammonia emissions during composting, but the biochar properties affect NH_3/NH_4^+ adsorption capacity and microbial growth, which may result in reducing or promoting ammonia emissions in the composting process [34]. In this study, observed emissions were in the range of 0–200 $\mu g \cdot g^{-1}$ d.m, with no specific pattern—no effects of biochar, dose, or temperature of composting were observed (Figure 4; Supplementary Materials Figure S4). Recorded emissions were much lower than those observed by He et al. who reported 400 μg·g^{-1} d.m. (powdered biochar) to even 800 μg·g^{-1} d.m. (granular biochar) in 10 days of pig manure/wheat straw composting and confirmed that powdered biochar is most suitable for controlling NH_3 emissions [1]. Additionally, the addition of compost biochars' usually has an effect on the significant increase in observed emissions. It was observed that low-temperature biochars (300 °C) mitigated NH_3 emissions better than high-temperature biochars (700 °C), possibly due to higher CEC [34]. The temperature used for production of biochars in the range of 550–650 °C could be not optimal for effective NH_3 mitigation. As an example, a high SSA is reportedly required for biochars to be effective composting amendments because they may enhance adsorption capacity and microbial activity [74]. High-temperature biochars (B600, B650) with higher SSAs reduced NH_3 emissions during composting slightly better than B550 (Supplementary Table S1). In addition, the pores on the biochar surface can become partially blocked during composting, which can promote anaerobic conditions, as evidenced by the observed H_2S emissions. Pore blockage can limit the potential of high SSA to mitigate emissions during composting as this interferes with the penetration of adsorbents and microorganisms into the composting matrix [75]. Furthermore, as the pyrolysis temperature increases, biochars become more aromatic, and toxicity increases (PAHs) [76]. These factors may decrease and change microbial activity (promote nitrification or denitrification) during composting, which has an effect on the observed greater emissions of ammonia. Other possible explanations of the low effectiveness of compost biochars for ammonia mitigation could be an increase in the content of nutrients and porosity which provide favorable conditions for microbial growth and hence accelerate the immobilization of ammonia into the microbial biomass [17].

3.2.3. Effect of Biochars on Nutrients and Heavy Metal Content during Initial Stages of Composting

C, H, N, S Changes

The elemental composition of the compost was evaluated by determining the content of C, H, N, and S in the biochar-amended feedstock after 10 days of incubation. In the case of carbon content in the post-incubated feedstock (Figure 5), the highest amounts were observed for samples with the addition of biochars produced at the temperature of 650 °C. It is widely documented that the pyrolysis temperature is crucial for the carbonization rate of biochar. High-temperature biochars are characterized by a polymerized, aromatic structure [35,77], which results in increased carbon content and was confirmed in our study. Although the effect of the condition of pyrolysis on compost biochars' C content is clear, the dose of biochar did not have a significant effect on the C content in feedstock among the tested treatments, since for most doses the differences in C content were insignificant, reaching very similar values (no significant statistical change was observed). The mixing of stable, well-carbonized biochar with easily decomposable compost has been suggested as a solution for the enhancement of process performance and prevention of carbon losses [78]. For example, Gao et al., who tested the effect of co-composting with 300–400 °C biochar, noted that the amendment significantly increased the C content of composted material, which rose to 32.0% from 24.8% [79]. In the research of Darby et al., the addition of biochar produced at 600 °C increased the C content of the co-composted material from 28.2% to 35.1% [38]. However, in our study, the significant effect of biochar on C content in composted feedstock was not confirmed, which may be an effect of the relatively short time of the experiment. Nevertheless, the results of C content stay in line with CO_2 emissions, as the highest carbon content was noted in treatments with lowest carbon dioxide emissions—incubated at 70 °C and amended with B650. It is noteworthy that the C content measured in our experiments is similar to findings of the aforementioned authors, reaching 35–40%.

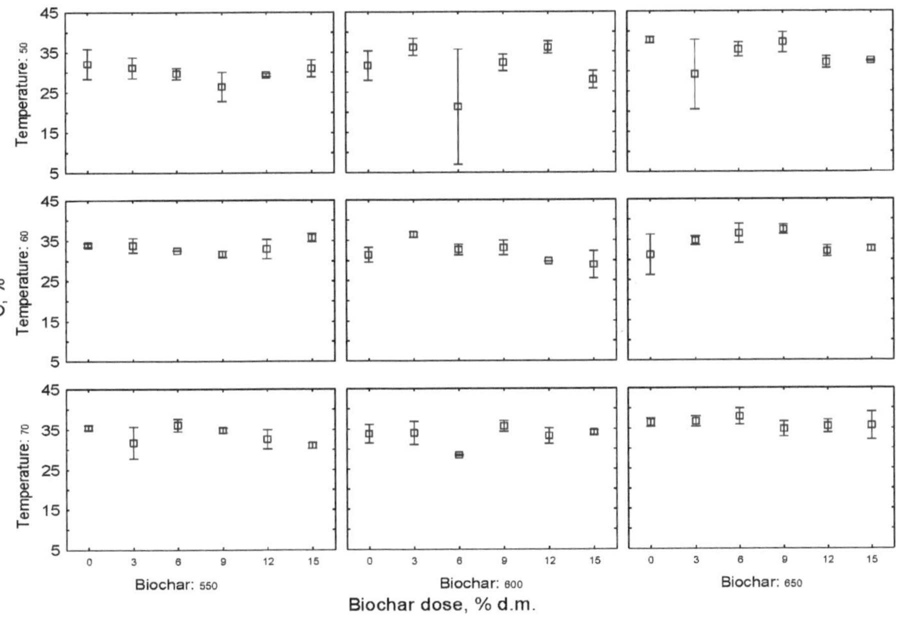

(**a**)

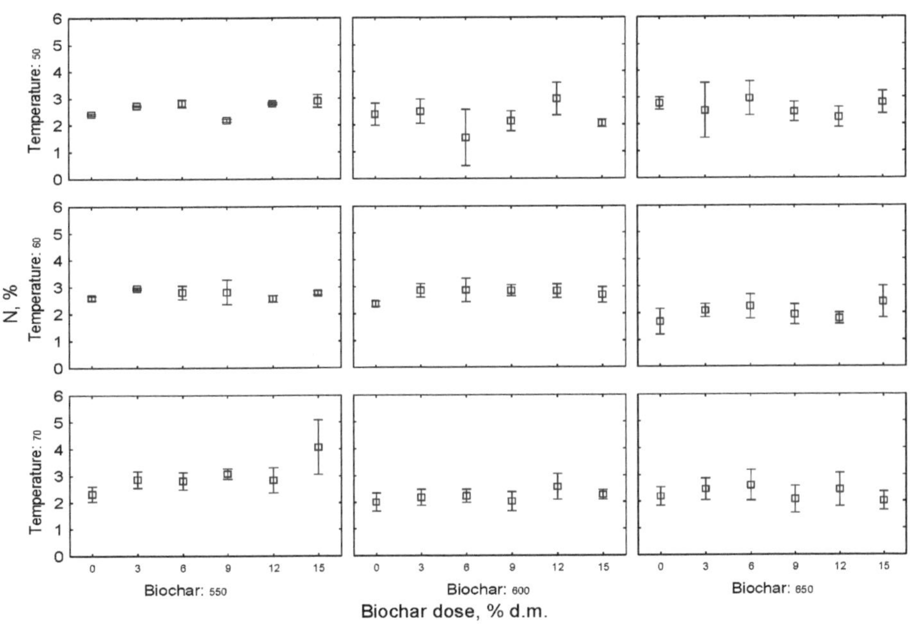

(**b**)

Figure 5. Effect of biochar dose for (**a**) C, % and (**b**) N, %, during the first 10 days of composting process at a laboratory scale; no statistical differences were observed.

Carbon and nitrogen are known as sources of energy for microbes, determining microbial activity in composted material. In general, biochar amendment does not directly supply the nitrogen and its impact on the compost quality results from an effect on nitrogen dynamics [80]. Biochars have the ability to absorb nitrogen ions, thus reducing the losses and enhancing the content of the element in composted material. Nitrogen content in analyzed samples varied between 2 and 3%, with no statistically significant effect of the biochar dose. Despite no statistical differences, it can be noted that in the majority of treatments, N content in samples without biochar amendments was noticeably lower than in treatments with 3–12% of biochar. As the incubation temperature of the material increases during the composting process, the percentage of the element decreases. Lower values of N were also seen in the case of biochar produced at 650 °C. Although some authors claimed that biochar-amended piles showed significantly higher N content in comparison to control treatments, in our study we did not clearly confirm this trend. Nevertheless, processes that have an impact on nitrogen dynamics are complex, and reports of enriching N content in the biochar-composted material involve longer incubation times, of up to 40 days [81,82]. On the other hand, presented observations support the theories that despite the role of biochar in the mitigation of carbon and nitrogen losses from composting processes, the final C/N ratio in the material is not significantly affected by the addition of biochar.

In the case of H, the content of the element in the analyzed material did not exceed a value of 5% (Supplementary Materials Figure S5a). The highest H values were observed in material composted with biochar produced at 650 °C and incubated at 60 °C, while the same biochar incubated at 70 °C showed the most stabilized values for all doses. From a statistical point of view, no correlation was observed between the parameters presented and the H content of the samples. For the S content, the greatest discrepancies were seen especially between the temperature of the pyrolysis process and the content of the element in the samples. It was observed that as the temperature of the pyrolysis process increased, the percentage of S decreased (Supplementary Materials Figure S5b). Increasing the pyrolysis temperature causes the loss of volatile, easily decomposable elements such as H and S; therefore, their content in biochar amendment was low and enrichment of the composted material was not expected [83]. The greatest differences between the doses of biochar in the material were shown in samples where biochar produced at 550 °C was used. The addition of biochar obtained by pyrolysis at 550 °C caused a decrease in the proportion of S, especially for the higher doses. As sulfur and hydrogen are not crucial nutrients for evaluating properties of the organic material, the literature lacks reports on the effect of biochar on H and S content during composting. Hagemann, who evaluated the impact of a 4.3% dose of three different biochars on manure composting, found a mixed effect on sulfur content depending on the biochar type, whereas the H content was not elaborated [72].

Nutrients and Heavy Metal Changes

The compositional makeup of organic matter that has undergone the process of composting can be analyzed to deduce its fertilizing potential and ascertain its suitability for application in soil. This evaluation takes into consideration the presence of both elemental, trace, and heavy metal components. This study elaborates on the total amounts of biologically significant elements like Ca, Mg, K, Na, P, and trace metals including Zn, Pb, Cd, Cu, Cr, Ni, and Hg (Figure 5). The presence of specific elements within a material can exert a significant influence on the properties of the soil into which it is administered. This influence may be advantageous or detrimental, depending on the nature of the elements present and the intended outcome of the application. Therefore, it is imperative to carefully evaluate the composition of any material prior to its use in soil management. Parameters such as the incubation temperature of the material, the temperature of biochar production, and the dose of biochar added to the feedstock were checked.

The addition of biochar to composted feedstock altered the total content of macronutrients: Ca, Na, Mg, and K. Considering the temperature of pyrolysis, the presence of nutrients in biochar-amended compost was the highest in the B550 treatments and decreased along with the increasing temperature of pyrolysis (Figure 6). The literature reports claim that a higher temperature of pyrolysis promotes the concentration of nutrients in biochar; however, our study did not confirm this trend [84]. The Ca, Na, Mg, and K content in tested biochars was higher than in compost before pyrolysis; however, B650 showed the lowest nutrient concentration, which was reflected in the compost–biochar composition. The variability in the content of nutrients is an effect of their different volatility and stability, affected by pyrolysis conditions [85]. Considering the biochar dose, some significant effects on composted feedstock were observed. In every treatment, the concentration of Na and Mg significantly increased, suggesting that biochars introduce additional elements and also prevent their losses in the composting process. For phosphorus and potassium, the effect was not that clear, as some significant decreases were observed in comparison to the unamended feedstock. For example, 3 and 9% of biochar reduced the content of phosphorus, similar to the observations of Hagemann et al., who explained this phenomenon by P losses during the composting process [86]. Significantly reduced amounts of total K in co-composted biochar is in line with the findings of Mujtaba et al. who examined fruit waste and vegetables as a feedstock [87]. In general, higher biochar doses in the co-composting process yield a material with elevated content of macronutrients, as an additional source of the elements is introduced, with the exception of K.

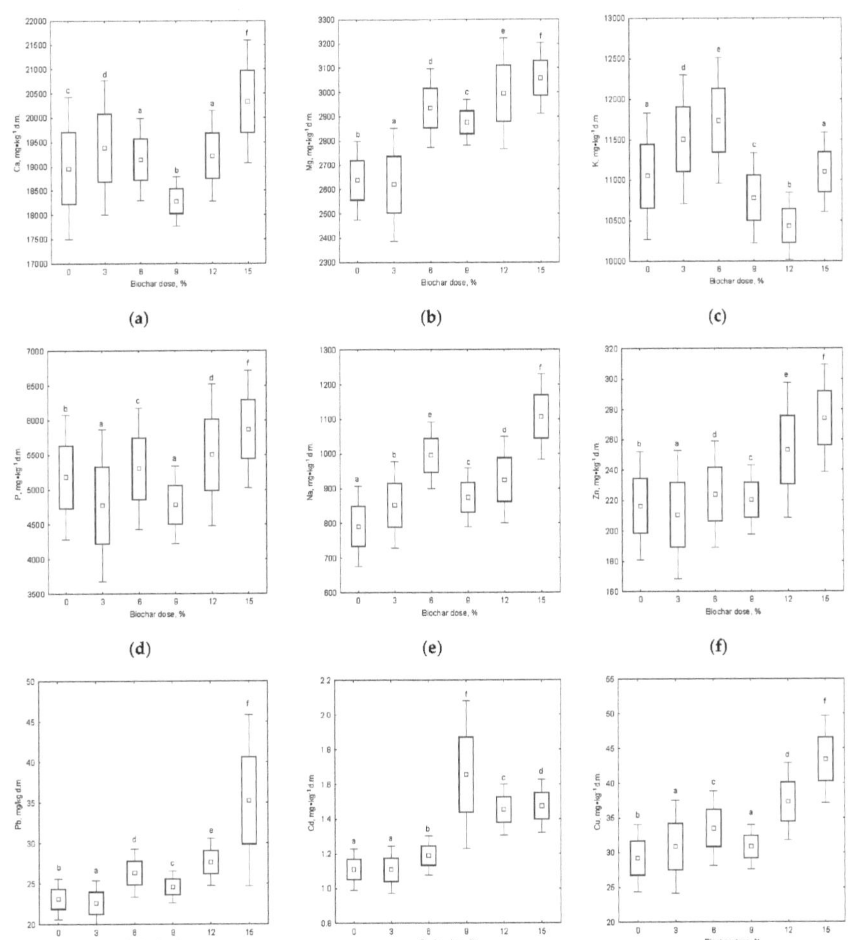

Figure 6. *Cont.*

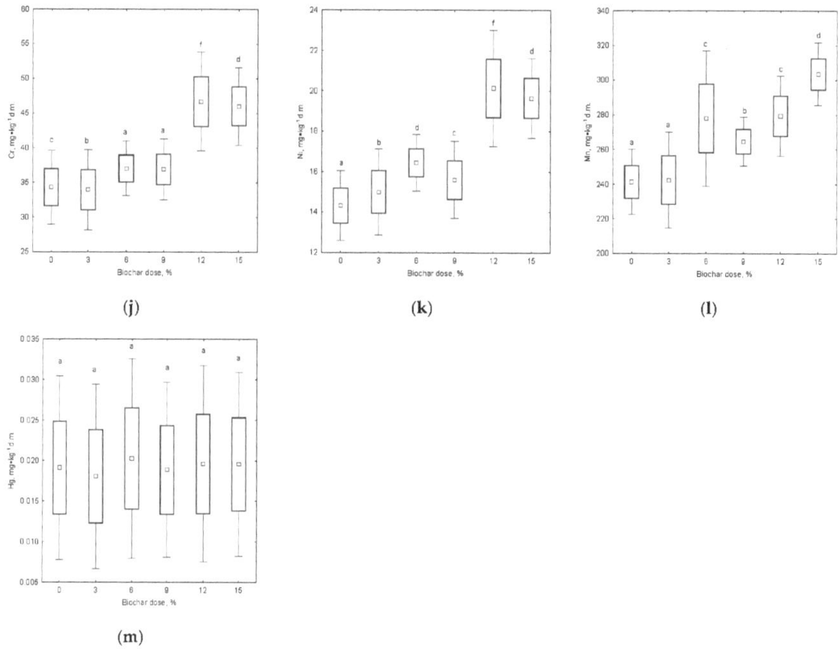

Figure 6. Effect of compost biochars' addition dose for nutrients and heavy metal content in substrate after 10 days of composting (**a**) calcium (Ca), (**b**) magnesium (Mg), (**c**) potassium (K), (**d**) phosphorus (P), (**e**) sodium (Na), (**f**) zinc (Zn), (**g**) lead (Pb), (**h**) cadmium (Cd), (**i**) copper (Cu), (**j**) chrome (Cr), (**k**) nickel (Ni), (**l**) manganese (Mn), and (**m**) mercury (Hg). Letters (a, b, c, d, e, f) indicate the homogeneity group according to Tukey's test at significance level $p < 0.05$.

Total trace metal content in organic amendments is one of the most important factors determining their suitability for application purposes. The results of selected trace metal contents in co-composted materials were much more varied than for other parameters studied. In general, considering the biochar type, at higher temperatures of pyrolysis the concentration of Cd, Cu, and Hg decreased; meanwhile, the content of Ni, Cr and Pb tended to be the highest in treatment with B600 (Figure 6). For Hg, no significant differences were observed between tested pyrolysis conditions. The results suggest that during pyrolysis, heavy metals were enriched in the biochars, as an effect of the better thermo-stability of metallic elements than other organic compounds present in the pyrolyzed biomass [88]. The fact that noted enrichment levels vary with tested metals may be an effect of metal speciation, influencing the temperature of decomposition [89].

Changes in metal concentrations during the composting process statistically differed between incubation temperatures. For all tested heavy metals, their mean content was the lowest after 10 days of co-composting at 70 °C, and this difference was statistically significant for Ni, Cr, and Cu. Fluctuations in trace metal content throughout the composting were observed by other authors [90]; however, the direction of the changes was often varied, depending on metal type. Composting and maturation processes decrease the heavy metal content and availability, and at higher temperatures, this effect was more significant [91].

Considering biochar doses on the content of potentially harmful metals in composted feedstock, it can be seen that a higher amount of introduced biochar amendment often resulted in an elevated metal content. This trend was clear for the 12% and 15% dose—a large amount of biochars, enriched in metals during pyrolysis process, led to enhanced concentrations in co-composted substrate. In the majority of cases, that enhancement was statistically significant, with the exception of Hg. Hagemann et al. who examined compost–biochar (with a dose of 4.3%), noted that the amendment did not significantly elevate the content of trace metals [84], and this is in agreement with our observations, where small doses (3–6%) of biochar were applied. Therefore, although small doses of biochars often do not have a statistically significant effect on HM content, higher doses can result in elevated metal concentrations, because of the additional pool of elements introduced with enriched, pyrolyzed materials [92]. This should be the subject of special attention when using compost–biochar in practice, due to the possibility of introducing harmful metals into the soil, their uptake by plants, and their inclusion in the food chain [93].

4. Conclusions

This study shows the effect of the addition of compost biochars to the composting process—the addition of compost biochars effectively decreases the CO_2 emissions and increases the nutrient content in the composting matrix. The effective use of compost biochars for gas mitigation during the intensive phase of composting demands the specific conditions of the process. It is presumed that the main mechanism of CO_2 mitigation was the addition of biochar with its inorganic content with a high composition of nutrients that enhance the microorganisms' activity. This effect should be examined in future studies. The following recommendations in compost biochars' application in compost can be formed:

- The use of compost biochars was the most effective in CO_2 mitigation—total reduction in emissions > 80% at 60 °C; lower reduction < 20% at 50 and 70 °C.
- The effect for CO, H_2S, and NH_3 mitigation was unclear; surprisingly, the reduction in CO emissions was observed at a temperature of 50 °C, with 30–60% effectiveness for B550, and 20–30% for B600; however, at 70 °C, an increase of >300% in CO emissions was observed; for H_2S, the most effective was incubation at 50 °C—a reduction > 50%, with no positive effect for 60 and 70 °C.
- No positive effect for NH_3 mitigation was observed, probably due to the high pH content of biochar.
- The addition of small doses (3–6% d.m.) of compost biochars reduced the observed emissions and significantly improved (the best treatment was B650, incubation at 60 °C, with 3% of the addition of biochar) the content of nutrients in composting matrix (P, K, Mg, Ca). In the same treatments, the content of trace metals was safe for the future use of compost in agriculture.

This study confirms that the use of pyrolysis for compost materials improves the properties of compost biochars, and has good potential for enhancing the composting process. This gives a new alternative niche for managing low quality compost and to become a part of the bioeconomic approach.

Supplementary Materials: The following supporting information can be downloaded at: https://www.mdpi.com/article/10.3390/ma17030563/s1, Figure S1: cumulative emission of CO_2 in 10 days of composting at different temperatures, and biochar type; Figure S2: cumulative emission of CO in 10 days of composting at different temperatures, and biochar type; Figure S3: cumulative emission of H_2S in 10 days of composting at different incubation temperatures, and biochar type; Figure S4: cumulative emission of NH_3 in 10 days of composting at different incubation temperatures, and biochar type; Figure S5: effect of biochar dose for (a) H, % and (b) S, %, during the first 10 days of composting process at laboratory scale; no statistical differences were observed according to Tukey's test at significance level $p < 0.05$; Figure S6: effect of incubation temperature for content of elements in substrate after 10 days of composting (a) calcium (Ca), (b) magnesium (Mg), (c) potassium (K), (d) phosphorus (P), (e) sodium (Na), (f) zinc (Zn), (g) lead (Pb), (h) cadmium (Cd), (i) copper (Cu), (j) chrome (Cr), (k) nickel (Ni), (l) manganese (Mn), (m) mercury (Hg); letters (a, b, c, d, e, f) indicate the homogeneity group according to Tukey's test at significance level $p < 0.05$; Figure S7: effect of biochar type (B550, B600, B650) for content of elements in substrate after 10 days of composting (a) calcium (Ca), (b) magnesium (Mg), (c) potassium (K), (d) phosphorus (P), (e) sodium (Na), (f) zinc (Zn), (g) lead (Pb), (h) cadmium (Cd), (i) copper (Cu), (j) chrome (Cr), (k) nickel (Ni), (l) manganese (Mn), (m) mercury (Hg); letters (a, b, c, d, e, f) indicate the homogeneity group according to Tukey's test at significance level $p < 0.05$; Table S1: total gas emissions after 10 days of the composting process at laboratory scale; Table S2: characteristics of material after 10 days of incubation—pH and LOI.

Author Contributions: Conceptualization, S.S.-D.; methodology, S.S.-D.; software, S.S.-D.; validation, S.S.-D., M.B. and J.R.; formal analysis, S.S.-D. and M.B.; investigation, S.S.-D., M.B., J.R. and E.S.; resources, S.S.-D. and E.S.; data curation, S.S.-D., M.B., J.R. and E.S.; writing—original draft preparation, S.S.-D., M.B. and J.R.; writing—review and editing, S.S.-D., M.B. and J.R.; visualization, S.S.-D. and J.R.; supervision, S.S.-D.; project administration, S.S.-D.; funding acquisition, S.S.-D. All authors have read and agreed to the published version of the manuscript.

Funding: The research leading to these results has received funding from the Norway Grants 2014–2021 through the National Centre for Research and Development. Project number: NOR/SGS/CompoChar/0090/2020.

Institutional Review Board Statement: Not applicable.

Informed Consent Statement: Not applicable.

Data Availability Statement: The data presented in this study are available in the Supplementary Materials.

Acknowledgments: The authors would like to thank the Best-Eko sp. z.o.o. company for the opportunity to collect compost and for cooperation. The authors would also like to thank Paweł Telega for his help in the research.

Conflicts of Interest: The authors declare no conflicts of interest.

References

1. He, X.; Yin, H.; Han, L.; Cui, R.; Fang, C.; Huang, G. Effects of Biochar Size and Type on Gaseous Emissions during Pig Manure/Wheat Straw Aerobic Composting: Insights into Multivariate-Microscale Characterization and Microbial Mechanism. *Bioresour. Technol.* **2019**, *271*, 375–382. [CrossRef]
2. Kweku, D.W.; Bismark, O.; Maxwell, A.; Desmond, K.A.; Danso, K.B.; Asante Oti-Mensah, E.; Quachie, A.T.; Adormaa, B.B.; Dopico, E. Greenhouse Effect: Greenhouse Gases and Their Impact on Global Warming. *J. Sci. Res. Rep.* **2017**, *17*, 1–9. [CrossRef]
3. Cao, Y.; Wang, X.; Bai, Z.; Chadwick, D.; Misselbrook, T.; Sommer, S.G.; Qin, W.; Ma, L. Mitigation of Ammonia, Nitrous Oxide and Methane Emissions during Solid Waste Composting with Different Additives: A Meta-Analysis. *J. Clean. Prod.* **2019**, *235*, 626–635. [CrossRef]
4. Guan, P.; Prasher, S.O.; Afzal, M.T.; George, S.; Ronholm, J.; Dhiman, J.; Patel, R.M. Removal of Escherichia Coli from Lake Water in a Biochar-Amended Biosand Filtering System. *Ecol. Eng.* **2020**, *150*, 105819. [CrossRef]
5. Ermolaev, E.; Jarvis, Å.; Sundberg, C.; Smårs, S.; Pell, M.; Jönsson, H. Nitrous Oxide and Methane Emissions from Food Waste Composting at Different Temperatures. *Waste Manag.* **2015**, *46*, 113–119. [CrossRef]
6. Yang, Y.; Kumar Awasthi, M.; Du, W.; Ren, X.; Lei, T.; Lv, J. Compost Supplementation with Nitrogen Loss and Greenhouse Gas Emissions during Pig Manure Composting. *Bioresour. Technol.* **2020**, *297*, 122435. [CrossRef]
7. Chen, X.; Zhao, Y.; Zeng, C.; Li, Y.; Zhu, L.; Wu, J.; Chen, J.; Wei, Z. Assessment Contributions of Physicochemical Properties and Bacterial Community to Mitigate the Bioavailability of Heavy Metals during Composting Based on Structural Equation Models. *Bioresour. Technol.* **2019**, *289*, 121657. [CrossRef] [PubMed]
8. Keiluweit, M.; Nico, P.S.; Johnson, M.; Kleber, M. Dynamic Molecular Structure of Plant Biomass-Derived Black Carbon (Biochar). *Environ. Sci. Technol.* **2010**, *44*, 1247–1253. [CrossRef] [PubMed]
9. Xu, X.; Cao, X.; Zhao, L.; Zhou, H.; Luo, Q. Interaction of Organic and Inorganic Fractions of Biochar with Pb(II) Ion: Further Elucidation of Mechanisms for Pb(II) Removal by Biochar. *RSC Adv.* **2014**, *4*, 44930–44937. [CrossRef]
10. Wang, H.; He, Z.; Lu, Z.; Zhou, J.; Van Nostrand, J.D.; Xu, X.; Zhang, Z. Genetic Linkage of Soil Carbon Pools and Microbial Functions in Subtropical Freshwater Wetlands in Response to Experimental Warming. *Appl. Environ. Microbiol.* **2012**, *78*, 7652–7661. [CrossRef] [PubMed]
11. Tang, L.; De Yang, G.; Zeng, G.M.; Cai, Y.; Li, S.S.; Zhou, Y.Y.; Pang, Y.; Liu, Y.Y.; Zhang, Y.; Luna, B. Synergistic Effect of Iron Doped Ordered Mesoporous Carbon on Adsorption-Coupled Reduction of Hexavalent Chromium and the Relative Mechanism Study. *Chem. Eng. J.* **2014**, *239*, 114–122. [CrossRef]
12. Albert, H.A.; Li, X.; Jeyakumar, P.; Wei, L.; Huang, L.; Huang, Q.; Kamran, M.; Shaheen, S.M.; Hou, D.; Rinklebe, J.; et al. Influence of Biochar and Soil Properties on Soil and Plant Tissue Concentrations of Cd and Pb: A Meta-Analysis. *Sci. Total Environ.* **2021**, *755*, 142582. [CrossRef]
13. Gupta, S.; Kua, H.W.; Low, C.Y. Use of Biochar as Carbon Sequestering Additive in Cement Mortar. *Cem. Concr. Compos.* **2018**, *87*, 110–129. [CrossRef]
14. Nan, H.; Yang, F.; Zhao, L.; Mašek, O.; Cao, X.; Xiao, Z. Interaction of Inherent Minerals with Carbon during Biomass Pyrolysis Weakens Biochar Carbon Sequestration Potential. *ACS Sustain. Chem. Eng.* **2019**, *7*, 1591–1599. [CrossRef]

15. Ghorbani, M.; Neugschwandtner, R.W.; Konvalina, P.; Asadi, H.; Kopecký, M.; Amirahmadi, E. Comparative Effects of Biochar and Compost Applications on Water Holding Capacity and Crop Yield of Rice under Evaporation Stress: A Two-Years Field Study. *Paddy Water Environ.* **2023**, *21*, 47–58. [CrossRef]
16. Li, Y.; Kumar Awasthi, M.; Sindhu, R.; Binod, P.; Zhang, Z.; Taherzadeh, M.J. Biochar Preparation and Evaluation of Its Effect in Composting Mechanism: A Review. *Bioresour. Technol.* **2023**, *384*, 129329. [CrossRef] [PubMed]
17. Awasthi, M.K.; Wang, M.; Chen, H.; Wang, Q.; Zhao, J.; Ren, X.; Li, D.S.; Awasthi, S.K.; Shen, F.; Li, R.; et al. Heterogeneity of Biochar Amendment to Improve the Carbon and Nitrogen Sequestration through Reduce the Greenhouse Gases Emissions during Sewage Sludge Composting. *Bioresour. Technol.* **2017**, *224*, 428–438. [CrossRef] [PubMed]
18. Yang, X.; Duan, P.; Cao, Y.; Wang, K.; Li, D. Mechanisms of Mitigating Nitrous Oxide Emission during Composting by Biochar and Calcium Carbonate Addition. *Bioresour. Technol.* **2023**, *388*, 129772. [CrossRef] [PubMed]
19. Behera, S.; Samal, K. Sustainable Approach to Manage Solid Waste through Biochar Assisted Composting. *Energy Nexus* **2022**, *7*, 100121. [CrossRef]
20. Beesley, L.; Moreno-Jiménez, E.; Gomez-Eyles, J.L. Effects of Biochar and Greenwaste Compost Amendments on Mobility, Bioavailability and Toxicity of Inorganic and Organic Contaminants in a Multi-Element Polluted Soil. *Environ. Pollut.* **2010**, *158*, 2282–2287. [CrossRef]
21. Tang, W.W.; Zeng, G.M.; Gong, J.L.; Liang, J.; Xu, P.; Zhang, C.; Huang, B. Bin Impact of Humic/Fulvic Acid on the Removal of Heavy Metals from Aqueous Solutions Using Nanomaterials: A Review. *Sci. Total Environ.* **2014**, *468–469*, 1014–1027. [CrossRef]
22. Beesley, L.; Moreno-Jiménez, E.; Gomez-Eyles, J.L.; Harris, E.; Robinson, B.; Sizmur, T. A Review of Biochars' Potential Role in the Remediation, Revegetation and Restoration of Contaminated Soils. *Environ. Pollut.* **2011**, *159*, 3269–3282. [CrossRef]
23. Castro-Herrera, D.; Prost, K.; Kim, D.G.; Yimer, F.; Tadesse, M.; Gebrehiwot, M.; Brüggemann, N. Biochar Addition Reduces Non-CO2 Greenhouse Gas Emissions during Composting of Human Excreta and Cattle Manure. *J. Environ. Qual.* **2023**, *52*, 814–828. [CrossRef]
24. Świechowski, K.; Liszewski, M.; Babelewski, P.; Koziel, J.A.; Białowiec, A. Oxytree Pruned Biomass Torrefaction: Mathematical Models of the Influence of Temperature and Residence Time on Fuel Properties Improvement. *Materials* **2019**, *12*, 2228. [CrossRef]
25. Rosik, J.; Łyczko, J.; Marzec, Ł.; Stegenta-Dąbrowska, S. Application of Composts' Biochar as Potential Sorbent to Reduce VOCs Emission during Kitchen Waste Storage. *Materials* **2023**, *16*, 6413. [CrossRef]
26. *PN-EN 14346:2011*; Waste Characteristics. Calculation of Dry Mass on the Basis of Dry Residue or Water Content. Polski Komitet Normalizacji: Warsaw, Poland, 2011.
27. Kozłowski, K.; Dach, J.; Czekała, W.; Malińska, K.; Świechowski, K.; Pulka, J.; Lewicki, A. Influence of the Parameters of Used Biochar on the Dark Fermentation Process. *Energies* **2023**, *16*, 7484. [CrossRef]
28. *PN-EN ISO 16948:2015-07*; Solid Biofuels. Total Carbon, Ammonia and Hydrogen Determination. Polski Komitet Normalizacji: Warsaw, Poland, 2015.
29. Zhao, S.X.; Ta, N.; Wang, X.D. Effect of Temperature on the Structural and Physicochemical Properties of Biochar with Apple Tree Branches as Feedstock Material. *Energies* **2017**, *10*, 1293. [CrossRef]
30. Bednik, M.; Medyńska-Juraszek, A.; Ćwieląg-Piasecka, I. Effect of Six Different Feedstocks on Biochar's Properties and Expected Stability. *Agronomy* **2022**, *12*, 1525. [CrossRef]
31. Karczewska, A. Metal Species Distribution in Top- and Sub-Soil in an Area Affected by Copper Smelter Emissions. *Appl. Geochem.* **1996**, *11*, 35–42. [CrossRef]
32. Medyńska-Juraszek, A.; Marcinkowska, K.; Gruszka, D.; Kluczek, K. The Effects of Rabbit-Manure-Derived Biochar Co-Application with Compost on the Availability and Heavy Metal Uptake by Green Leafy Vegetables. *Agronomy* **2022**, *12*, 2552. [CrossRef]
33. *PN-EN ISO 11885:2009*; Water Quality: Determination of Selected Elements by Inductively Excited Plasma Optical Emission Spectrometry. Polski Komitet Normalizacji: Warsaw, Poland, 2009.
34. Shin, Y.; Iwabuchi, K.; Itoh, T. Low-Temperature Biochars Are More Effective in Reducing Ammonia Emissions through Various Mechanisms during Manure Composting. *J. Mater. Cycles Waste Manag.* **2023**, *26*, 138–148. [CrossRef]
35. Tomczyk, A.; Sokołowska, Z.; Boguta, P. Biochar Physicochemical Properties: Pyrolysis Temperature and Feedstock Kind Effects. *Rev. Environ. Sci. Biotechnol.* **2020**, *19*, 191–215. [CrossRef]
36. Zhang, Z.; Duan, C.; Liu, Y.; Li, A.; Hu, X.; Chen, J.; Zhang, S.; Li, X.; Che, R.; Li, S.; et al. Green Waste and Sewage Sludge Feeding Ratio Alters Co-Composting Performance: Emphasis on the Role of Bacterial Community during Humification. *Bioresour. Technol.* **2023**, *380*, 129014. [CrossRef] [PubMed]
37. Su, L.; Chen, M.; Zhuo, G.; Ji, R.; Wang, S.; Zhang, L.; Zhang, M.; Li, H. Comparison of Biochar Materials Derived from Coconut Husks and Various Types of Livestock Manure, and Their Potential for Use in Removal of H2s from Biogas. *Sustainability* **2021**, *13*, 6262. [CrossRef]
38. Darby, I.; Xu, C.Y.; Wallace, H.M.; Joseph, S.; Pace, B.; Bai, S.H. Short-Term Dynamics of Carbon and Nitrogen Using Compost, Compost-Biochar Mixture and Organo-Mineral Biochar. *Environ. Sci. Pollut. Res.* **2016**, *23*, 11267–11278. [CrossRef] [PubMed]
39. Białowiec, A.; Pulka, J.; Styczyńska, M.; Koziel, J.A.; Kalka, J.; Jureczko, M.; Felis, E.; Manczarski, P. Is Biochar from the Torrefaction of Sewage Sludge Hazardous Waste? *Materials* **2020**, *13*, 3544. [CrossRef] [PubMed]
40. Luo, D.; Wang, L.; Nan, H.; Cao, Y.; Wang, H.; Kumar, T.V.; Wang, C. Phosphorus Adsorption by Functionalized Biochar: A Review. *Environ. Chem. Lett.* **2022**, *21*, 497–524. [CrossRef]

41. EUR-Lex—32019R1009—EN—EUR-Lex. Available online: https://eur-lex.europa.eu/legal-content/PL/TXT/?qid=1561534804127&uri=CELEX:32019R1009 (accessed on 16 November 2023).
42. Wang, H.; Wang, X.; Cui, Y.; Xue, Z.; Ba, Y. Slow Pyrolysis Polygeneration of Bamboo (Phyllostachys Pubescens): Product Yield Prediction and Biochar Formation Mechanism. *Bioresour. Technol.* **2018**, *263*, 444–449. [CrossRef]
43. Silva, A.C.; Rocha, P.; Antelo, J.; Valderrama, P.; López, R.; Geraldo, D.; Proença, M.F.; Pinheiro, J.P.; Fiol, S.; Bento, F. Comparison of a Variety of Physico-Chemical Techniques in the Chronological Characterization of a Compost from Municipal Wastes. *Process Saf. Environ. Prot.* **2022**, *164*, 781–793. [CrossRef]
44. Cole, E.J.; Zandvakili, O.R.; Xing, B.; Hashemi, M.; Herbert, S.; Mashayekhi, H.H. Dataset on the Effect of Hardwood Biochar on Soil Gravimetric Moisture Content and Nitrate Dynamics at Different Soil Depths with FTIR Analysis of Fresh and Aged Biochar. *Data Brief.* **2019**, *25*, 104073. [CrossRef]
45. Domingues, R.R.; Trugilho, P.F.; Silva, C.A.; De Melo, I.C.N.A.; Melo, L.C.A.; Magriotis, Z.M.; Sánchez-Monedero, M.A. Properties of Biochar Derived from Wood and High-Nutrient Biomasses with the Aim of Agronomic and Environmental Benefits. *PLoS ONE* **2017**, *12*, e0176884. [CrossRef]
46. Adegoke, T.O.; Ku, H.H. Temperature Response of Ammonia Emission from Sandy Loam Soil Amended with Manure Compost and Urea. *Environ. Technol. Innov.* **2023**, *31*, 103226. [CrossRef]
47. Sundberg, C.; Yu, D.; Franke-Whittle, I.; Kauppi, S.; Smårs, S.; Insam, H.; Romantschuk, M.; Jönsson, H. Effects of PH and Microbial Composition on Odour in Food Waste Composting. *Waste Manag.* **2013**, *33*, 204–211. [CrossRef] [PubMed]
48. Waqas, M.; Nizami, A.S.; Aburiazaiza, A.S.; Barakat, M.A.; Ismail, I.M.I.; Rashid, M.I. Optimization of Food Waste Compost with the Use of Biochar. *J. Environ. Manag.* **2018**, *216*, 70–81. [CrossRef]
49. Czekała, W.; Malińska, K.; Cáceres, R.; Janczak, D.; Dach, J.; Lewicki, A. Co-Composting of Poultry Manure Mixtures Amended with Biochar—The Effect of Biochar on Temperature and C-CO_2 Emission. *Bioresour. Technol.* **2016**, *200*, 921–927. [CrossRef]
50. Jia, X.; Wang, M.; Yuan, W.; Ju, X.; Yang, B. The Influence of Biochar Addition on Chicken Manure Composting and Associated Methane and Carbon Dioxide Emissions. *Bioresources* **2016**, *11*, 5255–5264. [CrossRef]
51. Jindo, K.; Sonoki, T.; Matsumoto, K.; Canellas, L.; Roig, A.; Sanchez-Monedero, M.A. Influence of Biochar Addition on the Humic Substances of Composting Manures. *Waste Manag.* **2016**, *49*, 545–552. [CrossRef] [PubMed]
52. Zhang, J.; Chen, G.; Sun, H.; Zhou, S.; Zou, G. Straw Biochar Hastens Organic Matter Degradation and Produces Nutrient-Rich Compost. *Bioresour. Technol.* **2016**, *200*, 876–883. [CrossRef] [PubMed]
53. Raclavská, H.; Růžičková, J.; Raclavský, K.; Juchelková, D.; Kucbel, M.; Švédová, B.; Slamová, K.; Kacprzak, M. Effect of Biochar Addition on the Improvement of the Quality Parameters of Compost Used for Land Reclamation. *Environ. Sci. Pollut. Res.* **2023**, *30*, 8563–8581. [CrossRef]
54. Ghanbarpour Mamaghani, Z.; Hawboldt, K.A.; MacQuarrie, S. Adsorption of CO_2 Using Biochar—Review of the Impact of Gas Mixtures and Water on Adsorption. *J. Environ. Chem. Eng.* **2023**, *11*, 109643. [CrossRef]
55. Mondal, M.K.; Balsora, H.K.; Varshney, P. Progress and Trends in CO_2 Capture/Separation Technologies: A Review. *Energy* **2012**, *46*, 431–441. [CrossRef]
56. Dissanayake, P.D.; Choi, S.W.; Igalavithana, A.D.; Yang, X.; Tsang, D.C.W.; Wang, C.H.; Kua, H.W.; Lee, K.B.; Ok, Y.S. Sustainable Gasification Biochar as a High Efficiency Adsorbent for CO_2 Capture: A Facile Method to Designer Biochar Fabrication. *Renew. Sustain. Energy Rev.* **2020**, *124*, 109785. [CrossRef]
57. Nguyen, M.K.; Lin, C.; Hoang, H.G.; Sanderson, P.; Dang, B.T.; Bui, X.T.; Nguyen, N.S.H.; Vo, D.V.N.; Tran, H.T. Evaluate the Role of Biochar during the Organic Waste Composting Process: A Critical Review. *Chemosphere* **2022**, *299*, 134488. [CrossRef]
58. Yan, H.; Yang, H.; Li, K.; Zhu, P.; Li, X.; Li, Q. Biochar Addition Modified Carbon Flux and Related Microbiota in Cow Manure Composting. *Waste Biomass Valorization* **2023**, *14*, 847–858. [CrossRef]
59. Wang, H.; Lu, Y.; Xu, J.; Liu, X.; Sheng, L. Effects of Additives on Nitrogen Transformation and Greenhouse Gases Emission of Co-Composting for Deer Manure and Corn Straw. *Environ. Sci. Pollut. Res.* **2021**, *28*, 13000–13020. [CrossRef] [PubMed]
60. Ottani, F.; Parenti, M.; Santunione, G.; Moscatelli, G.; Kahn, R.; Pedrazzi, S.; Allesina, G. Effects of Different Gasification Biochar Grain Size on Greenhouse Gases and Ammonia Emissions in Municipal Aerated Composting Processes. *J. Environ. Manag.* **2023**, *331*, 117257. [CrossRef] [PubMed]
61. Sobieraj, K.; Stegenta-Dąbrowska, S.; Zafiu, C.; Binner, E.; Białowiec, A. Carbon Monoxide Production during Bio-Waste Composting under Different Temperature and Aeration Regimes. *Materials* **2023**, *16*, 4551. [CrossRef]
62. Sánchez-García, M.; Alburquerque, J.A.; Sánchez-Monedero, M.A.; Roig, A.; Cayuela, M.L. Biochar Accelerates Organic Matter Degradation and Enhances N Mineralisation during Composting of Poultry Manure without a Relevant Impact on Gas Emissions. *Bioresour. Technol.* **2015**, *192*, 272–279. [CrossRef]
63. Stegenta-Dabrowska, S.; Sobieraj, K.; Koziel, J.A.; Bieniek, J.; Bialowiec, A. Kinetics of Biotic and Abiotic CO Production during the Initial Phase of Biowaste Composting. *Energies* **2020**, *13*, 5451. [CrossRef]
64. Hellebrand, H.J.; Schade, G.W. Carbon Monoxide from Composting Due to Thermal Oxidation of Biomass. *J. Environ. Qual.* **2008**, *37*, 592–598. [CrossRef]
65. Sobieraj, K.; Stegenta-Dąbrowska, S.; Luo, G.; Koziel, J.A.; Białowiec, A. Biological Treatment of Biowaste as an Innovative Source of CO—The Role of Composting Process. *Front. Bioeng. Biotechnol.* **2023**, *11*, 1126737. [CrossRef]

66. Chung, W.J.; Chang, S.W.; Chaudhary, D.K.; Shin, J.D.; Kim, H.; Karmegam, N.; Govarthanan, M.; Chandrasekaran, M.; Ravindran, B. Effect of Biochar Amendment on Compost Quality, Gaseous Emissions and Pathogen Reduction during in-Vessel Composting of Chicken Manure. *Chemosphere* **2021**, *283*, 131129. [CrossRef]
67. Yuan, J.H.; Xu, R.K.; Zhang, H. The Forms of Alkalis in the Biochar Produced from Crop Residues at Different Temperatures. *Bioresour. Technol.* **2011**, *102*, 3488–3497. [CrossRef] [PubMed]
68. Liu, Y.; Ma, R.; Li, D.; Qi, C.; Han, L.; Chen, M.; Fu, F.; Yuan, J.; Li, G. Effects of Calcium Magnesium Phosphate Fertilizer, Biochar and Spent Mushroom Substrate on Compost Maturity and Gaseous Emissions during Pig Manure Composting. *J. Environ. Manag.* **2020**, *267*, 110649. [CrossRef] [PubMed]
69. Ouyang, X.; Lin, H.; Hu, Z.; Zheng, Y.; Li, P.; Huang, W. Effect of Biochar Structure on H2S Emissions during Sludge Aerobic Composting: Insights into Microscale Characterization and Microbial Mechanism. *Biomass Convers. Biorefin.* **2022**, *1*, 1–14. [CrossRef]
70. Nguyen, M.K.; Lin, C.; Hoang, H.G.; Bui, X.T.; Ngo, H.H.; Le, V.G.; Tran, H.T. Investigation of Biochar Amendments on Odor Reduction and Their Characteristics during Food Waste Co-Composting. *Sci. Total Environ.* **2023**, *865*, 161128. [CrossRef] [PubMed]
71. Awasthi, M.K.; Zhang, Z.; Wang, Q.; Shen, F.; Li, R.; Li, D.S.; Ren, X.; Wang, M.; Chen, H.; Zhao, J. New Insight with the Effects of Biochar Amendment on Bacterial Diversity as Indicators of Biomarkers Support the Thermophilic Phase during Sewage Sludge Composting. *Bioresour. Technol.* **2017**, *238*, 589–601. [CrossRef] [PubMed]
72. Liu, N.; Zhou, J.; Han, L.; Ma, S.; Sun, X.; Huang, G. Role and Multi-Scale Characterization of Bamboo Biochar during Poultry Manure Aerobic Composting. *Bioresour. Technol.* **2017**, *241*, 190–199. [CrossRef] [PubMed]
73. Liu, W.; Huo, R.; Xu, J.; Liang, S.; Li, J.; Zhao, T.; Wang, S. Effects of Biochar on Nitrogen Transformation and Heavy Metals in Sludge Composting. *Bioresour. Technol.* **2017**, *235*, 43–49. [CrossRef]
74. Agyarko-Mintah, E.; Cowie, A.; Singh, B.P.; Joseph, S.; Van Zwieten, L.; Cowie, A.; Harden, S.; Smillie, R. Biochar Increases Nitrogen Retention and Lowers Greenhouse Gas Emissions When Added to Composting Poultry Litter. *Waste Manag.* **2017**, *61*, 138–149. [CrossRef]
75. Zhang, J.; Lu, M.; Wan, J.; Sun, Y.; Lan, H.; Deng, X. Effects of PH, Dissolved Humic Acid and Cu^{2+} on the Adsorption of Norfloxacin on Montmorillonite-Biochar Composite Derived from Wheat Straw. *Biochem. Eng. J.* **2018**, *130*, 104–112. [CrossRef]
76. Vijayanand, M.; Ramakrishnan, A.; Subramanian, R.; Issac, P.K.; Nasr, M.; Khoo, K.S.; Rajagopal, R.; Greff, B.; Wan Azelee, N.I.; Jeon, B.H.; et al. Polyaromatic Hydrocarbons (PAHs) in the Water Environment: A Review on Toxicity, Microbial Biodegradation, Systematic Biological Advancements, and Environmental Fate. *Environ. Res.* **2023**, *227*, 115716. [CrossRef]
77. Neeli, S.T.; Ramsurn, H. Synthesis and Formation Mechanism of Iron Nanoparticles in Graphitized Carbon Matrices Using Biochar from Biomass Model Compounds as a Support. *Carbon* **2018**, *134*, 480–490. [CrossRef]
78. Oldfield, T.L.; Sikirica, N.; Mondini, C.; López, G.; Kuikman, P.J.; Holden, N.M. Biochar, Compost and Biochar-Compost Blend as Options to Recover Nutrients and Sequester Carbon. *J. Environ. Manag.* **2018**, *218*, 465–476. [CrossRef]
79. Gao, S.; Harrison, B.P.; Thao, T.; Gonzales, M.L.; An, D.; Ghezzehei, T.A.; Diaz, G.; Ryals, R.A. Biochar Co-Compost Improves Nitrogen Retention and Reduces Carbon Emissions in a Winter Wheat Cropping System. *GCB Bioenergy* **2023**, *15*, 462–477. [CrossRef]
80. Schulz, H.; Dunst, G.; Glaser, B. Positive Effects of Composted Biochar on Plant Growth and Soil Fertility. *Agron. Sustain. Dev.* **2013**, *33*, 817–827. [CrossRef]
81. Samudrika, K.P.D.; Ariyawansha, R.T.K.; Basnayake, B.F.A.; Siriwardana, A.N. Optimization of Biochar Additions for Enriching Nitrogen in Active Phase Low-Temperature Composting. *Org. Agric.* **2020**, *10*, 449–463. [CrossRef]
82. Bello, A.; Deng, L.; Sheng, S.; Jiang, X.; Yang, W.; Meng, Q.; Wu, X.; Han, Y.; Zhu, H.; Xu, X. Biochar Reduces Nutrient Loss and Improves Microbial Biomass of Composted Cattle Manure and Maize Straw. *Biotechnol. Appl. Biochem.* **2020**, *67*, 799–811. [CrossRef]
83. Ippolito, J.A.; Spokas, K.A.; Novak, J.M.; Lentz, R.D.; Cantrell, K.B. Biochar Elemental Composition and Factors Infl Uencing Nutrient Retention. In *Biochar for Environmental Management: Science, Technology and Implementation*; Routledge: London, UK, 2015; Volume 139.
84. Nwajiaku, I.M.; Olanrewaju, J.S.; Sato, K.; Tokunari, T.; Kitano, S.; Masunaga, T. Change in Nutrient Composition of Biochar from Rice Husk and Sugarcane Bagasse at Varying Pyrolytic Temperatures. *Int. J. Recycl. Org. Waste Agric.* **2018**, *7*, 269–276. [CrossRef]
85. Hossain, M.K.; Strezov Vladimir, V.; Chan, K.Y.; Ziolkowski, A.; Nelson, P.F. Influence of Pyrolysis Temperature on Production and Nutrient Properties of Wastewater Sludge Biochar. *J. Environ. Manag.* **2011**, *92*, 223–228. [CrossRef] [PubMed]
86. Hagemann, N.; Subdiaga, E.; Orsetti, S.; de la Rosa, J.M.; Knicker, H.; Schmidt, H.P.; Kappler, A.; Behrens, S. Effect of Biochar Amendment on Compost Organic Matter Composition Following Aerobic Composting of Manure. *Sci. Total Environ.* **2018**, *613–614*, 20–29. [CrossRef] [PubMed]
87. Mujtaba, G.; Hayat, R.; Hussain, Q.; Ahmed, M. Physio-Chemical Characterization of Biochar, Compost and Co-Composted Biochar Derived from Green Waste. *Sustainability* **2021**, *13*, 4628. [CrossRef]
88. Yuan, H.; Lu, T.; Huang, H.; Zhao, D.; Kobayashi, N.; Chen, Y. Influence of Pyrolysis Temperature on Physical and Chemical Properties of Biochar Made from Sewage Sludge. *J. Anal. Appl. Pyrolysis* **2015**, *112*, 284–289. [CrossRef]
89. Yoshida, T.; Antal, M.J. Sewage Sludge Carbonization for Terra Preta Applications. *Energy Fuels* **2009**, *23*, 5454–5459. [CrossRef]

90. Hanc, A.; Szakova, J.; Svehla, P. Effect of Composting on the Mobility of Arsenic, Chromium and Nickel Contained in Kitchen and Garden Waste. *Bioresour. Technol.* **2012**, *126*, 444–452. [CrossRef]
91. Tibu, C.; Annang, T.Y.; Solomon, N.; Yirenya-Tawiah, D. Effect of the Composting Process on Physicochemical Properties and Concentration of Heavy Metals in Market Waste with Additive Materials in the Ga West Municipality, Ghana. *Int. J. Recycl. Org. Waste Agric.* **2019**, *8*, 393–403. [CrossRef]
92. De Figueiredo, C.C.; Chagas, J.K.M.; da Silva, J.; Paz-Ferreiro, J. Short-Term Effects of a Sewage Sludge Biochar Amendment on Total and Available Heavy Metal Content of a Tropical Soil. *Geoderma* **2019**, *344*, 31–39. [CrossRef]
93. Medyńska-Juraszek, A.; Bednik, M.; Chohura, P. Assessing the Influence of Compost and Biochar Amendments on the Mobility and Uptake of Heavy Metals by Green Leafy Vegetables. *Int. J. Environ. Res. Public Health* **2020**, *17*, 7861. [CrossRef] [PubMed]

materials

MDPI

Article

Investigation of the Pozzolanic Activity Improvement of Yellow Phosphorus Slag with Thermal Activation

Xinyue Liu [1,2], Xiaoming Liu [1,2,*], Zengqi Zhang [1,2,*], Chao Wei [2], Qingsen Zeng [2], Yantian Li [2] and Shanliang Ma [2]

[1] State Key Laboratory of Advanced Metallurgy, University of Science and Technology Beijing, Beijing 100083, China; b20200134@xs.ustb.edu.cn

[2] School of Metallurgical and Ecological Engineering, University of Science and Technology Beijing, Beijing 100083, China; weichao0810@126.com (C.W.); g20208335@xs.ustb.edu.cn (Q.Z.); liyantian2@126.com (Y.L.); g20208295@xs.ustb.edu.cn (S.M.)

* Correspondence: liuxm@ustb.edu.cn (X.L.); zhangzq@ustb.edu.cn (Z.Z.)

Abstract: Yellow phosphorus slag (YPS) is a byproduct from the production of yellow phosphorus. It has potential pozzolanic activity and can be used as a supplementary cementitious material. However, the early strength of cement mortar decreases significantly with increasing YPS dosage, which restricts the utilization of YPS in cement and concrete. This study aimed to increase the pozzolanic activity of YPS ash by thermal activation. The strength method, alkali dissolution method and polymerization degree method were used to evaluate the effect of thermal activation at different temperatures on the pozzolanic activity of YPS ash. The results showed that YPS ash calcined at 800 °C helps to enhance the early strength because the fluorine in cuspidine ($Ca_4Si_2O_7F_2$) is insoluble, reducing the retarding effect on the mortar. The higher late strength of YPS ash calcined at 100 °C was due to the low polymerization degree of [SiO_4]. The pozzolanic activity of YPS ash is positively correlated with the dissolution concentration of (Si + Al) and the compressive strength and negatively associated with the polymerization degree. This paper shows a possibility for the large-scale utilization of YPS.

Keywords: yellow phosphorus slag; thermal activation; pozzolanic activity; compressive strength; alkali dissolution; polymerization degree

1. Introduction

Yellow phosphorus slag (YPS) is a byproduct from the production of yellow phosphorus using phosphorus ore by the electric furnace method [1,2]. Phosphate ore reacts with coke to form phosphorus, and its impurities react with silica to obtain YPS [3]. Each ton of yellow phosphorus will produce approximately 8–12 tons of YPS [4]. China, as the largest yellow phosphorus producer, produces approximately 8 million tons of YPS annually, but the comprehensive utilization rate of YPS is lower than 50% [4]. Most YPS is stacking-treated by the factory, which not only takes up land resources but also damages the environment and affects human health by dissolving the phosphorus and fluorine components in rainwater [5,6]. Therefore, it is essential and urgent to find ways to dispose of YPS.

Meanwhile, cement is one of the three basic materials in the construction industry and has high carbon emissions and energy consumption in its production [7]. In recent decades, CO_2 emission reduction, environmental protection and energy conservation have been major concerns for the cement industry [8,9]. A large number of industrial solid wastes have been widely applied in cement and concrete as cementitious materials or mineral admixtures [10–12]. This not only reduces the cost of solid waste dam management and operation and cement production but also decreases the environmental pollution problems caused by solid waste storage [13].

The chemical composition of YPS varies depending on the raw materials and manufacturing process. Molten YPS is rapidly cooled by water quenching from high temperature to room temperature. CaO and SiO_2 are the main components of YPS, and the glassy-phase content can reach 85–90% in YPS [14]. Therefore, YPS has potential pozzolanic activity and can be applied to the field of building materials such as cement, concrete, supplementary cementitious materials, road base materials and backfill materials [15,16]. A small amount of YPS replacing silicate cement in concrete can decrease the dosage of silicate cement. YPS reacts with calcium hydroxide in cement and produces C-S-H gel during cementitious reaction, which can refine the microstructure of concrete, reduce harmful pores and improve the mechanical properties of concrete. Wang et al. [17] investigated the impact of YPS on the durability and mechanical properties of concrete. The results indicated that the activity of YPS was greater than that of fly ash. The high-volume electric furnace nickel slag-YPS concrete showed similar or even better mechanical properties and chloride ion permeability than high-volume fly ash concrete at later stages. Yang et al. [18] found that adding YPS significantly improved the flowability and reduced the autogenous shrinkage in the early stage of ultrahigh performance concrete. Although it decreases the early compressive strength of ultrahigh performance concrete, it still promotes long-term strength development.

The strength of cement mortar at an early age decreases significantly with increasing YPS dosage, which restricts the utilization of YPS in cement and concrete [19]. The early activity of YPS is comparatively poor, and the fluorine and phosphorus in YPS will delay the hardening and setting of cement [20]. The phosphorus is uniformly distributed in YPS. The distribution pattern of P is very similar to that of Si. P is mainly present in the solid solution of $3CaO{\cdot}P_2O_5$ and $2CaO{\cdot}SiO_2$ [21]. Therefore, it is vital to improve the pozzolanic activity of YPS and enhance the early strength of high-value YPS-based cementitious materials. Mechanical, chemical and thermal activation methods are applied to enhance the pozzolanic activity of YPS [22–26]. He et al. [27] noted that the slurry pH, activity index and uniformity of YPS particle sizes significantly increased with increasing YPS granularity, and the impact on the strength and setting time of cement paste was significantly reduced. Hu et al. [28] found that superfine YPS refined the microstructure of concrete, which helped to improve the resistance to chloride ion penetration and carbonation of concrete in the later stage. In addition, superfine YPS is helpful for the late development on the splitting tensile strength and compressive strength of concrete. Allahverdi et al. [29] reported that an improvement in the specific surface area of YPS with mechanical activation significantly reduced the amount of water absorption and the total open pore volume of mortars. The synergistic combination of mechanical activation and chemical activation can enhance the quality of high YPS cement. Additionally, chemically activated YPS-based composite cement has a superior resistance to sodium sulfate [30]. Zhang et al. [31] found that the application of YPS retarded the early hydration of cement, but increasing the curing temperature tended to decrease this retarding effect. Wang et al. [32] showed that adding YPS increased the late compressive strength of fly ash-based geopolymers. Increasing the curing temperature and NaOH concentration can improve the compressive strength of geopolymers. Heat treatment or high-temperature curing can increase strength development but may also reduce ultimate strength [33]. However, most of the scholars studied the effect of high-temperature curing on the properties of YPS-based cement. The effect of heat treatment on the pozzolanic activity and structural change in YPS has rarely been studied.

The aim of this study is to increase the pozzolanic activity of YPS by thermal activation. The strength method, alkali dissolution method and polymerization degree method were used to evaluate the effect of thermal activation at different temperatures on the pozzolanic activity of YPS. The Fourier-transform infrared spectrometry (FTIR) and X-ray diffraction (XRD) techniques were used to characterize the phases formed during the calcination process. This study explores the mechanism of YPS during calcination and lays the foundation for the large-scale utilization of YPS.

2. Materials and Methods

YPS was provided by Guizhou, China, and 42.5 grade cement was taken from the Hebei cement plant. The standard sand was obtained from the Xiamen sand plant. The NaOH (Aladdin) was an analytical grade chemical reagent. Table 1 shows the chemical compositions of the raw materials. The main components of both YPS and cement are CaO and SiO_2, and less Al_2O_3.

Table 1. The chemical compositions of raw materials (wt.%).

Oxides	CaO	SiO_2	Al_2O_3	Fe_2O_3	MgO	Na_2O	K_2O	SO_3	P_2O_5	F
YPS	44.72	34.84	6.00	1.19	1.94	0.35	1.43	2.36	3.18	3.24
Cement	63.48	20.70	5.00	3.59	2.76	0.14	0.82	2.59	0.075	0.096

First, YPS was dried in the oven and milled in a ball mill for 90 min. The specific surface area of the ground YPS measured by the FBT-9 full automation instrument testing specific surface was 421 m^2/kg, and Figure 1 shows its particle size distribution tested by a laser particle size analyzer (Mastersizer 2000, Malvern Instruments Ltd., Malvern, UK). The ground YPS particles are within the range of 0.5–18.7 μm with a mean of 7.30 μm.

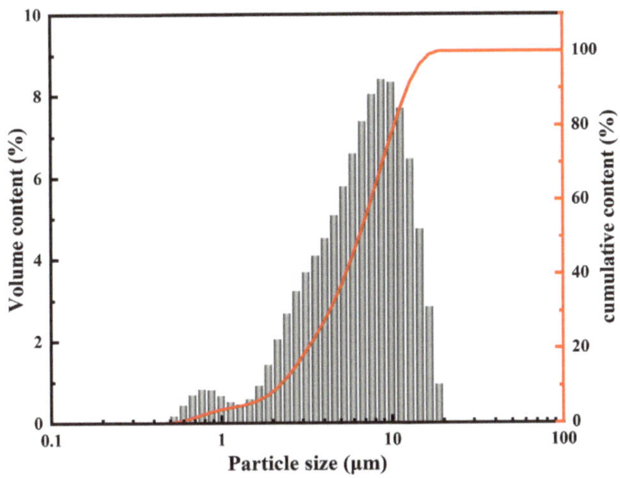

Figure 1. The particle size distribution of YPS.

Second, 300 g of ground YPS was placed in a muffle furnace, calcined at 100–1000 °C for 3 h and then cooled to room temperature. The heating speed of the muffle furnace was 5 °C/min. The YPS ash calcined at 800 °C and 1000 °C required grinding in a small ball mill for 20 min. YPS (n = 100–1000) represents the YPS ash calcined at different temperatures. Figure 2 shows the visual attributes of YPS ash. The color of YPS ash gradually changed from gray-white to flesh pink with the increase in calcination temperature. Finally, the pozzolanic activity of YPS ash was tested by the strength method, alkali dissolution method and polymerization degree method.

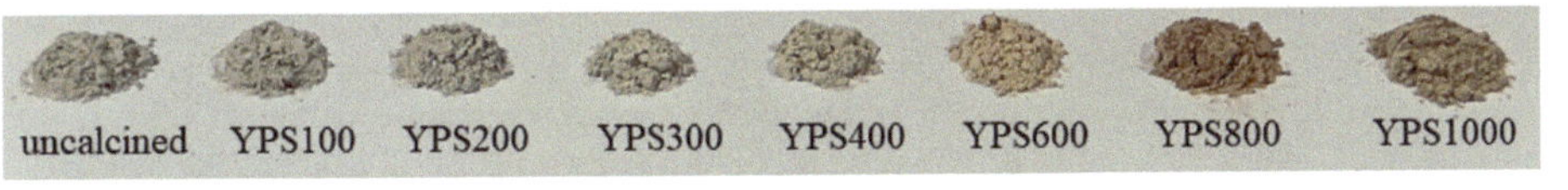

Figure 2. The visual attributes of YPS ash.

The strength method was carried out according to the Chinese standard "ground granulated electric furnace phosphorus slag powder used for cement and concrete" (GB/T 26751-2022) [34]. The activity of YPS ash was evaluated by comparing the compressive strength of cement mortar with or without the addition of YPS ash. The specific steps were as follows: 225 g of water, 135 g of YPS ash, 315 g of cement and 1350 g of standard sand were mixed and stirred and then injected into a 40 × 40 × 160 cm mold with vibration; the mortar was put into a curing box with a temperature of 20 ± 1 °C and humidity of 95% for 3, 7 and 28 d. The proportion of the control group was 450 g of cement and 1350 g of standard sand. Table 2 shows the mix proportions of each sample. A press (HYE-300-10, Beijing Sanyulutong Instrument Co., Ltd., Beijing, China) was used to test the compressive strength of the mortar samples. The pozzolanic activity index (A) of YPS ash was evaluated according to Equation (1):

$$\mathrm{A} = \frac{R_i}{R_0} \times 100, \tag{1}$$

where R_i (MPa) represents the compressive strength of the YPS sample and R_0 (MPa) represents the compressive strength of the control sample.

Table 2. The mix proportions of the sample.

Sample	YPS Ash/g	Cement/g	Standard Sand/g	Water/g	w/c Ratio
YPS sample	135	315	1350	225	0.5
Control sample	0	450	1350	225	0.5

The steps of the alkali dissolution method were as follows: 1 g of YPS ash was taken into 100 mL of 1 mol/L NaOH solution, sealed and put in a curing box at 20 ± 1 °C for 7 d; then, it was filtered to obtain the filtrate. An ICP-OES (Optima 7000DV, PerkinElmer Instrument Co., Ltd., Shanghai, China) was used to test the concentrations of Si and Al in the filtrate.

The polymerization degree method used Origin software to fit and calculate the peak area of Si(Al)Q^n at 800–1200 cm^{-1} of the FTIR spectra. Then, the polymerization degree of $[SiO_4]$ was evaluated using the concept of the relative bridging oxygen bond (RBO) according to Equation (2) [35]:

$$\mathrm{RBO} = \frac{1}{4}\left(1 \times \frac{Q^1}{\sum Q^n} + 2 \times \frac{Q^2}{\sum Q^n} + 3 \times \frac{Q^3}{\sum Q^n} + 4 \times \frac{Q^4}{\sum Q^n}\right) = \frac{1}{4} \times \frac{\sum n \times Q^n}{\sum Q^n} \tag{2}$$

The chemical compositions of the YPS ash and cement were tested by X-ray fluorescence spectrometry (Shimadzu XRF-1700 series, Shanghai, China). The phase composition of YPS ash at different calcination temperatures was analyzed by an X-ray diffractometer (D8 advance, Germany). The experimental conditions were Cu target, current 40 mA, voltage 40 kV, scanning speed of 10°/min and scanning range of 10–90°.

Infrared spectra of YPS ash calcined at different temperatures were analyzed by Fourier-transform infrared spectrometry (Nicolet IS10, Thermo Nicolet Corporation, Madison, GA, USA) using the KBr pellet technique. The specific steps were as follows: 200 mg of KBr and 2 mg of sample were mixed evenly, ground to less than 5 μm and made into transparent slices. The measured wavenumber range was 400–4000 cm^{-1}.

3. Results

3.1. XRD Analysis

YPS is a complex industrial waste whose chemical composition and mineral phase vary greatly depending on the process and the raw materials used in its production. The main chemical compositions of YPS used in this study are calcium, silicon and aluminum, as well as small amounts of phosphorus, fluorine, sulfur, magnesium, potassium and iron. The XRD spectra of YPS ash uncalcined and calcined at 100–1000 °C are shown in

Figure 3. The XRD spectra of uncalcined YPS and YPS100-600 are similar. None of them have significant diffraction peaks, and there is a major broad peak at 20–40°. This indicates that the YPS samples under these conditions are low chemically stable glassy substances, which is favorable for the pozzolanic reactions.

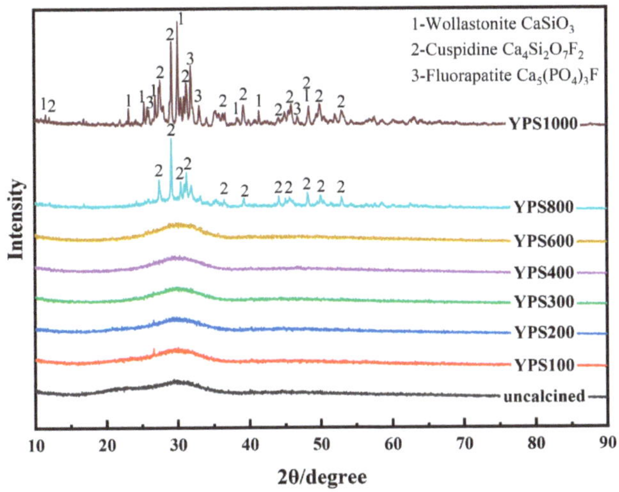

Figure 3. XRD patterns of YPS ash uncalcined and calcined at different temperatures.

The phase compositions of YPS ash change gradually with increasing temperature. In the XRD spectra of YPS800, the diffraction peaks of cuspidine ($Ca_4Si_2O_7F_2$) appeared with the presence of bun peaks, and the area of the bun peaks decreases. The XRD spectra of YPS1000 shows that the diffraction peaks of the different phases are sharp and clear, indicating that there is little amorphous content in YPS ash and that the crystallinity of the phases is high. As the calcination temperature is raised from 800 °C to 1000 °C, the wollastonite ($CaSiO_3$) and fluorapatite ($Ca_5(PO_4)_3F$) phase appears, and the diffraction peak of cuspidine in YPS ash is enhanced. The main phase compositions of YPS1000 are wollastonite, cuspidine and fluorapatite.

3.2. Pozzolanic Activity Evaluation by Compressive Strength

The pozzolanic activity of YPS ash can be calculated by the compressive strength of the YPS mortar. Figure 4 shows the compressive strength of YPS cement mortars and the pozzolanic activity index of YPS ash for 3, 7 and 28 d (the compressive strength of the control cement mortar is 37.1 MPa, 44.3 MPa and 49.8 MPa at 3, 7 and 28 d, respectively.). The compressive strength of the cement mortar varies with the calcination temperature, which proves that the calcination temperature has a significant effect on the cementitious properties of YPS ash. The compressive strength of YPS cement mortar probably increases and then reduces with the increasing calcination temperature. YPS800 has the highest 3 d compressive strength, with a value of 24.8 MPa, which is 55.0% higher than that of uncalcined YPS. This may be because the fluorine present in cuspidine is mainly insoluble and has little effect on the setting time of mortar [21]. However, the 28 d strength of YPS800 is relatively low, because the formation of crystalline phases makes the silica and aluminum in YPS800 difficult to dissolve at the later stage. This phenomenon illustrates that heat treatment can increase early strength but reduce ultimate strength. The 3 d compressive strength from YPS800 to YPS1000 decreases, which may be due to the generation of soluble $Ca_5(PO_4)_3F$ at 1000 °C, prolonging the setting time of the cement mortar and reducing its strength [21]. Compared to YPS uncalcined and YPS100-600, the 3 d compressive strength of YPS1000 is higher because a portion of fluorine in YPS1000 is present in the cuspidine, which reduces the retardation effect of fluorine on YPS1000 mortar. This also demonstrates that phosphorus and fluorine in the amorphous phase of YPS uncalcined and YPS100-600 are more easily soluble and affect the setting of YPS mortar. The 7 d compressive strength of the YPS300 mortar is the best, with a value of 31.5 MPa. The 28 d compressive strength of the YPS100 mortar is the highest, with a value of 55.8 MPa. This may be due to the gradual decrease in the amorphous substances in YPS with increasing calcination temperature.

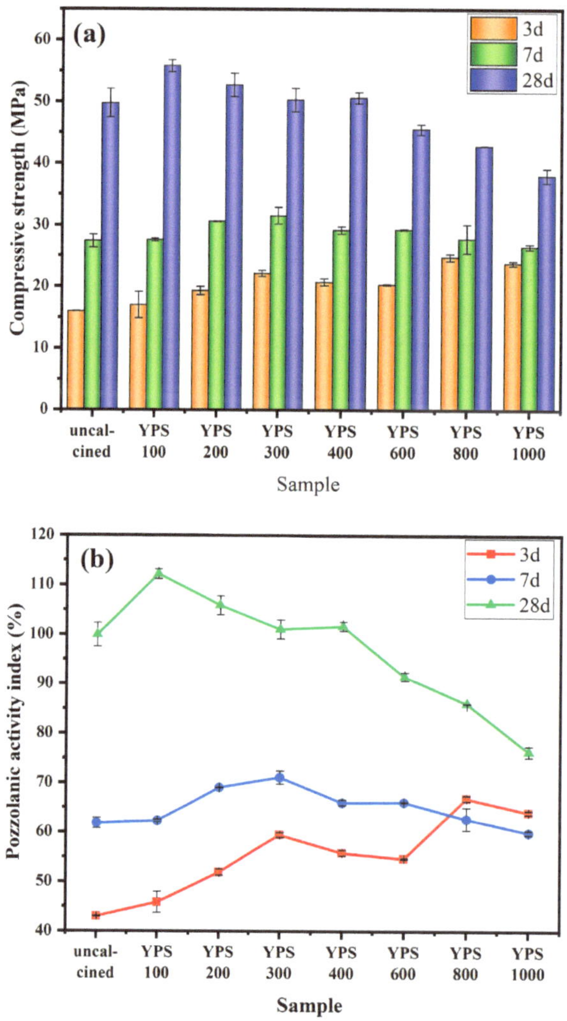

Figure 4. The (**a**) compressive strength of samples and (**b**) pozzolanic activity index of YPS ash uncalcined and calcined at different temperatures.

As can be seen from Figure 4b, the change in the pozzolanic activity index of YPS uncalcined and calcined at different temperatures is the same as the change in its compressive strength. The calcination of YPS can improve the pozzolanic activity index at 3 d, but it will reduce the pozzolanic activity index at 28 d, and the enhancement of the pozzolanic activity index at 7 d is less obvious. The calcination of YPS at 800 °C is favorable to enhance the compressive strength of YPS cement mortar at an early age, while YPS calcined at 100 °C is conducive to increasing the ultimate strength of YPS cement mortar.

3.3. Pozzolanic Activity Evaluation by Dissolution Concentrations of Si and Al

Reactive Si and Al dissolved in alkaline environments can take part in the hydration reaction of cement. Therefore, the pozzolanic activity of YPS ash can be evaluated by the alkaline dissolution method. The dissolution concentrations of Si and Al in a 1 mol/L NaOH solution of YPS uncalcined and calcined at different temperatures are shown in Table 3. The dissolution concentrations of Si and Al first increase and then decrease with increasing temperature. The dissolution concentrations of Si and Al in YPS100 are 41.01 mg/L and 30.08 mg/L, respectively. Its dissolution concentration of (Si + Al) is the highest, which is consistent with the 28 d compressive strength of YPS cement mortars. The lowest dissolution concentrations of Si, Al and (Si + Al) are 7.197 mg/L, 10.08 mg/L and 17.28 mg/L in YPS1000, respectively. And its 28 d compressive strength and pozzolanic activity index at 28 d are also the lowest.

Table 3. The dissolution concentrations of Si and Al in YPS ash uncalcined and calcined at different temperatures (mg/L).

Sample	Dissolution Concentration of Si	Dissolution Concentration of Al	Dissolution Concentration of (Si + Al)
uncalcined	40.83	25.01	65.84
YPS100	41.01	30.08	71.09
YPS200	35.71	31.79	67.50
YPS300	33.71	27.19	60.90
YPS400	28.17	23.84	52.01
YPS600	24.97	15.29	40.26
YPS800	15.79	14.16	29.95
YPS1000	7.197	10.08	17.28

The trend of the pozzolanic activity index for 28 d with temperature is basically consistent with that of the dissolution concentrations of Si and Al. Therefore, the relationship between the pozzolanic activity index of YPS ash at 28 d and the dissolution concentrations of Si and Al has been studied, as shown in Figure 5. The dissolution concentrations of Si, Al and (Si + Al) in YPS ash are positively related to its pozzolanic activity index for 28 d. The pozzolanic activity index for 28 d increases with increasing dissolution concentrations of Si, Al and (Si + Al). The linear fit R^2 about pozzolanic activity index and dissolution concentrations of Si, Al and (Si + Al) was progressively increasing with values of 0.87, 0.91 and 0.93, respectively. This shows that the dissolution concentrations of both Si and Al have an effect on the pozzolanic activity of YPS ash. The change rule of the pozzolanic activity index for 3 and 7 d and the dissolution concentration of Si and Al is a little different, which indicates that the early strength of YPS mortar is not only affected by the dissolution concentration of Si and Al but also by the role of soluble fluorine and phosphorus.

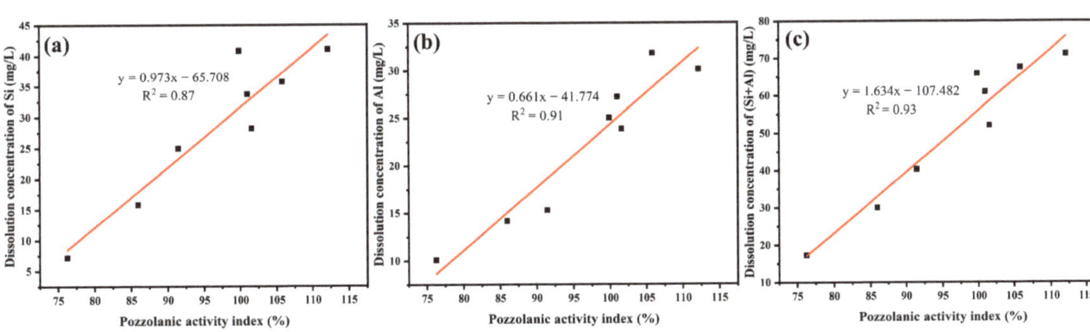

Figure 5. The relationship between the dissolution concentrations of (**a**) Si, (**b**) Al and (**c**) (Si + Al) and the pozzolanic activity index of YPS.

3.4. Pozzolanic Activity Evaluation by FTIR

The FTIR spectra of YPS ash uncalcined and calcined at 100–1000 °C are shown in Figure 6. The FTIR spectra of YPS ash uncalcined and calcined in the range of 100–600 °C are very similar and show analogous absorption bands. The FTIR spectra can be divided into three regions of 800–1200 cm^{-1}, 600–800 cm^{-1} and 400–600 cm^{-1}. The peaks of 800–1200 cm^{-1} correspond to the asymmetric tensile vibrations of Si-O-(Si, Al) linked to the tetrahedral of $[SiO_4]$ or $[AlO_4]^-$ [32]. The peaks at 600–800 cm^{-1} are the symmetric stretching vibrations of Si-O-Al or Si-O-Si in the tetrahedra of $[SiO_4]$ or $[AlO_4]^-$ [35]. The peaks in the 400–600 cm^{-1} range are the bending vibrations of Si-O-Si(Al) [29,35]. The positions and areas of the peaks of 800–1200 cm^{-1} vary with temperature, indicating that the bonds between Si, Al and O in YPS ash are combined or broken. The peaks of the Si-O-Si/Al bonds move to higher wavenumbers (949 cm^{-1}) with increasing temperature, indicating an improvement in the polymerization degree. The absorption peaks belong to PO_4^{3-} bands at 471 cm^{-1}, 560 cm^{-1} and 1027 cm^{-1}, which proves the formation of fluorapatite at 1000 °C [36]. This is consistent with the results of the XRD analysis.

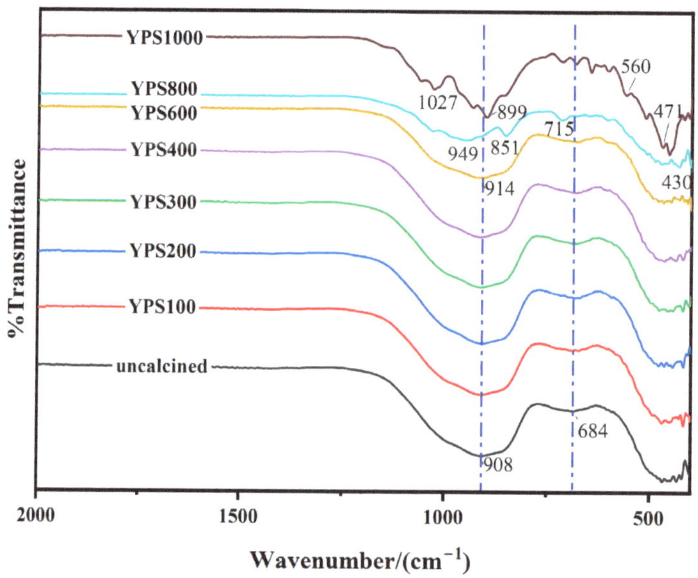

Figure 6. FTIR spectra of YPS ash uncalcined and calcined at different temperatures.

The five bands at 1200 cm^{-1}, 1050 cm^{-1}, 1000 cm^{-1}, 900 cm^{-1} and 850 cm^{-1} are assigned to Q^4 Si, Q^3 Si, Q^2 Si, Q^1 Si and Q^0 Si, respectively. The superscript n is the number of bridge oxygens attached to each Si. The polymerization degree of the silicate structure is higher as n increases. The peaks between 800 and 1200 cm^{-1} were separated and fitted with Origin software to calculate the RBO. Figure 7 shows the fitted peaks between 800 and 1200 cm^{-1} of YPS ash uncalcined and calcined at different temperatures. The relevant parameters of the peaks are shown in Table 4. The polymerization degree of YPS ash first decreases and then increases with increasing temperature. The high degree of polymerization indicates that silicon and aluminum are relatively stable. The lower degree of polymerization demonstrates that the Si-O-Si(Al) in YPS ash is broken, and the active silicon and active aluminum in YPS ash are increased [35]. The RBO of the YPS100 is the smallest, with a value of 0.3757. The RBO of the YPS1000 is the biggest, with a value of 0.4982. This result is consistent with the 28 d compressive strength and dissolution concentrations of Si and Al. In addition, the R^2 of the infrared fitting result decreases when the temperature is more than 800 °C. This is due to the appearance of a new crystalline phase, which can also be seen from the XRD results. Therefore, the RBO evaluation method is suitable for materials with no other influence in the range of 800–1200 cm^{-1}.

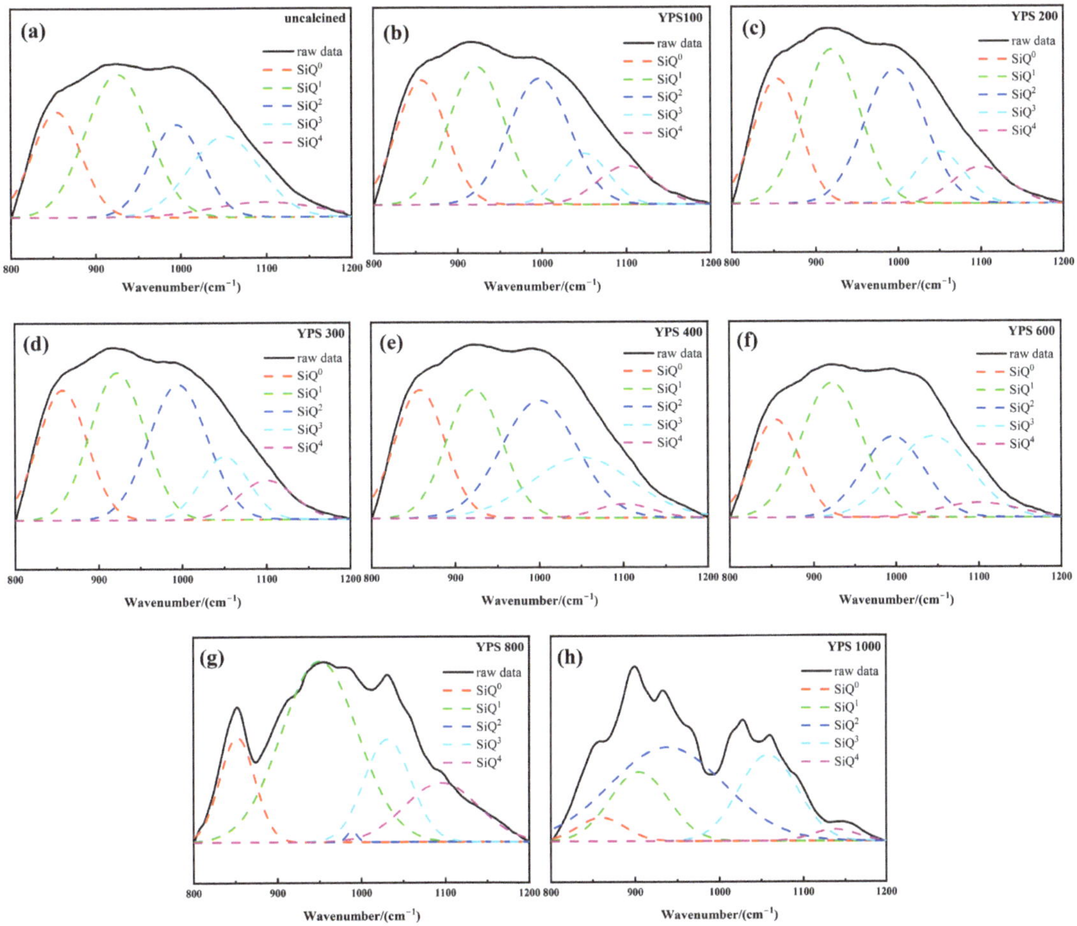

Figure 7. The fitted peaks in the range of 800–1200 cm^{-1} of YPS ash uncalcined and calcined at different temperatures: (**a**) uncalcined; (**b**) YPS100; (**c**) YPS200; (**d**) YPS300; (**e**) YPS400; (**f**) YPS600; (**g**) YPS800; (**h**) YPS1000.

Table 4. The relevant peaks parameters of YPS ash uncalcined and calcined at different temperatures.

Sample	Relative Content/%					RBO	R^2
	SiQ^0	SiQ^1	SiQ^2	SiQ^3	SiQ^4		
uncalcined	19.24	34.83	18.43	21.78	5.72	0.3998	0.996
YPS100	23.22	29.34	29.75	9.33	8.36	0.3757	0.996
YPS200	20.63	32.40	30.14	9.24	7.59	0.3769	0.997
YPS300	22.75	29.38	28.86	11.07	7.94	0.3802	0.997
YPS400	21.48	24.35	30.66	20.95	2.56	0.3969	0.996
YPS600	17.78	33.02	19.95	24.81	4.44	0.4128	0.997
YPS800	13.68	51.86	0.42	17.95	16.09	0.4273	0.983
YPS1000	5.14	19.10	49.46	23.94	2.36	0.4982	0.963

Figure 8 shows the relationship between the pozzolanic activity and RBO of YPS ash uncalcined and calcined at different temperatures. The RBO of YPS ash tends to decrease with increasing pozzolanic activity. Figure 9 shows the relationship between the RBO and the dissolution concentrations of Si, Al and (Si + Al) of YPS ash uncalcined and calcined at different temperatures. RBO has a linear relationship with the dissolution concentrations of Si and Al. The decrease in RBO promotes an increase in the dissolution concentration of Si and Al in the YPS ash. The linear fit R^2 about the RBO and pozzolanic activity is 0.87. The linear fits R^2 about the RBO and dissolution concentrations of Si, Al and (Si + Al) are 0.80, 0.79 and 0.83, respectively. The linear fits R^2 are poor compared with that of pozzolanic activity and dissolution concentrations of Si, Al and (Si + Al). This is mainly due to the fact that the infrared peaks in the range of 800–1200 cm^{-1} of YPS1000 are interfered with by other functional groups and the fitting effect of RBO is weakened. This shows that the dissolution concentrations of both Si and Al have an effect on the pozzolanic activity of YPS ash. The polymerization degree method for evaluating the pozzolanic activity of YPS ash takes less time and is more convenient compared to the strength and alkali dissolution methods, but it is susceptible to the influence of the physical phase.

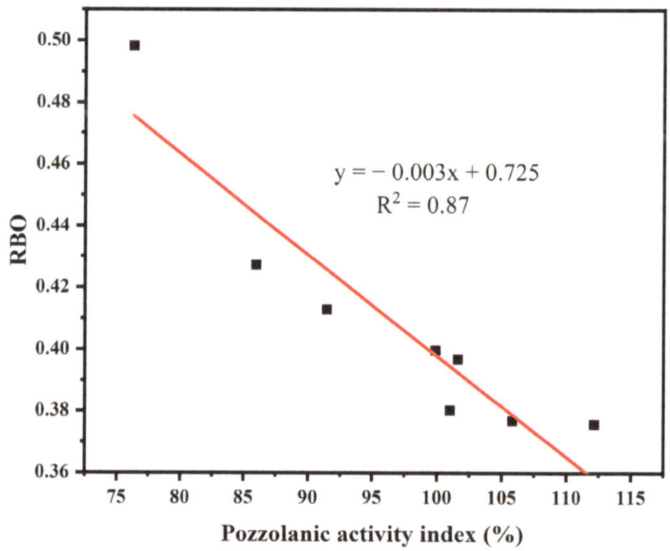

Figure 8. The relationship between RBO and the pozzolanic activity index of YPS ash.

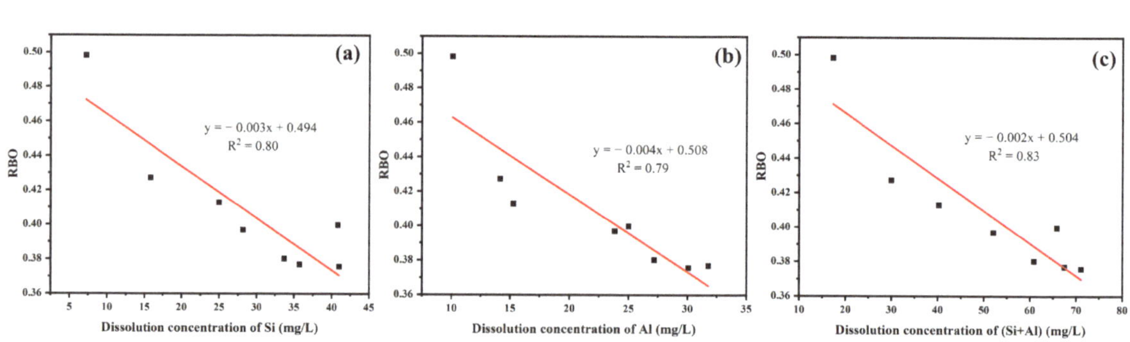

Figure 9. The relationship between RBO and the dissolution concentrations of (**a**) Si; (**b**) Al; and (**c**) (Si + Al).

4. Conclusions

In this paper, the phase and structural transformations of YPS ash during calcination at 100–1000 °C were investigated, and the changes in the pozzolanic activity of YPS ash were studied by the compressive strength method, alkaline dissolution method and polymerization degree method. This study provides insight into the gelling behavior of YPS ash.

The uncalcined YPS mainly contains amorphous aluminosilicates. As the calcination temperature increases to 800 °C, crystalline phases appear in the YPS ash. Cuspidine ($Ca_4Si_2O_7F_2$) is formed at 800 °C. The formation of fluorapatite ($Ca_5(PO_4)_3F$) and wollastonite ($CaSiO_3$) occurs at 1000 °C. The polymerization degree decreases and then increases with increasing calcination temperature. The polymerization degree is lowest at 100 °C. Due to the change in phase and polymerization degree, the calcination temperature has a great influence on the pozzolanic activity of YPS ash. It was found that YPS ash calcined at 800 °C favored the early strength of the sample, and YPS ash calcined at 100 °C had the highest 28 d compressive strength.

The relationships between the dissolution concentration of (Si + Al), compressive strength and polymerization degree and the pozzolanic activity of YPS ash were analyzed. The pozzolanic activity of YPS ash is positively correlated with the compressive strength and the dissolution concentration of (Si + Al) and negatively associated with the polymerization degree.

This study provides a possibility for the utilization of YPS in cement and concrete, which is conducive to the use of YPS in large dosages.

Author Contributions: Conceptualization, X.L. (Xinyue Liu), X.L. (Xiaoming Liu) and Z.Z.; validation, C.W. and Q.Z.; investigation, X.L. (Xinyue Liu), C.W., Q.Z. and Y.L.; data curation, X.L. (Xinyue Liu) and S.M.; writing—original draft preparation, X.L. (Xinyue Liu); writing—review and editing, X.L. (Xiaoming Liu) and Z.Z.; supervision, X.L. (Xiaoming Liu) and Z.Z. All authors have read and agreed to the published version of the manuscript.

Funding: This research was funded by the National Natural Science Foundation of China, grant number 52074035, the Young Elite Scientists Sponsorship Program by CAST, grant number 2022QNRC001, and the Fundamental Research Funds for the Central Universities, grant numbers 00007720 and 00007469.

Institutional Review Board Statement: Not applicable.

Informed Consent Statement: Not applicable.

Data Availability Statement: Data sharing is not applicable to this article.

Conflicts of Interest: The authors declare no conflict of interest.

References

1. Hassankhani-Majd, Z.; Anbia, M. Recovery of Valuable Materials from Phosphorus Slag Using Nitric Acid Leaching Followed by Precipitation Method. *Resour. Conserv. Recycl.* **2021**, *169*, 105547. [CrossRef]
2. Akcil, A.; Karshigina, Z.B.; Bochevskaya, Y.G.; Abisheva, Z.S. Conditions of Nitric Acid Treatment of Phosphorus Slag for Rems Recovery and Production of Precipitated Silicon Dioxide. *Complex Use Miner. Resour.* **2018**, *305*, 28–38. [CrossRef]
3. Abisheva, Z.S.; Karshigina, Z.B.; Bochevskaya, Y.G.; Akcil, A.; Sargelova, E.A.; Kvyatkovskaya, M.N.; Silachyov, I.Y. Recovery of Rare Earth Metals as Critical Raw Materials from Phosphorus Slag of Long-Term Storage. *Hydrometallurgy* **2017**, *173*, 271–282. [CrossRef]
4. Liu, X.; Liu, X.; Zhang, Z. Recycling and Comprehensive Utilization of Yellow Phosphorus Slag in Building Materials: A Review. *Constr. Build. Mater.* **2023**, *396*, 132384. [CrossRef]
5. Nguyen, V.H.; Nguyen, V.C.; Nguyen, T.C.; Tran, D.M.T.; Nguyen, T.T.T.; Vu, Q.T.; Nguyen, D.T.; Thai, H. Treatment of Yellow Phosphorus Slag and Reuse of It as an Absorbent of Chromium (VI) Ions and Methylene Blue. *J. Chem.* **2020**, *2020*, 1834829. [CrossRef]
6. Hassankhani-Majd, Z.; Anbia, M.; Tavoussi-Shirazi, D.S. A Zero-Waste Process for the Extraction of Highly Pure Nanoporous Silica Particles from Phosphorus Slag. *JOM* **2022**, *74*, 1002–1011. [CrossRef]
7. Naseroleslami, R.; Bakhshi, J.; Nemati Chari, M.; Yaghoobi, M.A.; Haji Mahdi, M. Properties of Cement Mortars Containing Phosphorous Slag. *J. Mater. Civ. Eng.* **2023**, *35*, 04022384. [CrossRef]

8. Pavlíková, M.; Rovnaníková, P.; Záleská, M.; Pavlík, Z. Diatomaceous Earth—Lightweight Pozzolanic Admixtures for Repair Mortars—Complex Chemical and Physical Assessment. *Materials* **2022**, *15*, 6881. [CrossRef] [PubMed]
9. Maghsoodloorad, H.; Allahverdi, A. Efflorescence Formation and Control in Alkali-Activated Phosphorus Slag Cement. *Int. J. Civ. Eng.* **2016**, *14*, 425–438. [CrossRef]
10. Mehdizadeh, H.; Shao, X.; Mo, K.H.; Ling, T.C. Enhancement of Early Age Cementitious Properties of Yellow Phosphorus Slag via CO_2 Aqueous Carbonation. *Cem. Concr. Compos.* **2022**, *133*, 104702. [CrossRef]
11. Mehdizadeh, H.; Kani, E. Modeling the Influence of Chemical Composition on Compressive Strength Behavior of Alkali-Activated Phosphorus Slag Cement Using Statistical Design. *Can. J. Civ. Eng.* **2018**, *45*, 1073–1083. [CrossRef]
12. Golewski, G. Evaluation of Morphology and Size of Cracks of the Interfacial Transition Zone (ITZ) in Concrete Containing Fly Ash (FA). *J. Hazard. Mater.* **2018**, *357*, 298–304. [CrossRef] [PubMed]
13. Allahverdi, A.; Mahinroosta, M.; Pilehvar, S. A Temperature–Age Model For Prediction of Compressive Strength of Chemically Activated High Phosphorus Slag Content Cement. *Int. J. Civ. Eng.* **2017**, *15*, 839–847. [CrossRef]
14. Shaikezhan, A.; Anuarova, A.D.; Antonovic, V. Cement Slurry from Electro-Phosphoric Slag. *Mag. Civ. Eng.* **2020**, *98*, 9806. [CrossRef]
15. Vafaei, M.; Allahverdi, A.; Dong, P.; Bassim, N.; Mahinroosta, M. Resistance of Red Clay Brick Waste/Phosphorus Slag-Based Geopolymer Mortar to Acid Solutions of Mild Concentration. *J. Build. Eng.* **2021**, *34*, 102066. [CrossRef]
16. Mahambetova, U.; Nuranbayeva, B.; Estemesov, Z.; Sadykov, P.; Mamyrbayev, O.; Oralbekova, D. Development And Research Of The Influence Of The Composition And Concentration Of Activators On The Strength Of Phosphorus Slag Binders. *East.-Eur. J. Enterp. Technol.* **2021**, *5*, 54–61. [CrossRef]
17. Wang, Q.; Huang, Z.; Wang, D. Influence of High-Volume Electric Furnace Nickel Slag and Phosphorous Slag on the Properties of Massive Concrete. *J. Therm. Anal. Calorim.* **2018**, *131*, 873–885. [CrossRef]
18. Yang, R.; Yu, R.; Shui, Z.; Gao, X.; Xiao, X.; Zhang, X.; Wang, Y.; He, Y. Low Carbon Design of an Ultra-High Performance Concrete (UHPC) Incorporating Phosphorous Slag. *J. Clean. Prod.* **2019**, *240*, 118157. [CrossRef]
19. Allahverdi, A.; Bahri Rasht Abadi, M.M. Resistance of Chemically Activated High Phosphorous Slag Content Cement against Frost-Salt Attack. *Cold Reg. Sci. Technol.* **2014**, *98*, 18–25. [CrossRef]
20. Maghsoodloorad, H.; Allahverdi, A. Developing Low-Cost Activators for Alkali-Activated Phosphorus Slag-Based Binders. *J. Mater. Civ. Eng.* **2017**, *29*, 04017006. [CrossRef]
21. Wang, Y.; Zhang, N.; Xiao, H.; Zhao, J.; Zhang, Y.; Liu, X. Structural Characterization of Phosphorous Slag Regarding Occurrence State of Phosphorus in Dicalcium Silicate. *Materials* **2022**, *15*, 7450. [CrossRef]
22. Vafaei, M.; Allahverdi, A. Strength Development and Acid Resistance of Geopolymer Based on Waste Clay Brick Powder and Phosphorous Slag. *Struct. Concr.* **2019**, *20*, 1596–1606. [CrossRef]
23. Mehdizadeh, H.; Najafi Kani, E.; Palomo Sanchez, A.; Fernandez-Jimenez, A. Rheology of Activated Phosphorus Slag with Lime and Alkaline Salts. *Cem. Concr. Res.* **2018**, *113*, 121–129. [CrossRef]
24. Salehi, A.; Najafi Kani, E. Green Cylindrical Mesoporous Adsorbent Based on Alkali-Activated Phosphorous Slag: Synthesis, Dye Removal, and RSM Modeling. *Adsorption* **2018**, *24*, 647–666. [CrossRef]
25. Mehdizadeh, H.; Najafi Kani, E. Rheology and Apparent Activation Energy of Alkali Activated Phosphorous Slag. *Constr. Build. Mater.* **2018**, *171*, 197–204. [CrossRef]
26. Maghsoodloorad, H.; Khalili, H.; Allahverdi, A. Alkali-Activated Phosphorous Slag Performance under Different Curing Conditions: Compressive Strength, Hydration Products, and Microstructure. *J. Mater. Civ. Eng.* **2017**, *30*, 04017253. [CrossRef]
27. He, X.; Ye, Q.; Yang, J.; Dai, F.; Su, Y.; Wang, Y.; Bohumír, S. Physico-Chemical Characteristics of Wet-Milled Ultrafine-Granulated Phosphorus Slag as a Supplementary Cementitious Material. *J. Wuhan Univ. Technol. Mater. Sci. Ed.* **2018**, *33*, 625–633. [CrossRef]
28. Hu, J. Comparison between the Effects of Superfine Steel Slag and Superfine Phosphorus Slag on the Long-Term Performances and Durability of Concrete. *J. Therm. Anal. Calorim.* **2017**, *128*, 1251–1263. [CrossRef]
29. Allahverdi, A.; Mahinroosta, M. Mechanical Activation of Chemically Activated High Phosphorous Slag Content Cement. *Powder Technol.* **2013**, *245*, 182–188. [CrossRef]
30. Allahverdi, A.; Akhondi, M.; Mahinroosta, M. Superior Sodium Sulfate Resistance of a Chemically Activated Phosphorus Slag–Based Composite Cement. *J. Mater. Civ. Eng.* **2017**, *29*, 04016231. [CrossRef]
31. Zhang, Z.; Wang, Q.; Yang, J. Hydration Mechanisms of Composite Binders Containing Phosphorus Slag at Different Temperatures. *Constr. Build. Mater.* **2017**, *147*, 720–732. [CrossRef]
32. Wang, Y.; Xiao, R.; Hu, W.; Jiang, X.; Zhang, X.; Huang, B. Effect of Granulated Phosphorus Slag on Physical, Mechanical and Microstructural Characteristics of Class F Fly Ash Based Geopolymer. *Constr. Build. Mater.* **2021**, *291*, 123287. [CrossRef]
33. Allahverdi, A.; Pilehvar, S.; Mahinroosta, M. Influence of Curing Conditions on the Mechanical and Physical Properties of Chemically-Activated Phosphorous Slag Cement. *Powder Technol.* **2016**, *288*, 132–139. [CrossRef]
34. Administration, M.; Standardization, A. *GB/T 26751-2022*; The Chinese Standard Ground Granulated Electric Furnace Phosphorus Slag Powder Used for Cement and Concrete, in Chinese Code GB/T 26751-2022. Chinese Standard Press: Beijing, China, 2022.

35. Wang, Y.; Liu, X.; Xie, Z.; Wang, H.; Zhang, W.; Xue, Y. Rapid Evaluation of the Pozzolanic Activity of Bayer Red Mud by a Polymerization Degree Method: Correlations with Alkali Dissolution of (Si + Al) and Strength. *Materials* **2021**, *14*, 5546. [CrossRef] [PubMed]
36. Liu, X.; Ren, Y.; Zhang, Z.; Liu, X.; Wang, Y. Harmless Treatment of Electrolytic Manganese Residue: Ammonia Nitrogen Recovery, Preparation of Struvite and Nonsintered Bricks. *Chem. Eng. J.* **2023**, *455*, 140739. [CrossRef]

MDPI AG
Grosspeteranlage 5
4052 Basel
Switzerland
Tel.: +41 61 683 77 34

Materials Editorial Office
E-mail: materials@mdpi.com
www.mdpi.com/journal/materials

www.ingramcontent.com/pod-product-compliance
Lightning Source LLC
Chambersburg PA
CBHW041401210926
51850CB00013B/3150